Modern Number Theory

Modern Number Theory

Editor: Rowan Payne

STATES
ACADEMIC PRESS
www.statesacademicpress.com

Published by States Academic Press
109 South 5th Street,
Brooklyn, NY 11249, USA
www.statesacademicpress.com

Modern Number Theory
Edited by Rowan Payne

International Standard Book Number: 978-1-63989-362-1 (Hardback)

Cataloging-in-Publication Data

Modern number theory / edited by Rowan Payne.
p. cm.
Includes bibliographical references and index.
ISBN 978-1-63989-362-1
1. Number theory. 2. Algebra. 3. Mathematics. I. Payne, Rowan.
QA241 .N86 2022
512.7--dc23

Contents

Preface

The main aim of this book is to educate learners and enhance their research focus by presenting diverse topics covering this vast field. This is an advanced book which compiles significant studies by distinguished experts in the area of analysis. This book addresses successive solutions to the challenges arising in the area of application, along with it; the book provides scope for future developments.

The branch of pure mathematics devoted primarily to the study of integers and integer-valued functions is known as number theory. The concepts of prime numbers as well as the properties of mathematical objects made out of integers such as rational numbers or generalizations of the integers, and algebraic integers are studied within this field. Some of the main subdivisions of number theory are elementary number theory, analytic number theory, algebraic number theory, Diophantine geometry, probabilistic number theory, arithmetic combinatorics, and computational number theory. Number theory has applications in discrete mathematics courses for computer science, numerical analysis, and cryptography. This book will help new researchers by foregrounding their knowledge in this branch. The various sub-fields of number theory along with technological progress that have future implications are glanced at herein. This book aims to equip students and experts with the advanced topics and upcoming concepts in this area.

It was a great honour to edit this book, though there were challenges, as it involved a lot of communication and networking between me and the editorial team. However, the end result was this all-inclusive book covering diverse themes in the field.

Finally, it is important to acknowledge the efforts of the contributors for their excellent chapters, through which a wide variety of issues have been addressed. I would also like to thank my colleagues for their valuable feedback during the making of this book.

Editor

1

A NOTE ON THE PERMUTATION BEHAVIOUR OF THE POLYNOMIAL $g_{n,q}$

Neranga Fernando
Department of Mathematics, Northeastern University, Boston, Massachusetts
w.fernando@northeastern.edu

Abstract
Let $q = 4$ and k be an even positive integer. In this note, we present a class of permutation polynomials over $\mathbb{F}_{q^{3k}}$. We also present a generalization.

1. Introduction

Let p be a prime and q a power of p. Let $\mathbb{F}_q$ denote the finite field with q elements, $\text{Char}\mathbb{F}_q$ denote the characteristic of $\mathbb{F}_q$, and e be a positive integer. A permutation polynomial (PP) of $\mathbb{F}_q$ is a polynomial $f \in \mathbb{F}_q[x]$ such that the mapping $x \mapsto f(x)$ is a permutation of $\mathbb{F}_q$. Permutation polynomials over finite fields have important applications in coding theory, cryptography, finite geometry, combinatorics and computer science, among other fields. Most studies on PPs have been over finite fields. However, several authors have studied PPs over finite commutative rings. We refer the reader to [9], [4] and references therein for further details on PPs over finite commutative rings.

For each integer $n \geq 0$, there is a polynomial $g_{n,q} \in \mathbb{F}_p[x]$ defined by the functional equation

$$\sum_{c \in \mathbb{F}_q} (x + c)^n = g_{n,q}(x^q - x). \tag{1}$$

The polynomial $g_{n,q}$ was introduced in [5], and its permutation property was studied in [2, 6]. The objective is to determine the triples $(n, e; q)$ of positive integers for which $g_{n,q}$ is a PP of $\mathbb{F}_{q^e}$. We call a triple of integers $(n, e; q)$ *desirable* if $g_{n,q}$ is a PP of $\mathbb{F}_{q^e}$.

Define

$$S_{m,q} = x + x^q + \cdots + x^{q^{m-1}} \in \mathbb{F}_p[x],$$

where m is a positive integer and $p = \text{Char}\mathbb{F}_q$. When q is fixed, we write $S_{m,q} = S_m$. Note that S_m is a particular type of linearized polynomial and $S_e = \text{Tr}_{q^e/q}$, where $\text{Tr}_{q^e/q}$ is the *trace* function from $\mathbb{F}_{q^e}$ to $\mathbb{F}_q$. The polynomial $g_{n,q}$ can be expressed in terms of polynomials of the form S_m when the integer n takes certain forms. PPs arising from polynomials $g_{n,q}$ that involve S_m were studied in [3].

Desirable triples obtained from computer searches were presented in [1, 2, 6]. Table 1 in the Appendix of the present paper contains PPs over $\mathbb{F}_{q^e}$ defined by the polynomial $g_{n,q}$ when $q = 4$ and $e = 6$. In [1], Hou and the author of the present paper discovered several new classes of PPs which explained several entries of Table 3 in [2]. But, as a whole, determining the permutation behaviour of the polynomial $g_{n,q}$ appears to be a difficult problem. We refer the reader to [6] for more background of the polynomial $g_{n,q}$, and to [1] for a complete table of desirable triples when $q = 4$.

PPs over $\mathbb{F}_{q^{3k}}$, where q is a power of 2, have appeared in the literature; see [2, 7, 10]. In [7], Hou confirmed a class of PPs over $\mathbb{F}_{4^{3k}}$ that was conjectured in [2]. In [10], Yuan and Ding generalized Hou's result and obtained several more classes of PPs of the form $L(x)+S_{2k}^a+S_{2k}^b$ over $\mathbb{F}_{q^{3k}}$, where q is a power of 2. In this paper, we present a new class of PPs over $\mathbb{F}_{q^{3k}}$ defined by the polynomial $g_{n,q}$, where $q = 4$ and $k > 0$ is an even integer, which explains another entry, $n = 65921$, of Table 1 in the Appendix.

The note is organized as follows. In Section 2, we present the main result of the paper, Theorem 1, which gives a new class of PPs over $\mathbb{F}_{4^{3k}}$. In Section 3, we explain an entry of Table 1 in the Appendix using the results obtained in Section 2. In Section 4, we present a generalization. In Section 2 and Section 3, we assume that $q = 4$ and use $e = 3k$ for some even k unless otherwise specified.

2. A Class of Permutation Polynomials Over $\mathbb{F}_{q^{3k}}$

We recall two facts:

Fact 1. ([8, Theorem 7.7]) *Let $f \in \mathbb{F}_q[x]$. Then f is a PP of $\mathbb{F}_q$ if and only if for all $a \in \mathbb{F}_q^*$, $\sum_{x\in\mathbb{F}_q} \zeta_p^{\mathrm{Tr}_{q/p}(af(x))} = 0$, where $p = \mathrm{Char}\mathbb{F}_q$ and $\zeta_p = e^{2\pi i/p}$.*

Fact 2. ([2, Lemma 6.13]) *Let p be a prime and $f\colon \mathbb{F}_p^n \to \mathbb{F}_p$ a function. If there exists a $y \in \mathbb{F}_p^n$ such that $f(x+y) - f(x)$ is a nonzero constant for all $x \in \mathbb{F}_p^n$, then $\sum_{x\in\mathbb{F}_p^n} \zeta_p^{f(x)} = 0$.*

Lemma 1. *Let k be a positive integer, q a power of 2, $b \in \mathbb{F}_{q^{3k}}^*$ and $y \in \mathbb{F}_{q^k}^*$. Then $Tr_{q^{3k}/2}\big(b\,(y + y^q + \cdots + y^{q^{k-2}})\big) = Tr_{q^{3k}/2}\big(y\,(b^{q^{3k}} + b^{q^{3k-1}} + \cdots + b^{q^{2k+2}})\big)$.*

Proof. We write $\mathrm{Tr} = \mathrm{Tr}_{q^{3k}/2}$. We have

$$\begin{aligned}\mathrm{Tr}\big(b\,(y + y^q + \cdots + y^{q^{k-2}})\big) &= \mathrm{Tr}\big(by + by^q + \cdots + by^{q^{k-2}}\big)\\ &= \mathrm{Tr}\big(by\big) + \mathrm{Tr}\big(by^q\big) + \cdots + \mathrm{Tr}\big(by^{q^{k-2}}\big)\\ &= \mathrm{Tr}\big(b^{q^{3k}}y^{q^{3k}}\big) + \mathrm{Tr}\big(b^{q^{3k-1}}y^{q^{3k}}\big) + \cdots + \mathrm{Tr}\big(b^{q^{2k+2}}y^{q^{3k}}\big)\\ &= \mathrm{Tr}\big(b^{q^{3k}}y\big) + \mathrm{Tr}\big(b^{q^{3k-1}}y\big) + \cdots + \mathrm{Tr}\big(b^{q^{2k+2}}y\big)\\ &= \mathrm{Tr}\big(yb^{q^{3k}} + yb^{q^{3k-1}} + \cdots + yb^{q^{2k+2}}\big)\\ &= \mathrm{Tr}\big(y\,(b^{q^{3k}} + b^{q^{3k-1}} + \cdots + b^{q^{2k+2}})\big).\end{aligned}$$

□

Lemma 2. *Let k be a positive integer, q a power of 2, $x \in \mathbb{F}_{q^{3k}}$ and $c \in \mathbb{F}_{q^{3k}} \setminus \mathbb{F}_{q^k}$. Then $Tr_{q^{3k}/2}\big(c(x + x^q + \cdots + x^{q^{2k-1}})\big) = Tr_{q^{3k}/2}\big(x(c^{q^{3k}} + c^{q^{3k-1}} + \cdots + c^{q^{k+1}})\big)$.*

Proof. We write $\mathrm{Tr} = \mathrm{Tr}_{q^{3k}/2}$. We have

$$\begin{aligned}\mathrm{Tr}\big(c(x + x^q + \cdots + x^{q^{2k-1}})\big) &= \mathrm{Tr}\big(cx + cx^q + \cdots + cx^{q^{2k-1}}\big)\\ &= \mathrm{Tr}\big(cx\big) + \mathrm{Tr}\big(cx^q\big) + \cdots + \mathrm{Tr}\big(cx^{q^{2k-1}}\big)\\ &= \mathrm{Tr}\big(c^{q^{3k}}x^{q^{3k}}\big) + \mathrm{Tr}\big(c^{q^{3k-1}}x^{q^{3k}}\big) + \cdots + \mathrm{Tr}\big(c^{q^{k+1}}x^{q^{3k}}\big)\\ &= \mathrm{Tr}\big(c^{q^{3k}}x\big) + \mathrm{Tr}\big(c^{q^{3k-1}}x\big) + \cdots + \mathrm{Tr}\big(c^{q^{k+1}}x\big)\\ &= \mathrm{Tr}\big(xc^{q^{3k}} + xc^{q^{3k-1}} + \cdots + xc^{q^{k+1}}\big)\\ &= \mathrm{Tr}\big(x\,(c^{q^{3k}} + c^{q^{3k-1}} + \cdots + c^{q^{k+1}})\big).\end{aligned}$$

□

The following theorem is the main theorem of this paper.

Theorem 1. *Let $q = 4$, $e = 3k$, where $k > 0$ is an even integer. Then $g = S_{k+1}^2 + S_{2k}^{q^k+1}$ is a PP of $\mathbb{F}_{q^e}$.*

Proof. By Fact 1, it suffices to show that

$$\sum_{x \in \mathbb{F}_{q^e}} (-1)^{\mathrm{Tr}_{q^e/2}(ag(x))} = 0$$

for all $0 \neq a \in \mathbb{F}_{q^e}$. We write $\mathrm{Tr} = \mathrm{Tr}_{q^e/2}$. We consider two cases, where $\mathrm{Tr}_{q^e/q^k}(a) \neq 0$ and $\mathrm{Tr}_{q^e/q^k}(a) = 0$.

`Case 1.` Assume $\mathrm{Tr}_{q^e/q^k}(a) \neq 0$. By Fact 2, it suffices to show that there exists $y \in \mathbb{F}_{q^e}$ such that $\mathrm{Tr}\big(ag(x+y) - ag(x)\big)$ is a nonzero constant for all $x \in \mathbb{F}_{q^e}$.

Let $y \in \mathbb{F}_{q^k}^*$. Then for $x \in \mathbb{F}_{q^e}$, we have

$$\begin{aligned}
&\operatorname{Tr}\big(ag(x+y) - ag(x)\big)\\
&= \operatorname{Tr}\big[a\big(S_{k+1}^2(x+y) + S_{2k}^{q^k+1}(x+y) - S_{k+1}^2(x) - S_{2k}^{q^k+1}(x)\big)\big]\\
&= \operatorname{Tr}(a\, S_{k+1}^2(y)) \qquad\qquad (S_{2k}(y) = 0 \text{ since } y \in \mathbb{F}_{q^k})\\
&= \operatorname{Tr}(a\,(y + y^q + \cdots + y^{q^{k-1}} + y^{q^k})^2)\\
&= \operatorname{Tr}(a\,(y + y^q + \cdots + y^{q^{k-1}} + y)^2)\\
&= \operatorname{Tr}(a\,(y^q + y^{q^2} + \cdots + y^{q^{k-1}})^2)\\
&= \operatorname{Tr}(a\,(y + y^q + \cdots + y^{q^{k-2}})^{2q})\\
&= \operatorname{Tr}(a\, S_{k-1}^{2q}(y))\\
&= \operatorname{Tr}(a^{2q^{e-2}}\, S_{k-1}(y))\\
&= \operatorname{Tr}(b\, S_{k-1}(y)) \qquad\qquad (a^{2q^{e-2}} = b, \text{ i.e. } a = b^{2q})\\
&= \operatorname{Tr}\big(b\,(y + y^q + \cdots + y^{q^{k-2}})\big)\\
&= \operatorname{Tr}\big(y\,(b^{q^{3k}} + b^{q^{3k-1}} + \cdots + b^{q^{2k+2}})\big) \qquad\qquad \text{(Lemma 1)}\\
&= \operatorname{Tr}_{q^k/2}\big(y\,(c^{q^{2k+2}} + \cdots + c^{q^{3k}})\big), \qquad\qquad \text{where } c = \operatorname{Tr}_{q^e/q^k}(b) \neq 0.
\end{aligned}$$

Now consider $\gcd(\mathrm{x}^{q^{2k+2}} + \mathrm{x}^{q^{2k+3}} + \cdots + \mathrm{x}^{q^{3k}},\ \mathrm{x}^{q^k} + \mathrm{x})$. By [8, Theorem 3.62], we only need to consider $\gcd(\mathrm{x}^{2k+2} + \mathrm{x}^{2k+3} + \cdots + \mathrm{x}^{3k},\ \mathrm{x}^k + 1)$.

Since k is even, $\gcd(\mathrm{x}^{2k+2}+\mathrm{x}^{2k+3}+\cdots+\mathrm{x}^{3k},\ \mathrm{x}^k+1) = \gcd(1+\mathrm{x}+\cdots+\mathrm{x}^{k-2},\ \mathrm{x}^k+1) = 1$, which implies $c^{q^{2k+2}} + \cdots + c^{q^{3k}} \neq 0$, and hence there exists $y \in \mathbb{F}_{q^k}$ such that $\operatorname{Tr}\big(ag(x+y) - ag(x)\big) = \operatorname{Tr}_{q^k/2}\big(y\,(c^{q^{2k+2}} + \cdots + c^{q^{3k}})\big)$ is a nonzero constant, which is 1.

`Case 2.` Assume $\operatorname{Tr}_{q^e/q^k}(a) = 0$. Then $a = b + b^{q^k}$ for some $b \in \mathbb{F}_{q^e}$. Since $a \neq 0$, we have $b \notin \mathbb{F}_{q^k}$. For $x \in \mathbb{F}_{q^e}$, we have

$$\begin{aligned}
\operatorname{Tr}\big(ag(x)\big) &= \operatorname{Tr}\big(bg(x) + b^{q^k} g(x)\big)\\
&= \operatorname{Tr}\big(bg(x)\big) + \operatorname{Tr}\big(b^{q^k} g(x)\big)\\
&= \operatorname{Tr}\big(bg(x)\big) + \operatorname{Tr}\big(bg(x)^{q^{2k}}\big)\\
&= \operatorname{Tr}\big(bg(x) + bg(x)^{q^{2k}}\big)\\
&= \operatorname{Tr}\big(b(g(x) + g(x)^{q^{2k}})\big).
\end{aligned}$$

Note that $S_{2k}^{q^k} + S_{2k}^{q^{2k}} \equiv S_{2k} \pmod{\mathrm{x}^{q^e} - \mathrm{x}}$ and

$$S_{k+1}^2 + S_{k+1}^{2q^{2k}} \equiv x^2 + x^{2q^k} + S_{2k}^2 + S_{2k}^{2q^k} \pmod{\mathrm{x}^{q^e} - \mathrm{x}}.$$

Then we have

$$\begin{aligned} g(\mathrm{x}) + g(\mathrm{x})^{q^{2k}} &\equiv S_{k+1}^2 + S_{2k}^{q^k+1} + S_{k+1}^{2q^{2k}} + S_{2k}^{q^{2k}+1} \pmod{\mathrm{x}^{q^e} - \mathrm{x}} \\ &= S_{k+1}^2 + S_{k+1}^{2q^{2k}} + S_{2k} \cdot (S_{2k}^{q^k} + S_{2k}^{q^{2k}}) \\ &\equiv x^2 + x^{2q^k} + S_{2k}^2 + S_{2k}^{2q^k} + S_{2k}^2 \pmod{\mathrm{x}^{q^e} - \mathrm{x}} \\ &= (\mathrm{x} + \mathrm{x}^{q^k} + S_{2k}^{q^k})^2 \\ &= (S_{3k} + S_k^q)^2 \\ &\equiv (S_{2k}^{q^{k+1}})^2 \pmod{\mathrm{x}^{q^e} - \mathrm{x}}. \end{aligned} \tag{2}$$

So for $x \in \mathbb{F}_{q^e}$,

$$\begin{aligned} \mathrm{Tr}\big(ag(x)\big) &= \mathrm{Tr}\big(bS_{2k}(x)^{2q^{k+1}}\big) \\ &= \mathrm{Tr}\big(cS_{2k}(x)\big) && (b = c^{2q^{k+1}}) \\ &= \mathrm{Tr}\big(c(x + x^q + \cdots + x^{q^{2k-1}})\big) \\ &= \mathrm{Tr}\big(x(c^{q^{3k}} + c^{q^{3k-1}} + \cdots + c^{q^{k+1}})\big) && \text{(Lemma 2)}. \end{aligned}$$

Now consider $\gcd(\mathrm{x}^{q^{k+1}} + \mathrm{x}^{q^{k+2}} + \cdots + \mathrm{x}^{q^{3k}},\ \mathrm{x}^{q^{3k}} + \mathrm{x})$. By [8, Theorem 3.62], we only need to consider $\gcd(\mathrm{x}^{k+1} + \mathrm{x}^{k+2} + \cdots + \mathrm{x}^{3k},\ \mathrm{x}^{3k} + 1)$.

Since $\gcd(\mathrm{x}^{k+1} + \mathrm{x}^{k+2} + \cdots + \mathrm{x}^{3k},\ \mathrm{x}^{3k} + 1) = \gcd(1 + \mathrm{x} + \cdots + \mathrm{x}^{2k-1},\ \mathrm{x}^{3k} + 1) = \mathrm{x}^k + 1$, we see that for $z \in \mathbb{F}_{q^{3k}}$, we have $z^{q^{3k}} + z^{q^{3k-1}} + \cdots + z^{q^{k+1}} = 0$ if and only if $z \in \mathbb{F}_{q^k}$. Since $c \notin \mathbb{F}_{q^k}$, we have $c^{q^{3k}} + c^{q^{3k-1}} + \cdots + c^{q^{k+1}} \neq 0$. Therefore

$$\sum_{x \in \mathbb{F}_{q^e}} (-1)^{\mathrm{Tr}(ag(x))} = \sum_{x \in \mathbb{F}_{q^e}} (-1)^{\mathrm{Tr}(x(c^{q^{3k}} + c^{q^{3k-1}} + \cdots + c^{q^{k+1}}))} = 0. \qquad \square$$

3. Polynomial $g_{n,q}$

We first recall a fact:

Fact 3. ([2, Theorem 6.1]) *Let $q \geq 4$ be even, and let*

$$n = 1 + q^{a_1} + q^{b_1} + \cdots + q^{a_{q/2}} + q^{b_{q/2}},$$

where $a_i, b_i \geq 0$ are integers. Then

$$g_{n,q} = \sum_i S_{a_i} S_{b_i} + \sum_{i<j} (S_{a_i} + S_{b_i})(S_{a_j} + S_{b_j}).$$

Corollary 1. *Let $q = 4$, $e = 6$, and $n = 65921 = 1 + 2q^3 + q^4 + q^8$. Then*

$$g_{n,q} \equiv S_3^2 + S_4^{q^2+1} \pmod{\mathrm{x}^{q^e} - \mathrm{x}},$$

and $g_{n,q}$ is a PP of $\mathbb{F}_{q^e}$.

Proof. We write g_n for $g_{n,q}$. Let $a_1 = b_1 = 3, a_2 = 4$ and $b_2 = 8$ in Fact 3. Then we have

$$\begin{aligned} g_n &= S_3^2 + S_4 S_8 \\ &\equiv S_3^2 + S_4(S_6 + S_2) \pmod{\mathrm{x}^{q^e} - \mathrm{x}} \\ &= S_3^2 + S_4 S_4^{q^2} \\ &= S_3^2 + S_4^{q^2+1}. \end{aligned}$$

Let $k = 2$ in Theorem 1. Then it follows from Theorem 1 that g_n is a PP of $\mathbb{F}_{q^e}$. $\square$

Remark 1. Let k be odd and consider $g = S_{k+1}^2 + S_{2k}^{q^k+1}$. Note that g consists of even number of powers of x since S_{k+1} and S_{2k} consist of even number of powers of x. Thus the polynomial g defined in Theorem 1 is never a PP over $\mathbb{F}_{4^{3k}}$ when k is odd since $g(0) = g(1)$.

4. A Generalization

The following theorem is a generalization of Theorem 1.

Theorem 2. *Let q be a power of 2. Let $L \in \mathbb{F}_{q^{3k}}[x]$ be a 2-linearized polynomial such that*

(i) L permutes $\mathbb{F}_{q^k}$, and

(ii) $L + L^{q^{2k}} \equiv S_{2k}^2 + S_{2k}^{2q^{k+1}} \pmod{\mathrm{x}^{q^{3k}} - \mathrm{x}}$.

Then $L + S_{2k}^{q^k+1}$ is a PP of $\mathbb{F}_{q^{3k}}$.

Proof. Let $g = L + S_{2k}^{q^k+1}$. As in Case 1 in Theorem 1, assume $\mathrm{Tr}_{q^{3k}/q^k}(a) \neq 0$. By Fact 2, it suffices to show that there exists $y \in \mathbb{F}_{q^{3k}}$ such that $\mathrm{Tr}_{q^{3k}/2}\big(ag(x+y) - ag(x)\big)$ is a nonzero constant for all $x \in \mathbb{F}_{q^{3k}}$. We write $\mathrm{Tr} = \mathrm{Tr}_{q^{3k}/2}$.

Let $y \in \mathbb{F}_{q^k}^*$. Then for $x \in \mathbb{F}_{q^{3k}}$, we have

$$\begin{aligned} &\mathrm{Tr}\big(ag(x+y) - ag(x)\big) \\ &= \mathrm{Tr}\big[a\big(L(x+y) + S_{2k}^{q^k+1}(x+y) - L(x) - S_{2k}^{q^k+1}(x)\big)\big] \\ &= \mathrm{Tr}(a\,L(y)) \qquad (S_{2k}(y) = 0 \text{ since } y \in \mathbb{F}_{q^k}) \\ &= \mathrm{Tr}_{q^k/2}\big(L(y)\,\mathrm{Tr}_{q^{3k}/q^k}(a)\big). \end{aligned}$$

Since L permutes $\mathbb{F}_{q^k}$ and $y \in \mathbb{F}_{q^k}^*$, $L(y) \neq 0$. Thus we choose $y \in \mathbb{F}_{q^k}$ such that $\mathrm{Tr}_{q^k/2}\big(L(y)\,\mathrm{Tr}_{q^{3k}/q^k}(a)\big) \neq 0$. In Case 2 in Theorem 1, (2) still holds because of condition 2. $\square$

Note. Let $q = 4$, k be even and $L = S_{k+1}^2$. Then $L = x^2 \circ S_{k+1}$. Since $\gcd(2, q^k - 1) = 1$, x^2 permutes $\mathbb{F}_{q^k}$. For $x \in \mathbb{F}_{q^k}$, we have $S_{k+1} = x^q + x^{q^2} + \cdots + x^{q^{k-1}}$. Note that when k is odd S_{k+1} does not permute $\mathbb{F}_{q^k}$ since $S_{k+1}(0) = S_{k+1}(1)$.

Now consider $\gcd(\mathrm{x}^q + \mathrm{x}^{q^2} + \cdots + \mathrm{x}^{q^{k-1}}, \ \mathrm{x}^{q^k} + 1)$. By [8, Theorem 3.62], we only need to consider $\gcd(\mathrm{x}+\mathrm{x}^2+\cdots+\mathrm{x}^{k-1}, \ \mathrm{x}^k+1)$. Since $\gcd(\mathrm{x}+\mathrm{x}^2+\cdots+\mathrm{x}^{k-1}, \ \mathrm{x}^k+1) = 1$, S_{k+1} permutes $\mathbb{F}_{q^k}$. Thus L satisfies condition (i) in Theorem 2.

For (ii), we have

$$
\begin{aligned}
L + L^{q^{2k}} &= S_{k+1}^2 + \left(S_{k+1}^2\right)^{q^{2k}} \\
&= (x + x^q + x^{q^2} + \cdots + x^{q^k})^2 + (x^2 + x^{2q} + x^{2q^2} + \cdots + x^{2q^k})^{q^{2k}} \\
&= (x + x^q + x^{q^2} + \cdots + x^{q^k})^2 + (x^{q^{2k}} + x^{q^{2k+1}} + x^{q^{2k+2}} + \cdots + x^{q^{3k}})^2 \\
&= (x + x^q + x^{q^2} + \cdots + x^{q^k} + x^{q^{2k}} + x^{q^{2k+1}} + x^{q^{2k+2}} + \cdots + x^{q^{3k}})^2 \\
&\equiv S_{2k}^2 + S_{2k}^{2q^{k+1}} \pmod{x^{q^{3k}} - x}.
\end{aligned}
$$

Acknowledgement. The author is grateful to the referee and the editor for useful comments and suggestions.

References

[1] N. Fernando and X. Hou, From r-linearized polynomial equations to r^m-linearized polynomial equations, *Finite Fields Appl.* **37** (2016), 14 – 27.

[2] N. Fernando, X. Hou, and S. D. Lappano, A new approach to permutation polynomials over finite fields, II, *Finite Fields Appl.* **22** (2013), 122 – 158.

[3] N. Fernando, X. Hou, and S. D. Lappano, Permutation polynomials over finite fields involving $x + x^q + \cdots + x^{q^{a-1}}$, *Discrete Math.* **315** (2014), 173 – 184.

[4] D. Görcsös, G. Horváth, and A. Mészáros, Permutation polynomials over finite rings, *Finite Fields Appl.* **49** (2018), 198 – 211.

[5] X. Hou, Two classes of permutation polynomials over finite fields, *J. Combin. Theory Ser. A* **118** (2011), 448 – 454.

[6] X. Hou, A new approach to permutation polynomials over finite fields, *Finite Fields Appl.* **18** (2012), 492 – 521.

[7] X. Hou, Proof of a conjecture on permutation polynomials over finite fields, *Finite Fields Appl.* **24** (2013) 192 – 195.

[8] R. Lidl and H. Niederreiter, *Finite Fields*, 2nd ed., Cambridge Univ. Press, Cambridge, 1997.

[9] R. L. Rivest, Permutation polynomials modulo 2^w, *Finite Fields Appl.* **7** (2001), no. 2, 287 – 292.

[10] P. Yuan and C. Ding, Permutation polynomials of the form $L(x) + S_{2k}^a + S_{2k}^b$ over $\mathbb{F}_{q^{3k}}$, *Finite Fields Appl.* **29** (2014), 106 – 117.

Appendix

The following table is a part of [1, Table 1] of desirable triples $(n, e; q)$ with $q = 4$, $e = 6$ and $w_4(n) > 4$, where $w_4(n)$ is the base 4 weight of n.

Table 1: Desirable triples $(n, 6; 4)$, $w_4(n) > 4$

e	n	base 4 digits of n	reference
6	4361	1,2,0,0,1,0,1	[2, Theorem 6.12]
6	6161	1,0,1,0,0,2,1	[2, Theorem 6.12]
6	6401	1,0,0,0,1,2,1	[2, Theorem 6.12]
6	8227	3,0,2,0,0,0,2	[2, Theorem 6.10]
6	8707	3,0,0,0,2,0,2	[2, Theorem 6.11]
6	12299	3,2,0,0,0,0,3	[2, Theorem 6.6]
6	12307	3,0,1,0,0,0,3	[2, Theorem 6.8]
6	14339	3,0,0,0,0,2,3	[2, Theorem 6.6]
6	37121	1,0,0,0,1,0,1,2	[2, Theorem 6.12]
6	65801	1,2,0,0,1,0,0,0,1	[2, Corollary 6.16]
6	65921	1,0,0,2,1,0,0,0,1	Theorem 1
6	66307	3,0,0,0,3,0,0,0,1	[7, Theorem 1.1]
6	135209	1,2,2,0,0,0,1,0,2	
6	135217	1,0,3,0,0,0,1,0,2	[1, Proposition 3.4]
6	135457	1,0,2,0,1,0,1,0,2	[1, Proposition 3.7]
6	137249	1,0,2,0,0,2,1,0,2	
6	8388607	3,3,3,3,3,3,3,3,3,3,3,1	[6, Proposition 3.1]

2

AUGMENTED HILBERT SERIES OF NUMERICAL SEMIGROUPS

Jeske Glenn
Mathematics Department, University of Oregon, Eugene, Oregon
jeskeg@uoregon.edu

Christopher O'Neill
Mathematics Department, San Diego State University, San Diego, California
cdoneill@sdsu.edu

Vadim Ponomarenko
Mathematics Department, San Diego State University, San Diego, California
vponomarenko@sdsu.edu

Benjamin Sepanski
Mathematics Department, Baylor University, Waco, Texas
ben_sepanski@baylor.edu

Abstract

A numerical semigroup S is a subset of the non-negative integers containing 0 that is closed under addition. The Hilbert series of S (a formal power series equal to the sum of terms t^n over all $n \in S$) can be expressed as a rational function in t whose numerator is characterized in terms of the topology of a simplicial complex determined by membership in S. In this paper, we obtain analogous rational expressions for the related power series whose coefficient of t^n equals $f(n)$ for one of several semigroup-theoretic invariants $f : S \to \mathbb{R}$ known to be eventually quasipolynomial.

1. Introduction

A *numerical semigroup* is a subset $S \subset \mathbb{Z}_{\geq 0}$ containing 0 that is closed under addition and has finite complement, and a *factorization* of an element $n \in S$ is an expression of n as a sum of generators of S. A clear trend that has emerged in the study of numerical semigroups is the eventually quasipolynomial behavior of arithmetic invariants derived from their factorization structure [14]. More specifically, each of these invariants (which we call *S-invariants*) is a function assigning to each element $n \in S$ a value determined by the possible factorizations of n in S. This includes invariants from discrete optimization such as maximum and minimum factorization length [2], distinct factorization length count [11], and maximum and minimum 0-norm [1], as well as more semigroup-theoretic invariants like the delta set [7], ω-primality [12], and the catenary degree [6], each of which agrees with a quasipolynomial for large input.

When (eventually) quasipolynomial functions arise in combinatorial settings, there are several potential ways to study them: (i) directly, using tools specific to the setting in question; (ii) via combinatorial commutative algebra, using Hilbert functions of graded modules; and (iii) via rational generating functions. Approaches (ii) and (iii) were largely pioneered by Stanley [17], among others, and carry with them powerful algebraic tools. The eventually quasipolynomial behavior of each semigroup invariant mentioned above was initially examined using standard semigroup-theoretic tools, and more recently an approach using Hilbert functions was developed [11]. The goal of this paper is to initiate the use of approach (iii) in studying S-invariants.

To date, rational generating functions have been used to study several aspects of numerical semigroups [4, 8], primarily using the *Hilbert series*

$$\mathcal{H}(S;t) = \sum_{n \in S} t^n = \frac{\mathcal{K}(S;t)}{(1-t^{n_1})\cdots(1-t^{n_k})}$$

associated to each numerical semigroup $S = \langle n_1, \ldots, n_k \rangle$. A natural consequence of the Hilbert syzygy theorem from commutative algebra [10] states that the numerator $\mathcal{K}(S;t)$ in the second expression above is a polynomial in t whose coefficients are obtained from the graded Betti numbers of the defining toric ideal of S. An alternative characterization of the coefficients of $\mathcal{K}(S;t)$ (stated formally in Theorem 2) uses the topology of a simplicial complex determined by membership in S [5]. One of the key selling points of the latter characterization is that it is given entirely in terms of the underlying semigroup S, without the theoretical overhead often necessary when incorporating commutative algebra techniques.

The primary goal of this paper is to obtain analogous rational expressions for various *augmented Hilbert series*, which we define to be series of the form

$$\mathcal{H}_f(S;t) = \sum_{n \in S} f(n)t^n$$

where f is some S-invariant admitting eventually quasipolynomial behavior. We give two such expressions: (i) when $f(n)$ counts the number of distinct factorization lengths of n (Proposition 1) and (ii) when $f(n)$ is the maximum or minimum factorization length of n (Theorem 3). Examples 3 and 4 illustrate the need for distinct rational forms for these invariants. We also specify how to obtain the *dissonance point* of each quasipolynomial function f (i.e., the optimal bound on the start of quasipolynomiality) from the numerator of its rational generating function (Theorem 4). Lastly, we examine these rational expressions under the operation of gluing numerical semigroups (Section 5) and give a closed form for each rational expression in the special case when S has 2 generators (Section 6).

2. Background

Definition 1. A *numerical semigroup* S is a cofinite, additive subsemigroup of $\mathbb{Z}_{\geq 0}$. When we write $S = \langle n_1, \ldots, n_k \rangle$ in terms of generators, we assume $n_1 < \cdots < n_k$. The *Frobenius number* of S is the largest integer $\mathsf{F}(S)$ lying in the complement of S. A *factorization* of $n \in S$ is an expression

$$n = a_1 n_1 + \cdots + a_k n_k$$

of n as a sum of generators of S, and the *length* of a factorization is the sum $a_1 + \cdots + a_k$. The *set of factorizations* of $n \in S$ is

$$\mathsf{Z}_S(n) = \{a \in \mathbb{Z}_{\geq 0}^k : n = a_1 n_1 + \cdots + a_k n_k\}$$

and the *length set* of n is the set

$$\mathsf{L}_S(n) = \{a_1 + \cdots + a_k : a \in \mathsf{Z}_S(n)\}$$

of all possible factorization lengths of n. The *maximum and minimum factorization length functions*, and the *length denumerant* function, are defined as

$$\mathsf{M}_S(n) = \max \mathsf{L}_S(n) \qquad \mathsf{m}_S(n) = \min \mathsf{L}_S(n) \qquad \text{and} \qquad \mathsf{l}_S(n) = |\mathsf{L}_S(n)|,$$

respectively. The *Apéry set* of an element $n \in S$ is the set

$$\mathrm{Ap}(S; n) = \{m \in S : m - n \notin S\}.$$

It can be easily shown that $|\mathrm{Ap}(S; n)| = n$ for any $n \in S$.

Notation 1. Unless otherwise stated, thoughout the paper, $S = \langle n_1, \ldots, n_k \rangle$ denotes a numerical semigroup with a fixed generating set $n_1 < \cdots < n_k$.

Definition 2. A function $f : \mathbb{Z} \to \mathbb{R}$ is an S*-invariant* if $f(n) = 0$ for all $n \notin S$.

Definition 3. A function $f : \mathbb{Z} \to \mathbb{R}$ is an r*-quasipolynomial* of degree α if

$$f(n) = a_\alpha(n) n^\alpha + \cdots + a_1(n) n + a_0(n)$$

for periodic functions $a_0, \ldots, a_\alpha$, whose periods all divide r, with a_α not identically 0. We say f is *eventually quasipolynomial* if the above equality holds for all $n \gg 0$.

Theorem 1 ([2, 11]). *For sufficiently large* $n \in S$,

$$\mathsf{M}_S(n + n_1) = \mathsf{M}_S(n) + 1, \qquad \mathsf{m}_S(n + n_k) = \mathsf{m}_S(n) + 1, \\ \text{and} \qquad \mathsf{l}_S(n + n_1 n_k) = \mathsf{l}_S(n) + \tfrac{1}{d}(n_k - n_1),$$

where $d = \gcd\{n_i - n_{i-1} : i = 2, \ldots, d\}$. *In particular, the* S*-invariants* M_S*,* m_S*, and* l_S *are each eventually quasilinear.*

Definition 4. The *Hilbert series* of S is the formal power series

$$\mathcal{H}(S;t) = \sum_{n \in S} t^n \in \mathbb{Z}[\![t]\!].$$

Given $n \in S$, the *squarefree divisor complex* Δ_n is a simplicial complex on the ground set $[k] = \{1, \ldots, k\}$ where $F \in \Delta_n$ if $n - n_F \in S$, where $n_F = \sum_{i \in F} n_i$. The *Euler characteristic* of a simplicial complex Δ is the alternating sum

$$\chi(\Delta) = \sum_{F \in \Delta_n} (-1)^{|F|}.$$

Remark 1. The definition of Euler characteristic above differs slightly from the usual topological definition, but has the advantage that $\chi(\Delta) = 0$ for any contractible simplicial complex Δ.

Theorem 2 ([5]). *The Hilbert series of* S *can be written as*

$$\mathcal{H}(S;t) = \sum_{n \in S} t^n = \frac{\sum_{a \in \mathrm{Ap}(S;n_1)} t^a}{1 - t^{n_1}} = \frac{\sum_{m \in S} \chi(\Delta_m) t^m}{(1 - t^{n_1}) \cdots (1 - t^{n_k})},$$

where both numerators have finitely many terms.

Example 1. For $S = \langle 6, 9, 20 \rangle$, Theorem 2 yields

$$\mathcal{H}(S;t) = \frac{1 + t^9 + t^{20} + t^{29} + t^{40} + t^{49}}{1 - t^6} = \frac{1 - t^{18} - t^{60} + t^{78}}{(1 - t^6)(1 - t^9)(1 - t^{20})}.$$

Here, $\mathrm{Ap}(S;6) = \{0, 49, 20, 9, 40, 29\}$ and each entry is distinct modulo 6. The elements

$$18 = 3 \cdot 6 = 2 \cdot 9 \qquad \text{and} \qquad 60 = 4 \cdot 6 + 4 \cdot 9 = 3 \cdot 20$$

are, respectively, the first element that can be factored using 6's and 9's and the first element that can be factored using 6's, 9's, and 20's. In particular, these two elements encode minimal relations between the generators of S, viewed as minimal "trades" from one factorization to another. Moreover,

$$78 = 7 \cdot 6 + 4 \cdot 9 = 2 \cdot 9 + 3 \cdot 20$$

is the first element in which two distinct sequences of trades between factorizations are possible: one can perform the exchange $3 \cdot 6 \rightsquigarrow 2 \cdot 9$ followed by $4 \cdot 6 + 4 \cdot 9 \rightsquigarrow 3 \cdot 20$, or these trades can be applied in the reverse order. This represents a "relation between minimal relations". These properties are encoded in the element's respective squarefree divisor complexes, since Δ_{18} and Δ_{60} are each disconnected, and Δ_{78} is connected but has nontrivial 1-dimensional homology.

Remark 2. One remarkable aspect of Theorem 2 is that simple algebraic manipulation of the rational expression of the Hilbert series reveals additional structural information about the underlying semigroup. Indeed, cancelling all common factors in Example 1 yields $\mathsf{P}_S(t)/(1-t)$ (see [8]), where

$$\begin{aligned}\mathsf{P}_S(t) = 1 - t &+ t^6 - t^7 + t^9 - t^{10} + t^{12} - t^{13} + t^{15} - t^{16} + t^{18} \\ &- t^{19} + t^{20} - t^{22} + t^{24} - t^{25} + t^{26} - t^{28} + t^{29} - t^{31} + t^{32} \\ &- t^{34} + t^{35} - t^{37} + t^{38} - t^{43} + t^{44}\end{aligned}$$

has significantly more terms than the numerator of either form in Theorem 2. This is not a coincidence: since they represent the same power series, the fewer terms that appear in a particular expression, the more information each term must encode.

3. Numerators of Augmented Hilbert Series

In this section, we formally introduce augmented Hilbert series of a general semigroup invariant f (Definition 5), present two rational expressions in the spirit of Theorem 2 (Proposition 1 and Theorem 3), and illustrate and compare their use when f is one of the S-invariants appearing in Theorem 1 (Examples 3 and 4).

Definition 5. Fix an S-invariant f. The *augmented Hilbert series* of S with respect to f is the formal power series

$$\mathcal{H}_f(S;t) = \sum_{n \in S} f(n) t^n.$$

Given $n \in S$, the *weighted Euler characteristic of* Δ_n is defined as

$$\chi_f(\Delta_n) = \sum_{F \in \Delta_n} (-1)^{|F|} f(n - n_F),$$

and the *augmented Euler characteristic of* Δ_n is defined as

$$\widehat{\chi}_f(\Delta_n) = \sum_{F \in \Delta} (-1)^{|F|} (f(n - n_F) + |F|).$$

Example 2. Let $S = \langle 6, 9, 20 \rangle$. The complex Δ_{138} is given in Figure 1, and each face F is labeled with the value $\mathsf{M}_S(138 - n_F)$. Together with $\mathsf{M}_S(138) = 23$ as the label for the empty face, we obtain

$$\chi_{\mathsf{M}_S}(\Delta_{138}) = \widehat{\chi}_{\mathsf{M}_S}(\Delta_{138}) = 0,$$

in part because the label of each face containing the vertex 6 matches its label on the face obtained by deleting 6.

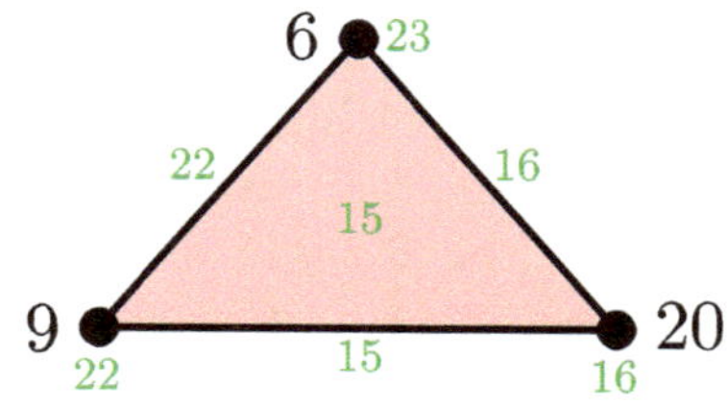

Figure 1: The complex Δ_{138} in $S = \langle 6, 9, 20\rangle$ in Example 2.

Proposition 1. *Fix an S-invariant f. For any fixed $p \in \mathbb{Z}_{\geq 1}$, we have*

$$\mathcal{H}_f(S;t) = \frac{\sum_{n\in S}(f(n) - 2f(n-p) + f(n-2p))t^n}{(1-t^p)^2} = \frac{\sum_{n\in S}\chi_f(\Delta_n)t^n}{\prod_{i=1}^k(1-t^{n_i})}.$$

Proof. Clearing respective denominators yields

$$\begin{aligned}(1-t^p)^2\sum_{n\in S} f(n)t^n &= \sum_{n\in S} f(n)t^n - \sum_{n\in S} 2f(n)t^{n+p} + \sum_{n\in S} f(n)t^{n+2p}\\ &= \sum_{n\in S}(f(n) - 2f(n-p) + f(n-2p))t^n,\end{aligned}$$

which proves the first equality, and the second equality follows from

$$\begin{aligned}\sum_{m\in S}\chi_f(\Delta_m)t^m &= \sum_{m\in S}\sum_{\substack{F\subseteq[k]\\ m-n_F\in S}}(-1)^{|F|}f(m-n_F)t^m = \sum_{A\subseteq[k]}\sum_{n\in S}(-1)^{|A|}f(n)t^{n+n_A}\\ &= \Big(\sum_{A\subseteq[k]}(-1)^{|A|}t^{n_A}\Big)\Big(\sum_{n\in S} f(n)t^n\Big) = \Big(\prod_{i=1}^k(1-t^{n_i})\Big)\Big(\sum_{n\in S} f(n)t^n\Big),\end{aligned}$$

where the second step uses the substitution $m = n + n_A$. □

Example 3. For $S = \langle 9, 10, 23\rangle$, we have

$$\begin{aligned}\sum_{n\in S}\chi_{\mathsf{M}_S}(n)t^n = t^9 + t^{10} + t^{18} + t^{20} + t^{23} + t^{27} + t^{30} + t^{36} + t^{40} + t^{45}\\ - t^{46} - 3t^{50} + t^{54} - t^{55} - t^{56} - t^{59} - 4t^{63} - t^{64} - t^{66}\\ - t^{68} + 2t^{73} - t^{76} - t^{77} + 3t^{86} - t^{90} + t^{113},\end{aligned}$$

whereas

$$\sum_{n\in S}\widehat{\chi}_{\mathsf{M}_S}(n)t^n = -2t^{46} - 4t^{50} - 5t^{63} + 5t^{73} + 6t^{86} - t^{90} + t^{113}.$$

This difference in number of terms occurs in nearly every example of $\mathcal{H}_{\mathsf{M}_S}(S;t)$ the authors have computed, and illustrates the primary reason for Theorem 3: filtering many of the extraneous terms from the first expression above. At play here is the philosophy discussed in Remark 2, namely that expressions with fewer terms necessarily encode more combinatorial information per term.

Example 4. For $S = \langle 9, 10, 23 \rangle$ as in Example 3, the polynomials

$$\sum_{n \in S} \chi_{\mathsf{l}_S}(n) t^n = 1 - t^{140}$$

and

$$\sum_{n \in S} \widehat{\chi}_{\mathsf{l}_S}(n) t^n = 1 - t^9 - t^{10} - t^{18} - t^{20} - t^{23} - t^{27} - t^{30} - t^{36} - t^{40} - t^{45} - t^{46} - t^{50} - t^{54} + t^{55} + t^{56} + t^{59} - t^{63} + t^{64} + t^{66} + t^{68} + 3t^{73} + t^{76} + t^{77} + 3t^{86} - t^{140}$$

also differ greatly in the number of terms, but in the opposite direction. This is in part because $\mathsf{L}(0) = \{0\}$ for every numerical semigroup S, as the lack of a constant term in $\mathcal{H}_{\mathsf{l}_S}(S;t)$ adds many erroneous terms in the numerator of Proposition 1 that the constant term 1 in $\mathcal{H}_{\mathsf{l}_S}(S;t)$ avoids. Additionally, this example illustrates that examining S-invariants via generating functions will sometimes require specialized expressions, rather than a "one-size-fits-all" characterization.

Notation 2. In what follows, we make heavy use of the power series

$$z(t) = \prod_{i=1}^{k} \frac{1}{1 - t^{n_i}} \qquad \text{and} \qquad \lambda(t) = \sum_{i=1}^{k} \frac{t^{n_i}}{1 - t^{n_i}},$$

the second of which often occurs in the form $z(t)\lambda(t)$ (for instance, in Theorem 3). The coefficient of t^n in the series $z(t)$ (usually notated as $\mathcal{H}_{\partial}(S;t)$ in the literature) equals the number of factorizations of $n \in S$ (known as the *denumerant* of n), while the coefficients of $z(t)\lambda(t)$ are described in Lemma 1.

Lemma 1. *The power series $z(t)\lambda(t)$ is given by*

$$z(t)\lambda(t) = \sum_{n \in S} \ell(n) t^n,$$

where $\ell(n)$ denotes the sum of the lengths of every factorization of $n \in S$.

Proof. Rewrite $z(t)\lambda(t)$ as

$$z(t)\lambda(t) = \sum_{i=1}^{k} \frac{t^{n_i}}{(1 - t^{n_i})^2} \left(\prod_{j \neq i} \frac{1}{1 - t^{n_j}} \right) = \sum_{i=1}^{k} \left(\sum_{a_i \geq 0} a_i t^{a_i n_i} \right) \left(\prod_{j \neq i} \left(\sum_{a_j \geq 0} t^{a_j n_j} \right) \right).$$

In the final expression above, when expanding the product inside the outermost sum, the term t^n appears once for each factorization of n in S, with coefficient equal to the number of copies of n_i appearing in that factorization. As such,

$$z(t)\lambda(t) = \sum_{i=1}^{k} \sum_{n \in S} \sum_{a \in \mathsf{Z}(n)} a_i t^n = \sum_{n \in S} \sum_{a \in \mathsf{Z}(n)} |a| t^n = \sum_{n \in S} \ell(n) t^n,$$

as desired. □

Theorem 3. *Fix an S-invariant f. The augmented Hilbert series of f is given by*

$$\begin{aligned}\mathcal{H}_f(S;t) &= z(t)\lambda(t)\sum_{n\in S}\chi(\Delta_n)t^n + z(t)\sum_{n\in S}\widehat{\chi}_f(\Delta_n)t^n \\ &= \lambda(t)\mathcal{H}(S;t) + \frac{\sum_{n\in S}\widehat{\chi}_f(\Delta_n)t^n}{(1-t^{n_1})\cdots(1-t^{n_k})}.\end{aligned}$$

Proof. Multiplying both sides by the denominator of $z(t)$, Proposition 1 implies

$$\left(\prod_{i=1}^{k}(1-t^{n_i})\right)\sum_{n\in S} f(n)t^n = \sum_{m\in S}\sum_{F\in\Delta_m}(-1)^{|F|}f(m-n_F)t^m.$$

In the second term on the right-hand side of the claimed equality, we have

$$\begin{aligned}\sum_{m\in S}\widehat{\chi}_f(\Delta_m)t^m &= \sum_{m\in S}\sum_{F\in\Delta_m}(-1)^{|F|}(f(m-n_F)+|F|)t^m \\ &= \sum_{m\in S}\sum_{F\in\Delta_m}(-1)^{|F|}f(m-n_F)t^m + \sum_{m\in S}\sum_{F\in\Delta_m}(-1)^{|F|}|F|t^m,\end{aligned}$$

so it suffices to show that

$$\left(\sum_{i=1}^{k}\frac{t^{n_i}}{1-t^{n_i}}\right)\sum_{m\in S}\sum_{G\in\Delta_m}(-1)^{|G|}t^m + \sum_{m\in S}\sum_{F\in\Delta_m}(-1)^{|F|}|F|t^m = 0.$$

Indeed, multiplying the first part by $\prod_{j=1}^{k}(1-t^{n_j})$ yields

$$\begin{aligned}\sum_{i=1}^{k}t^{n_i}\left(\prod_{j\neq i}^{k}(1-t^{n_j})\right)\sum_{m\in S}\sum_{G\in\Delta_m}(-1)^{|G|}t^m &= \sum_{m\in S}\sum_{i=1}^{k}\sum_{\substack{A\subseteq[k]\\ i\in A}}(-1)^{|A|-1}\sum_{G\in\Delta_m}(-1)^{|G|}t^{m+n_A} \\ &= \sum_{m\in S}\sum_{A\subseteq[k]}(-1)^{|A|-1}|A|\sum_{G\in\Delta_m}(-1)^{|G|}t^{m+n_A} \\ &= \sum_{m\in S}\sum_{F\in\Delta_m}(-1)^{|F|-1}|F|\sum_{G\in\Delta_m}(-1)^{|G|}t^m\end{aligned}$$

and multiplying the second part by the same factor yields

$$\begin{aligned}\left(\prod_{j=1}^{k}(1-t^{n_j})\right)\sum_{m\in S}\sum_{F\in\Delta_m}(-1)^{|F|}|F|t^m &= \sum_{m\in S}\sum_{A\subseteq[k]}(-1)^{|A|}\sum_{F\in\Delta_m}(-1)^{|F|}t^{m+n_A} \\ &= \sum_{m\in S}\sum_{G\in\Delta_m}(-1)^{|G|}\sum_{F\in\Delta_m}(-1)^{|F|}t^m\end{aligned}$$

which completes the proof. □

4. The Dissonance Point

The numerator of each rational expression in Proposition 1 and Theorem 3 has finite degree when f is any S-invariant listed in Theorem 1. This follows from Theorem 1 and general facts from the theory of generating function [18], but we prove this fact in Proposition 2 using weighted and augmented Euler characteristics, as a demonstration of their utility.

The other main result of this section is Theorem 4, which demonstrates that when $f = \mathsf{M}_S$ or $f = \mathsf{m}_S$, we can recover from the degree of $\sum_{n \in S} \widehat{\chi}_f(\Delta_n)t^n$ the minimum integer input after which f becomes truly quasipolynomial. Note that by Proposition 1, this fact is immediate if the coefficients $\chi_f(\Delta_n)$ are used in place of $\widehat{\chi}_f(\Delta_n)$ for *any* eventually quasipolynomial function f.

Definition 6. Fix an S-invariant f that agrees with a quasipolynomial function $g : \mathbb{Z} \to \mathbb{R}$ for sufficiently large input values. The *dissonance point of* f is the largest integer $n \geq \mathsf{F}(S)$ such that $f(n) \neq g(n)$. We say the semigroup S is f*-harmonic* if $f(n) = g(n)$ for every $n \in S$.

Example 5. Let $S = \langle 9, 10, 23\rangle$ from Example 3. The dissonance point of M_S is 71, since $\mathsf{Z}(71) = \{(2, 3, 1)\}$ but $\mathsf{Z}(80) = \{(3, 3, 1), (0, 8, 0)\}$, so

$$8 = \mathsf{M}_S(80) > \mathsf{M}_S(71) + 1 = 7.$$

In particular, the longest factorization of 80 does not have any copies of the first generator. Generally, longer factorizations will involve more small generators than large generators, but even though $(3, 3, 1)$ has more copies of the smallest generator, it has enough larger generators to afford $(0, 8, 0)$ higher efficiency. This is exacerbated by the fact that 9 and 10 are close together, while 23 is significantly larger than both.

On the other hand, $S = \langle 6, 9, 20\rangle$ is M_S-harmonic, since

$$\mathsf{M}_S(n + 6) = \mathsf{M}_S(n) + 1$$

for every $n \in S$ by Theorem 1 and exhaustive computation for small n using, for instance, the `GAP` package `numericalsgps` [9].

Proposition 2. *If f is one of the S-invariants appearing in Theorem 1, then $\sum_{n \geq 0} \chi_f(\Delta_n)t^n$ and $\sum_{n \geq 0} \widehat{\chi}_f(\Delta_n)t^n$ have finitely many terms.*

Proof. We must show

$$\chi_f(\Delta_n) = \widehat{\chi}_f(\Delta_n) = 0$$

for all sufficiently large n. If $f = \mathsf{m}_S$, then f satisfies $f(n + n_k) = f(n) + 1$ for sufficiently large n, so provided that $n > \mathsf{F}(S) + n_{[k]}$ also holds, we have

$$\begin{aligned}\chi_f(\Delta_n) &= \sum_{F\subseteq[k]} (-1)^{|F|} f(n-n_F) \\ &= \sum_{F\subseteq[k-1]} (-1)^{|F|} f(n-n_F) + \sum_{F\subseteq[k-1]} (-1)^{|F|+1} f(n-n_F-n_k) \\ &= \sum_{F\subseteq[k-1]} (-1)^{|F|} (f(n-n_F) - f(n-n_F-n_k)) = \sum_{F\subseteq[k-1]} (-1)^{|F|} = 0.\end{aligned}$$

Additionally,

$$\widehat{\chi}_f(\Delta_n) - \chi_f(\Delta_n) = \sum_{F\subseteq[k]} (-1)^{|F|}|F| = 0,$$

which proves $\widehat{\chi}_f(\Delta_n) = 0$. Replacing n_k with n_1 throughout the above argument proves the same equalities hold for $f = \mathsf{M}_S$, leaving only the case $f = \mathsf{l}_S$. By Theorem 1, we have $f(n) = \frac{1}{d}(\mathsf{M}(n) - \mathsf{m}(n)) - l_0(n)$ for large n, where l_0 is some $n_1 n_k$-periodic function. As such,

$$\sum_{n\in S} f(n)t^n = \frac{1}{d}\Big(\sum_{n\in S} \mathsf{M}(n)t^n - \sum_{n\in S} \mathsf{m}(n)t^n\Big) - \sum_{n\in S} l_0(n)t^n,$$

and by Proposition 1, each power series on the right-hand side is rational with denominator dividing $\prod_{i=1}^{k}(1-t^{n_i})$. This proves $\chi_f(\Delta_n) = 0$ for large n. Just as above, $\widehat{\chi}_f(\Delta_n) = 0$ then readily follows for large n, so the proof is complete. □

Theorem 4. *If $f = \mathsf{M}_S$ or $f = \mathsf{m}_S$, then the dissonance point of f is $d - n_{[k]}$, where*

$$d = \deg\Big(\sum_{n\geq 0} \widehat{\chi}_f(\Delta_n)t^n\Big).$$

Proof. Suppose $f = \mathsf{m}_S$, and let $m \in S$ denote the largest element of S such that $\mathsf{m}(m - n_k) + 1 \neq \mathsf{m}(m)$. Clearly $m \leq d - n_{[k]}$, since each $n > m + n_{[k]}$ must have $\widehat{\chi}_f(\Delta_n) = 0$ by the proof of Proposition 2. Moreover,

$$\widehat{\chi}_f(\Delta_d) = \sum_{F\subseteq[k]} (-1)^{|F|}\big(f(d-n_F) + |F|\big) = (-1)^k\big(1 + f(d-n_{[k]}) - f(d-n_{[k-1]})\big)$$

is nonzero, proving the claim when $f = \mathsf{m}_S$. The case $f = \mathsf{M}_S$ follows analogously. □

5. Augmented Hilbert Series of Gluings

Gluing (Definition 7) is a method of combining two numerical semigroups S_1 and S_2 to obtain a numerical semigroup $S = d_1S_1 + d_2S_2$ whose factorization structure can be expressed explicitly in terms of the factorizations of S_1 and S_2 [16]. Several families of numerical semigroups of interest in the literature (e.g., complete intersection, supersymmetric, telescopic) are described in terms of gluings. Moreover, the Hilbert series of S can be concisely expressed as

$$\mathcal{H}(S;t) = (1 - t^{d_1 d_2})\mathcal{H}(S_1; t^{d_1})\mathcal{H}(S_2; t^{d_2})$$

in terms of the Hilbert series of S_1 and S_2 (see [8]).

One might hope that a similar relation can be obtained for augmented Hilbert series, but unfortunately this is not the case. In fact, even gluing two harmonic numerical semigroups need not yield a harmonic numerical semigroup; see Example 6. However, if the gluing is sufficiently well-behaved (see Definition 8), then an expression for the augmented Hilbert series of S can be obtained (Theorem 5).

Remark 3. All results and definitions in this section are stated in terms of the maximum factorization length S-invariant M_S, but analogous results (with analogous proofs) also hold for the minimum factorization length S-invariant m_S.

Definition 7. Fix numerical semigroups S_1 and S_2, and elements $d_1 \in S_2$ and $d_2 \in S_1$ that are not minimal generators of their respective semigroups. We say $S = d_1 S_1 + d_2 S_2$ is a *gluing of* S_1 *and* S_2 if $\gcd(d_1, d_2) = 1$.

Example 6. Let $S_1 = \langle 6, 10, 15 \rangle$ and $S_2 = \langle 5, 7 \rangle$, and let

$$S = 23S_1 + 27S_2 = \langle 138, 230, 345, 135, 162 \rangle.$$

Both S_1 and S_2 are M_S-harmonic (and supersymmetric, one of the most well-behaved families of numerical semigroups under gluing), but the glued numerical semigroup S fails to satisfy $\mathsf{M}_S(n + n_1) = \mathsf{M}_S(n) + 1$ for each n in the set

$$\left\{ \begin{array}{c} 831,\ 969,\ 993, 1061, 1131, 1155, 1199, 1223, 1291, 1293, \\ 1317, 1361, 1385, 1429, 1453, 1455, 1479, 1523, 1547, 1591, \\ 1615, 1617, 1685, 1709, 1753, 1777, 1847, 1915, 1939,\ 2077 \end{array} \right\}$$

(this can be verified using the `GAP` package `numericalsgps` [9]). The primary issue is that the images of the smallest generators of S_1 and S_2 are relatively close in S, a property that was observed by the second author when writing [2] to correlate with a large dissonance point for maximum factorization length.

Definition 8. Resume notation from Definition 7. We say S is a M_S*-harmonic gluing* if every $n \in S$ satifies $\mathsf{M}_S(n) = \mathsf{M}_{S_1}(n') + \mathsf{M}_{S_2}(n'')$, where $n = d_1 n' + d_2 n''$ for $n' \in S_1$, $n'' \in S_2$, and n' maximal among all such expressions. Note that this property is dependent on the order of S_1 and S_2. We define an m_S*-harmonic gluing* analogously, where the expression $n = d_1 n' + d_2 n''$ is chosen so that n'' is maximal.

Theorem 5. *If $S = d_1 S_1 + d_2 S_2$ is an M_S-harmonic gluing, then*

$$\mathcal{H}_{\mathsf{M}_S}(S;t) = \mathcal{H}(S_1;t^{d_1})\Big(\sum_{n\in A_2} \mathsf{M}_{S_2}(n)(t^{d_2})^n\Big) + \mathcal{H}_{\mathsf{M}_{S_1}}(S_1;t^{d_1})\Big(\sum_{n\in A_2}(t^{d_2})^n\Big),$$

where $A_2 = \mathrm{Ap}(S_2;d_1)$, and if S is an m_S-harmonic gluing, then

$$\mathcal{H}_{\mathsf{m}_S}(S;t) = \Big(\sum_{n\in A_1} \mathsf{m}_{S_1}(n)(t^{d_1})^n\Big)\mathcal{H}(S_2;t^{d_2}) + \Big(\sum_{n\in A_1}(t^{d_1})^n\Big)\mathcal{H}_{\mathsf{m}_{S_2}}(S_2;t^{d_2}),$$

where $A_1 = \mathrm{Ap}(S_1;d_2)$

Proof. The key is that whenever $n = d_1n' + d_2n'' \in S$ with $n' \in S_1$ and $n'' \in S_2$, we have n' maximal among all such expressions for n if and only if $n'' \in \mathrm{Ap}(S_2;d_1)$. Indeed, if $n'' - d_1 \in S_2$, then we can write $n = d_1(n' + d_2) + d_2(n'' - d_1)$, and the converse holds since $\gcd(d_1, d_2) = 1$. This implies the coefficient of t^n obtained from expanding the right-hand side of the first equality is $\mathsf{M}_{S_1}(n') + \mathsf{M}_{S_2}(n'')$, so the harmonic assumption on S proves the first equality. An analogous argument proves the second equality. □

6. Numerical Semigroups With 2 Generators

In this section, we restrict our attention to the case $S = \langle n_1, n_2\rangle$.

Theorem 6. *If $S = \langle n_1, n_2\rangle$, then*

$$\sum_{n\in S} \widehat{\chi}_{\mathsf{M}_S}(\Delta_n)t^n = -n_1t^{n_1n_2} \qquad \textit{and} \qquad \sum_{n\in S} \widehat{\chi}_{\mathsf{m}_S}(\Delta_n)t^n = -n_2t^{n_1n_2}.$$

Proof. It suffices to prove the first equality, as the second follows analogously. We use the well-known fact that

$$\mathrm{Ap}(S;n_1) = \{0, n_2, \ldots, (n_1 - 1)n_2\},$$

every element of which is uniquely factorable, and that $\mathsf{M}_S(n + n_1) = \mathsf{M}_S(n) + 1$ for every $n \in S$ [16]. As such,

$$\begin{aligned}\widehat{\chi}_{\mathsf{M}_S}(\Delta_{n_1n_2}) &= \mathsf{M}_S(n_1n_2) - (\mathsf{M}_S(n_1n_2 - n_1) + 1) - (\mathsf{M}_S(n_1n_2 - n_2) + 1)\\ &= n_2 - (n_2 - 1 + 1) - (n_1 - 1 + 1) = -n_1.\end{aligned}$$

For all other elements $n \neq n_1n_2$, the complex Δ_n is either (i) the a single vertex 1, (ii) the single vertex 2, or (iii) the full simplex $2^{[2]}$. In each case, one readily checks that $\widehat{\chi}_{\mathsf{M}_S}(\Delta_n) = 0$, thereby completing the proof. □

Remark 4. It is known that for $S = \langle n_1, n_2 \rangle$, no two factorizations of a given element $n \in S$ have the same length, so

$$\sum_{n \in S} \mathsf{l}_S(n) t^n = \sum_{n \in S} |Z_S(n)| t^n = z(t) = \frac{1}{(1 - t^{n_1})(1 - t^{n_2})}$$

Remark 5. The disparity between $\chi_{\mathsf{M}_S}(\Delta_n)$ and $\widehat{\chi}_{\mathsf{M}_S}(\Delta)$ is perhaps most exemplified in the case $S = \langle n_1, n_2 \rangle$. Indeed, for $S = \langle 9, 11 \rangle$, we have $\sum_{n \in S} \widehat{\chi}_{\mathsf{M}_S}(\Delta_n) t^n = -9t^{99}$ by Theorem 6, whereas

$$\begin{aligned}\sum_{n \in S} \chi_{\mathsf{M}_S}(\Delta_n) t^n = t^9 &+ t^{11} + t^{18} + t^{22} + t^{27} + t^{33} + t^{36} + t^{44} + t^{45} + t^{54} \\ &+ t^{55} + t^{63} + t^{66} + t^{72} + t^{77} + t^{81} + t^{88} + t^{90} - 7t^{99}\end{aligned}$$

has one additional term for each element of $\mathrm{Ap}(S; n_1)$ and $\mathrm{Ap}(S; n_2)$.

7. Future Work

The ω-primality invariant ω_S, a semigroup-theoretic measure of nonunique factorization [13], is also known to be eventually quasilinear over numerical semigroups. More precisely, for all sufficiently large $n \in S$,

$$\omega(n + n_1) = \omega(n) + 1.$$

Additionally, it is known [3] that the domain of ω_S can be naturally extended to the quotient group $\mathbb{Z}$, i.e., $\omega_S : \mathbb{Z} \to \mathbb{Z}_{\geq 0}$, in such a way that sufficiently negative input values yield 0. In many cases, after the domain is extended in this way, the lower bound on n after which quasilinearity holds for ω_S can be significantly lowered.

Problem 1. Find rational expressions for the power series $\sum_{n \in S} \omega_S(n) t^n$ and its extension $\sum_{n \in \mathbb{Z}} \omega_S(n) t^n$ in the style of Proposition 1 or Theorem 3.

There are eventually quasipolynomial S-invariants that arise naturally in studying numerical semigroups whose period does not divide the product $n_1 \cdots n_k$. For example, writing $\ell^\infty(a)$ for the component-wise maximum of $a \in \mathbb{Z}_{\geq 0}^k$, it is not hard to show

$$n \mapsto \min\{\ell^\infty(a) : a \in \mathsf{Z}(n)\}$$

is eventually quasilinear in n with period dividing $n_1 + \cdots + n_k$. As this often does not divide the product $n_1 \cdots n_k$, the rational expressions in Proposition 1 and Theorem 3 will not have numerators with finite degree.

Problem 2. Develop an analogue of Proposition 1 and Theorem 3 for S-invariants whose periods do not divide the product $n_1 \cdots n_k$.

Given $n \in S$, define the simplicial complex ∇_n with vertex set $\mathsf{Z}(n)$ where $F \subset \mathsf{Z}(n)$ is a face of ∇_n whenever there is some generator appearing in every factorization in F. The complex ∇_n is topologically equivalent to Δ_n (this was first observed in [15]), and thus is sometimes used in place of Δ_n when examining Hilbert series of numerical semigroups via Theorem 2.

Problem 3. Find labelings of the simplicial complex ∇_n so that the weighted and augmented Euler characteristic matches those of Δ_n.

Acknowledgements. Much of this work was completed as part of the 2017 San Diego State Mathematics Research Experience for Undergraduates, NSF grant 1061366.

References

[1] I. Aliev, J. De Loera, C. O'Neill, and T. Oertel, Sparse solutions of linear Diophantine equations, *SIAM J. Appl. Algebra Geom.* **1** (2017), no. 1, 239–253.

[2] T. Barron, C. O'Neill, and R. Pelayo, On the set of elasticities in numerical monoids, *Semigroup Forum* **94** (2017), no. 1, 37–50.

[3] T. Barron, C. O'Neill, and R. Pelayo, On dynamic algorithms for factorization invariants in numerical monoids, *Math. Comp.* **86** (2017), 2429–2447.

[4] M. Beck and S. Robins, *Computing the Continuous Discretely: Integer-Point Enumeration in Polyhedra*, Undergraduate Texts in Mathematics, Springer, New York, 2007.

[5] W. Bruns and J. Herzog, Semigroup rings and simplicial complexes, *J. Pure Appl. Algebra* **122** (1997), no. 3, 185–208.

[6] S. Chapman, M. Corrales, A. Miller, C. Miller, and D. Patel, The catenary and tame degrees on a numerical monoid are eventually periodic, *J. Aust. Math. Soc.* **97** (2014), no. 3, 289–300.

[7] S. Chapman, R. Hoyer, and N. Kaplan, Delta sets of numerical monoids are eventually periodic, *Aequationes Math.* **77** (2009), no. 3, 273–279.

[8] E. Ciolan, P. García-Sánchez, and P. Moree, Cyclotomic numerical semigroups, *SIAM J. Discrete Math.* **30** (2016), no. 2, 650–668.

[9] M. Delgado, P. García-Sánchez, and J. Morais, *NumericalSgps, A package for numerical semigroups*, Version 1.1.10 (2018), (Refereed GAP package), `https://gap-packages.github.io/numericalsgps/`.

[10] E. Miller and B. Sturmfels, *Combinatorial commutative algebra*, Graduate Texts in Mathematics, 227, Springer-Verlag, New York, 2005.

[11] C. O'Neill, On factorization invariants and Hilbert functions, *J. Pure Appl. Algebra* **221** (2017), no. 12, 3069–3088.

[12] C. O'Neill and R. Pelayo, On the linearity of ω-primality in numerical monoids, *J. Pure Appl. Algebra* **218** (2014), no. 9, 1620–1627.

[13] C. O'Neill and R. Pelayo, How do you measure primality?, *Amer. Math. Monthly* **122** (2015), no. 2, 121–137.

[14] C. O'Neill and R. Pelayo, Factorization invariants in numerical monoids, *Contemp. Math.* **685** (2017), 231–249.

[15] I. Ojeda and A. Vigneron-Tenorio, Simplicial complexes and minimal free resolution of monomial algebras, *J. Pure Appl. Algebra* **214** (2010), no. 6, 850–861.

[16] J. Rosales and P. García-Sánchez, *Numerical semigroups*, Developments in Mathematics, 20, Springer-Verlag, New York, 2009.

[17] R. Stanley, *Combinatorics and Commutative Algebra, 2nd Edition*, Progress in Mathematics, 41, Birkhäuser Boston, Inc., Boston, MA, 1996.

[18] R. Stanley, *Enumerative Combinatorics, Volume 1, 2nd Edition*, Cambridge Studies in Advanced Mathematics, 49, Cambridge University Press, Cambridge, 2012.

3

PARITY SEQUENCES OF THE 3X+1 MAP ON THE 2-ADIC INTEGERS AND EUCLIDEAN EMBEDDING

Olivier Rozier
Institut de Physique du Globe de Paris, Paris, France
rozier@ipgp.fr

Abstract

In this paper, we consider the one-to-one correspondence between a 2-adic integer and its parity sequence under iteration of the so-called "$3x + 1$" map. First, we prove a new formula for the inverse transform. Next, we briefly review what is known about the induced automorphism and study its dynamics on the 2-adic integers. We find that it is ergodic on many small odd invariant sets, and that it has two odd cycles of period 2 in addition to its two odd fixed points. Finally, a plane embedding is presented, for which we establish affine self-similarity by using functional equations.

1. Introduction

It is an unsolved problem [15, 19] to prove that the repeated iteration of the famous "$3x + 1$" map acting on the positive integers and defined by

$$T(n) = \begin{cases} \frac{3n+1}{2} & \text{if } n \text{ is odd,} \\ \frac{n}{2} & \text{otherwise,} \end{cases} \tag{1}$$

always leads to the value 1, whatever the starting integer of the sequence. And it is not even known whether the orbits $(n, T(n), T(T(n)), \ldots)$ are bounded for all n, nor is it known if there exists any non-trivial cycle. This problem, whose origin remains unclear (cf. History and Background section in [19, p. 5]), has received a great variety of names like the $3x+1$ problem, the Collatz conjecture, the Syracuse problem, Hasse's algorithm, *la conjecture tchèque*, etc.

Its intrinsic hardness is frequently attributed to the unpredictability of the successive parities of the iterates in most sequences, until 1 is reached [1, 10]. Therefore it seems relevant to study the relationship between the integers and their parity vectors (see Definition 1), so as to address the question of the existence and nature of some underlying structure.

Definition 1. For any two positive integers j and n, we call $V_j(n)$ the *parity vector* of n, of length j,

$$V_j(n) = \left(n, T(n), \ldots, T^{j-1}(n)\right) \mod 2, \tag{2}$$

where T^k denotes the k-th iterate of T.

This notion was introduced independently by Everett and Terras, and named the *parity vector* by Lagarias [17].

It was quite easy to state [13, 28] that any two integers have the same parity vector of length j if and only if they belong to the same congruence class modulo 2^j. From this property, we derive that each function V_j sends with a one-to-one correspondence any set of 2^j consecutive integers to the set of all parity vectors of length j. There is consequently an infinite class of integers producing exactly any finite sequence of parities under iteration of T.

2. Two Formulae for the Inverse Transform

In this part, we may freely extend the definition of the functions T and V_j to the ring $\mathbb{Z}$ of rational integers, as in [17].

The transformation of an integer n into its parity vector $V_j(n)$ is straightforward by applying the map T repeatedly j times. Conversely, one may use the forthcoming Lemma 1 to obtain all the integers that have any given parity vector. In fact, it is a well-known expression with various formulations [7, 23, 28] and generalizations [6, 22], further studied by Matthews in [21].

Definition 2. Let j be a positive integer. We say that a vector $S = (s_0, \ldots, s_{j-1})$ of length j is a finite *binary sequence* if $s_k = 0$ or 1 for all $0 \leq k \leq j-1$. We further define the *partial sum* functions σ_k applying on S by

$$\sigma_k(S) = \sum_{i=0}^{k} s_i \quad \text{for each } k \leq j-1. \tag{3}$$

The above functions σ_k are essentially the same as the functions pop_k introduced in [6] and used in a similar way. They frequently appear in various forms within the literature on the $3x+1$ problem.

Lemma 1. (First formulation of the inverse transform) *Let S be a finite binary sequence $(s_0, s_1, \ldots, s_{j-1})$ of length j. The set of integers n for which $V_j(n) = S$ is given by the congruence class*

$$n \equiv -\sum_{k=0}^{j-1} s_k\, 2^k\, 3^{-\sigma_k(S)} \pmod{2^j}. \tag{4}$$

Proof. Suppose that $V_j(n) = S$. Then equation (4) follows from the formula

$$2^j\, T^j(n) = 3^{\sigma_{j-1}(S)} \quad \left. n + \sum_{k=0}^{j-1} s_k\, 2^k\, 3^{-\sigma_k(S)} \right),$$

which is easy to state by induction on j (see [28]). □

Example 1. By Lemma 1, the odd integers n leading to sequences where every odd term is followed by exactly two even terms on the first j iterations of the map T are such that

$$n \equiv -\sum_{k=0}^{\lfloor \frac{j-1}{3} \rfloor} 8^k\, 3^{-(k+1)} \equiv -\frac{1}{3}\, \frac{\left(\frac{8}{3}\right)^{\lfloor \frac{j+2}{3} \rfloor} - 1}{\frac{8}{3} - 1} \equiv \frac{1}{5} \pmod{2^j}.$$

For the increasing lengths $j = 3, 6, 9, \ldots$, the smallest positive values of n are $5, 13, 205, \ldots$ respectively.

The discovery of a second formulation of the inverse transform came after studying the particular case of sequences where all terms but one are odd [26] (see also Example 2). It can be stated in different ways and we give a very short proof using a conjugate function. While this function already appears in the literature (e.g., [27, p. 26]), the resulting formula in Theorem 1 seems new.

Definition 3. Let us consider the function

$$\begin{aligned} U : \mathbb{Z} &\longrightarrow \mathbb{Z} \\ n &\longmapsto \begin{cases} \frac{n+1}{2} & \text{if } n \text{ is odd,} \\ \frac{3n}{2} & \text{otherwise.} \end{cases} \end{aligned}$$

The conjugacy relationship

$$U(n+1) = T(n) + 1 \tag{5}$$

holds for all n. Furthermore, for any binary sequence $S = (s_0, s_1, \ldots, s_{j-1})$ of length j and any integer n such that $U^k(n) \equiv s_k \pmod 2$ for each k, one has

$$n \equiv -\sum_{k=0}^{j-1} s_k\, 2^k\, 3^{\sigma_k(S)-k-1} \pmod{2^j}. \tag{6}$$

The latter congruence is derived from the equation

$$2^j\, U^j(n) = 3^{j-\sigma_{j-1}(S)} \quad \left. n + \sum_{k=0}^{j-1} s_k\, 2^k\, 3^{\sigma_k(S)-k-1} \right),$$

which may be easily proved by induction on j, exactly as in Lemma 1.

We are now able to provide a second formulation of the inverse of the functions V_j, which has little difference with the previous one. It turns out to be practical for sequences of T iterations that contain many odd terms, because the corresponding terms in formula (7) vanish.

Theorem 1. (Second formulation of the inverse transform) *Let S be a finite binary sequence $(s_0, s_1, \dots, s_{j-1})$. The set of integers n for which $V_j(n) = S$ is given by the congruence class*

$$n \equiv -1 - \sum_{k=0}^{j-1} (1 - s_k)\, 2^k\, 3^{-\sigma_k(S)} \pmod{2^j}. \tag{7}$$

Proof. Let n be such that $V_j(n) = S$, and consider the binary sequence

$$S_U = \left(U^k(n+1) \mod 2\right)_{k=0}^{j-1}.$$

The conjugacy (5) gives $U^k(n+1) = T^k(n) + 1$, so that $S_U = (1 - s_0, \dots, 1 - s_{j-1})$ and $\sigma_k(S_U) = k + 1 - \sigma_k(S)$ for every k.

It suffices to write the inverse formula (6) applied to $n + 1$,

$$n + 1 \equiv -\sum_{k=0}^{j-1} (1 - s_k)\, 2^k\, 3^{\sigma_k(S_U)-k-1} \pmod{2^j},$$

to conclude the proof. □

Example 2. Let j be a positive integer. Suppose we want to find the integers n for which the parity vector $V_j(n)$ contains exactly once the value 0. Then we can write that

$$n \equiv -1 - \left(\frac{2}{3}\right)^k \pmod{2^j}$$

where k is the only integer lower than j such that $T^k(n)$ is even. See [26, §6] for a brief study of those integers in $\mathbb{Z}^+$.

In fact, we obtain from Lemma 1 and Theorem 1 infinitely many formulations by considering linear combinations of (4) and (7). For example, a simple addition gives

$$2n + 1 \equiv -\sum_{k=0}^{j-1} 2^k\, 3^{-\sigma_k(V_j(n))} \pmod{2^j} \quad \text{for all integers } j > 0 \text{ and } n.$$

On the other hand, subtracting the second formulation (7) from the first formulation (4) yields the non-trivial congruence in Corollary 1.

Corollary 1. *Let j be a positive integer. For any finite binary sequence $S = (s_0, s_1, \ldots, s_{j-1})$,*

$$\sum_{k=0}^{j-1} (-1)^{s_k}\, 2^k\, 3^{-\sigma_k(S)} \equiv -1 \pmod{2^j}. \tag{8}$$

Proof. Let n be a positive integer such that $V_j(n) = S$. Subtracting each side of (7) from the corresponding side of (4) gives the desired result, by writing $(-1)^{s_k} = (1 - s_k) - s_k$ for any k. □

This corollary can also be proved directly by induction on j, then leading to an alternate proof of Theorem 1, which is left to the reader.

3. Ultrametric Extension

3.1. The Space of 2-adic Integers

Following Hasses's generalization of the $3x+1$ problem, it was suggested [21, 22] to extend the definition of the map T to the ring $\mathbb{Z}_2$ of 2-adic integers, that is, numbers of the form $\sum_{k=0}^{\infty} a_k 2^k$ with $a_k = 0$ or 1 for all k. The standard shorthand notation $(\ldots a_2 a_1 a_0)_2$ from right to left[1] may be used for the sake of conciseness, and the parentheses are most often omitted. A periodic expansion is usually indicated by an upper bar. For example, one may write

$$(\ldots 010101)_2 = \overline{01}_2 = \sum_{k=0}^{\infty} 2^{2k} = -\frac{1}{3}.$$

Recall that all rational numbers with an odd denominator have an eventually periodic expansion in $\mathbb{Z}_2$.

A metric can be derived from the 2-adic norm

$$\left| \sum_{k=0}^{\infty} a_k 2^k \right|_2 = 2^{-l} \quad \text{with } l = \min\{k \geq 0 : a_k \neq 0\}, \quad \text{and } |0|_2 = 0.$$

The space $\mathbb{Z}_2$ is then said to be *ultrametric*, due to the strong triangle inequality

$$|x + z|_2 \leq \max\left(|x + y|_2, |y + z|_2\right)$$

for all x, y and z. Therefore it is not Euclidean.

When needed, we apply the usual Haar measure on $\mathbb{Z}_2$, here noted μ, such that $\mu(\mathbb{Z}_2) = 1$, and refer to it as the 2-adic measure.

[1] Some authors prefer to write the 2-adic "digits" from left to right.

The function T remains well-defined on $\mathbb{Z}_2$, where it is known to be continuous and measure-preserving [21]. As was observed many times [1, 18, 23], iterating T on $\mathbb{Z}_2$ leads to a much greater variety of behaviors, due to its ergodic [21] and strongly mixing [17] dynamics, and interesting properties thus arise.

3.2. Parity Sequences

Let first introduce the notion of *parity sequence*.

Definition 4. For every 2-adic integer x, the infinite binary sequence

$$V_\infty(x) = \big(x, T(x), T^2(x), \ldots\big) \mod 2 \tag{9}$$

is called the *parity sequence* of x.

It is remarkable, as mentioned in [17], that the V_∞ function is a one-to-one and onto transform from $\mathbb{Z}_2$ to $\{0,1\}^\infty$. Every infinite binary sequence is the parity sequence, via T iteration, of exactly one 2-adic integer. As a consequence, there exist 2-adic cycles of every period. A complete list of the 23 cycles of period at most 6 is given in [18]. Since eventually periodic sequences have density zero in $\{0,1\}^\infty$, we infer that almost all orbits in $\mathbb{Z}_2$ do not contain a cycle.

From Lemma 1 and Theorem 1, one immediately derives two formulae to express the inverse transform V_∞^{-1}.

Corollary 2. *Let S be an infinite binary sequence $(s_0, s_1, s_2, \ldots)$. The 2-adic integer x such that $V_\infty(x) = S$ is given by any of the 2-adically convergent expansions*

$$x = -\sum_{k=0}^{\infty} s_k\, 2^k\, 3^{-\sigma_k(S)} \tag{10}$$

and

$$x = -1 - \sum_{k=0}^{\infty} (1 - s_k)\, 2^k\, 3^{-\sigma_k(S)}. \tag{11}$$

where σ_k denotes the partial sum function $\sigma_k(S) = \sum_{i=0}^{k} s_i$.

Example 3. Consider the binary sequence $S = (s_0, s_1, s_2, \ldots)$ where $s_k = 1$ for all k. Applying the inverse transform (11), we get $V_\infty^{-1}(S) = -1 = \ldots 111111_2$.

The question whether the inverse formula (10) leads to a convergent series when evaluated in the set of real numbers has been investigated in various papers [12, 20, 21]. Note that the sum of the series is negative or zero, when it exists. Interestingly, both series on the right hand side of (10) and (11) are expected to be divergent (resp. convergent) for the parity sequences of positive (resp. negative) rational integers.

One may also remark that the equation (10) is quite similar to the expression of the real function θ mentioned at the end of Coquet's paper [9, §7], which is convergent and turns out to be fractal.

We do not further discuss this issue in the present paper.

3.3. Automorphism

It is convenient to encode every parity sequence as a 2-adic integer, so as to give it a rational value when it is eventually periodic, as done by Lagarias in [17]. This yields an automorphism in $\mathbb{Z}_2$.

Definition 5. Let Q denote the function

$$\begin{aligned} Q : \mathbb{Z}_2 &\longrightarrow \mathbb{Z}_2 \\ x &\longmapsto \textstyle\sum_{k=0}^{\infty} s_k 2^k \end{aligned}$$

where $(s_0, s_1, s_2, \ldots)$ is the parity sequence of x, as defined by (9).

The function Q is a one-to-one and onto morphism [6, 17]. It is also non-expanding[2] with respect to the 2-adic norm, since it satisfies the 1-Lipschitz condition

$$|Q(x) - Q(y)|_2 \leq |x - y|_2 \quad \text{for all } x \text{ and } y,$$

or equivalently,

$$x \equiv y \pmod{2^n} \text{ implies } Q(x) \equiv Q(y) \pmod{2^n}. \tag{12}$$

The fact that Q is one-to-one further implies (see [2]) the reciprocal

$$x \equiv y \pmod{2^n} \text{ if and only if } Q(x) \equiv Q(y) \pmod{2^n}, \tag{13}$$

which makes Q a 2-adic isometry [6].

For convenience purposes, we prefer to use the simple notation Q, as in [1], rather than the original notation Q_∞. Its inverse Q^{-1}, called the $3x + 1$ *conjugacy map* and denoted by Φ in [6, 23], is known [5] to conjugate the map T with the shift map $\mathcal{S}$ whose definition follows. For the sake of clarity, we rephrase all known and conjectured properties related to Φ in terms of the function Q.

Definition 6. Let the *shift map* $\mathcal{S}$ denote the function

$$\begin{aligned} \mathcal{S} : \mathbb{Z}_2 &\longrightarrow \mathbb{Z}_2 \\ x &\longmapsto \begin{cases} \frac{x-1}{2} & \text{for x odd,} \\ \frac{x}{2} & \text{otherwise.} \end{cases} \end{aligned}$$

The conjugacy

$$T = Q^{-1} \circ \mathcal{S} \circ Q. \tag{14}$$

holds.

[2] In [6], the functions having the property (12) are said to be *solenoidal*.

In the context of the original $3x + 1$ problem, it is crucial to determine which values of Q are rational. Indeed, we have the well-known statements holding for all 2-adic integers x:

- the orbit $\left(x, T(x), T^2(x), \ldots\right)$ is eventually periodic if and only if $Q(x)$ is rational [1]. This is an immediate consequence of the conjugacy (14) as noted by Monks [23];
- if $Q(x)$ is rational, then x is rational [5] (see also [1], Theorem 5).

We know that all the cycles within the dynamics of T on $\mathbb{Z}_2$ are rational [18]. Lagarias' Periodicity Conjecture [17] asserts that the reciprocal of the second statement above also holds. This would imply that every rational point in $\mathbb{Z}_2$ is preperiodic.

Conjecture 1. *(Periodicity Conjecture)* For any 2-adic integer x, $Q(x)$ is rational if and only if x is rational.

Furthermore, the function Q allows us to formulate the $3x+1$ problem differently [5, 17].

Conjecture 2. ($3x + 1$ *Problem*)

$$Q\left(\mathbb{Z}^+\right) \subset \frac{1}{3}\mathbb{Z}, \quad \text{or equivalently,} \quad \mathbb{Z}^+ \subset Q^{-1}\left(\frac{1}{3}\mathbb{Z}\right).$$

It asserts that every positive integer has an eventually periodic parity sequence of period 2, ending with an infinite alternation of 0 and 1 (the case of a fixed parity is trivially excluded), which only occurs when some iterate reaches the trivial cycle (1,2). Note that the reverse inclusion in Conjecture 2 does not hold, since $Q^{-1}(1) = -1/3$, by formula (10).

3.4. Functional Equations

The semiring $\mathbb{N}$ of natural integers is completely generated by all finite compositions of the functions $x \longmapsto 2x$ and $x \longmapsto 2x+1$ starting from 0, thus reversing the action of the shift map $\mathcal{S}$. Therefore it is tempting to search for functional equations that express $Q(2x)$ and $Q(2x+1)$ from $Q(x)$. Such equations exist for any x in $\mathbb{Z}_2$ or in a subset of $\mathbb{Z}_2$, as shown in Theorem 2. It turns out that equation (18) is a sort of 2-adic extension of previous results by Andaloro [4] and Garner [14]. We also establish similar equations for the inverse transform Q^{-1} (see [12] for a generalization).

Theorem 2. *The functions Q^{-1} and Q are solution to the functional equations*

$$Q^{-1}(2x) = 2Q^{-1}(x), \tag{15}$$

$$Q^{-1}(2x+1) = \frac{2Q^{-1}(x) - 1}{3}, \tag{16}$$

$$Q(2x) = 2Q(x) \tag{17}$$

for all 2-adic integers x. Moreover

$$Q(2x+1) = 2Q(x) - 2^k + 1 \tag{18}$$

for $x \equiv -1 - (-2)^{k-2} \pmod{2^k}$ and $k \geq 2$.

Proof of (15) *and* (16). First, one may rewrite equation (14) as

$$T \circ Q^{-1} = Q^{-1} \circ \mathcal{S}. \tag{19}$$

Take a 2-adic integer x. Putting together (19) with the fact that

$$\mathcal{S}(2x+1) = \mathcal{S}(2x) = x$$

and that

$$Q^{-1}(y) \equiv y \pmod 2 \quad \text{for all } y,$$

we obtain

$$Q^{-1}(x) = Q^{-1} \circ \mathcal{S}(2x) = T \circ Q^{-1}(2x) = \frac{Q^{-1}(2x)}{2}$$

and

$$Q^{-1}(x) = Q^{-1} \circ \mathcal{S}(2x+1) = T \circ Q^{-1}(2x+1) = \frac{3Q^{-1}(2x+1)+1}{2}.$$

□

Proof of (17). Replacing x by $Q(x)$ in (15) gives $Q^{-1}(2Q(x)) = 2x$, leading to $2Q(x) = Q(2x)$. □

Proof of (18). Let $k \geq 2$ and let x, y be 2-adic integers such that $x = -1-(-2)^{k-2}+2^k y$. Starting from x and applying repeatedly the map T, it is easily seen that the first $k-3$ iterates are odd, while the next one is even: $T^{k-2}(x) = -1 - (-3)^{k-2} + 3^{k-2}(4y) \equiv 2 \pmod 4$. Setting $T^{k-2}(x) = 2 + 4z$, we get $T^{k-1}(x) = 1 + 2z$ and $T^k(x) = 2 + 3z$. Since x has the parity vector $V_k(x) = (1, 1, \ldots, 1, 0, 1)$, one may write

$$Q(x) = 1 + 2 + \ldots + 2^{k-3} + 2^{k-1} + 2^k Q(2+3z)$$

for $k \geq 3$. In the case $k = 2$, the above expression simplifies to $Q(x) = 2+4Q(2+3z)$.

On the other hand, starting from $2x+1$ and applying $k-1$ times the map T, we get after $(k-2)$ odd iterates the even value $T^{k-1}(2x+1) = -1+(-3)^{k-1}+3^{k-1}(4y) = 3T^{k-2}(x) + 2 = 8 + 12z$. The next two iterates are $T^k(2x+1) = 4 + 6z$ and $T^{k+1}(2x+1) = 2 + 3z$. It follows that $2x+1$ has the parity vector $V_{k+1}(2x+1) = (1, 1, \ldots, 1, 0, 0)$, so that

$$Q(2x+1) = 1 + 2 + \ldots + 2^{k-2} + 2^{k+1}Q(2+3z) = 2Q(x) - 2^k + 1.$$

□

One may ask whether Theorem 2 provides a general algorithm to calculate in a finite number of steps the exact value of the function Q applied to an arbitrary positive integer. Unfortunately, the answer appears to be negative, since equation (18) only applies to a subset of $\mathbb{N}$ of density $2^{-2} + 2^{-3} + 2^{-4} + \ldots = 1/2$.

The case $k = 2$ can be restated as

$$Q(4x+1) = 4Q(x) - 3 \quad \text{for } x \equiv 1 \pmod 2 \tag{20}$$

by replacing x by $2x$ in (18) and using (17). One may further combine (18) and (20) to produce the functional equations

$$Q(8x+5) = 4Q(2x+1) - 3 = 8Q(x) - 2^{k+2} + 1$$

for $x \equiv -1 - (-2)^{k-2} \pmod{2^k}$ and $k \geq 2$.

The function Q satisfies many other functional equations that are not combinations of (17) and (18) like

$$Q(3x+1) = Q(x) - 1 \quad \text{for } x \equiv 1 \pmod 2. \tag{21}$$

Such equations are always related to the phenomenon of coalescence within the dynamics of T. For example, the equation (21) derives directly from the equality $T(3x+1) = T(x)$ for $x \equiv 1 \pmod 2$; see [4, 14] for other examples of generic coalescences.

3.5. Ergodicity

The ergodic dynamics of the $3x+1$ map T on $\mathbb{Z}_2$ is quite well understood, and paradoxically, it does not provide any indication on the validity of the $3x+1$ Conjecture, as is discussed in [1].

Nevertheless, in view of the Periodicity Conjecture, it could be helpful to better specify the dynamics of Q, which appears to be more complicated.

In what follows, we refer to [2] for the ergodicity of a measure-preserving function on the 2-adic integers.

Since Q is isometric, it induces in the finite set $\mathbb{Z}/2^n\mathbb{Z}$ a permutation Q_n whose behavior is easier to study.

Definition 7. For all integers $n \geq 0$, let Q_n denote the function

$$\begin{aligned} Q_n : \mathbb{Z}/2^n\mathbb{Z} &\longrightarrow \mathbb{Z}/2^n\mathbb{Z} \\ x &\longmapsto Q(x) \mod 2^n. \end{aligned}$$

In [17], Lagarias showed that the order of Q_n is always a power of 2, and the following theorem was finally stated in [6].

Theorem 3. (Bernstein, Lagarias) *For every positive integer n, the length of any cycle in Q_n is a power of 2. Moreover, Q_n is a permutation of order 2^{n-4} for $n \geq 6$.*

When lifting from $\mathbb{Z}/2^n\mathbb{Z}$ to $\mathbb{Z}/2^{n+1}\mathbb{Z}$, it is known that any cycle of Q_n either splits into two cycles whose period is unchanged, or undergoes a period-doubling.

Definition 8. Let $m \geq k \geq 0$ and let $C = (c_1, \ldots, c_{2^k})$ be a cycle of the permutation Q_m of length 2^k. We say that C has an *ever-doubling period* if, for all $n \geq m$, the elements $c_1, \ldots, c_{2^k}$ of C are all included in a single cycle of Q_n of length 2^{n-m+k}.

The second part of Theorem 3 is based on the fact that the cycle (5,17) of Q_5 has an ever-doubling period (see [6]).

Now we can use this result to study the dynamics of Q and Q^{-1} on the topological space $\mathbb{Z}_2$. To this aim, we need the notion of 2-adic *ball*.

Definition 9. For any $y \in \mathbb{Z}_2$ and $r \geq 0$, let $B(y, r)$ denote the (closed) *ball*

$$B(y, r) = \{x \in \mathbb{Z}_2 : |x - y|_2 \leq r\}$$

with center y and radius r. Equivalently, one has

$$B(y, 2^{-k}) = \{x \in \mathbb{Z}_2 : x \equiv y \pmod{2^k}\}$$

for every integer $k \geq 0$, and its 2-adic measure is given by its radius:

$$\mu\left(B(y, 2^{-k})\right) = 2^{-k}.$$

Recall that the function Q is measure-preserving [6, 17] and, unlike the map T, is not ergodic on $\mathbb{Z}_2$, since it preserves the parity.

One may at first observe that all forward and backward orbits remain close to the initial point, and that the 2-adic distance is even smaller when the number of iterations is highly divisible by 2. This fact is illustrated in the table below that gives some of the iterates of the 2-adic integer $\overline{0110}1_2 = 1/5$.

j	$Q^j\left(\frac{1}{5}\right)$	$\left\lvert Q^j\left(\frac{1}{5}\right) - \frac{1}{5}\right\rvert_2$	$Q^{-j}\left(\frac{1}{5}\right)$	$\left\lvert Q^{-j}\left(\frac{1}{5}\right) - \frac{1}{5}\right\rvert_2$
1	$-\frac{1}{7} = \overline{001}_2$	2^{-2}	$\frac{13}{21} = \overline{001100}1_2$	2^{-2}
2	$\frac{17}{5} = \overline{0011}101_2$	2^{-4}	$-\frac{1}{11} = \overline{000101110}1_2$	2^{-4}
3	$\frac{1863}{31} = \ldots 1001_2$	2^{-2}	$\frac{373}{781} = \ldots 1001_2$	2^{-2}
4	$\ldots 00001101_2$	2^{-6}	$\ldots 10001101_2$	2^{-6}

The previous observation is due to the congruences (22) and (23), which follow from Theorem 3.

Corollary 3. *For all 2-adic integers x and all $k \geq 2$,*

$$Q^2(x) \equiv x \pmod{2^4}, \tag{22}$$

$$Q^{2^k}(x) \equiv x \pmod{2^{k+4}}, \tag{23}$$

or equivalently, $Q^2(x) \in B(x, 2^{-4})$ and $Q^{2^k}(x) \in B(x, 2^{-k-4})$.

Proof. Take an integer $k \geq 2$. Theorem 3 implies that the length of every cycle of Q_{k+4} divides 2^k, from which we infer the congruence (23).

Likewise, the permutation Q_4 has ten fixed points and three cycles of length 2, which are (1,5), (2,10), and (9,13). Hence, the equation (22). □

Consequently, whatever the 2-adic integer x, its forward and backward orbits under iteration of Q have elements arbitrarily close to x.

Corollary 4. *For all 2-adic integers x,*

$$\lim_{k\to\infty} Q^{2^k}(x) = \lim_{k\to\infty} Q^{-2^k}(x) = x.$$

Though the dynamics of Q are not truly ergodic on $\mathbb{Z}_2$, this may occur on some invariant subsets.

Let us recall a known criterion for the ergodicity of non-expanding[3] functions, by Anashin (Proposition 4.1 in [2]; see also [3]).

Theorem 4. (Anashin) *A non-expanding function $F : \mathbb{Z}_2 \to \mathbb{Z}_2$ is ergodic if and only if F induces modulo 2^n a permutation with a single cycle for all positive integers n.*

The next theorem shows that Q is ergodic in a neighborhood of each cycle of Q_m having an ever-doubling period, for any $m \geq 0$.

Theorem 5. *Let Q_m denote the permutation induced by Q in $\mathbb{Z}/2^m\mathbb{Z}$. For all $m \geq k \geq 0$ and all cycles $C = (c_1, \ldots, c_{2^k})$ of Q_m having an ever-doubling period, the restriction of Q to $B(c_1, 2^{-m}) \cup \ldots \cup B(c_{2^k}, 2^{-m})$ is ergodic.*

Proof. Let $n \geq m$. Put $K = B(c_1, 2^{-m}) \cup \ldots \cup B(c_{2^k}, 2^{-m})$ and $K_n = K \mod 2^n$. Since Q is isometric, the sets K and K_n are left invariant by Q and Q_n respectively.

Let C_n be the cycle of the permutation Q_n that contains all the elements of C. Its length is equal to 2^{n-m+k}. Moreover, it is included in K_n whose cardinality is equal to 2^{n-m+k}. Therefore the restriction of Q_n to the set K_n is a permutation with a single cycle C_n.

From Theorem 4, we deduce that the restriction of Q to the set K is ergodic. For completeness, it is not difficult to find a suitable bijection $F : \mathbb{Z}_2 \to K$ for which the conjugate function $F^{-1} \circ Q \circ F$ acting on $\mathbb{Z}_2$ is non-expanding and ergodic. For example, one may use the function

$$\begin{aligned} F : \mathbb{Z}_2 &\longrightarrow K \\ x &\longmapsto c_{i+1} + 2^{m-k}(x - i), \quad \text{where } i = x \mod 2^k, \end{aligned}$$

which is one-to-one and onto. □

[3] In [2], the term *compatible* is used instead of non-expanding for the same meaning.

m	k	C	$\mu(\omega(C))$
5	1	(5,17)	2^{-4}
6	2	(9, 29, 25, 13)	2^{-4}
6	2	(41, 61, 57, 45)	2^{-4}
8	2	(27, 251, 219, 59)	2^{-6}
8	2	(91, 187, 155, 123)	2^{-6}

Table 1: Odd cycles C of Q_m of length 2^k and having an ever-doubling period for $0 \leq m - k \leq 6$. The last column gives the 2-adic measure, equal to 2^{k-m}, of their ω-limit sets for the function Q.

k	N_k	$N_k \times 2^{-k}$	k	N_k	$N_k \times 2^{-k}$
1	0	0.000	9	11	0.021
2	0	0.000	10	29	0.028
3	0	0.000	11	54	0.026
4	3	0.187	12	91	0.022
5	0	0.000	13	118	0.014
6	2	0.031	14	213	0.013
7	10	0.078	15	282	0.008
8	11	0.042	16	436	0.006

Table 2: Numbers N_k of odd ergodic sets of 2-adic measure 2^{-k}, $k \leq 16$.

Definition 10. Whenever Q is ergodic on a (closed) invariant set with positive measure, we call it an *ergodic set*. Moreover, we call *ergodic domain* the union of all the ergodic sets.

The fact that Q is bijective further implies that Q^{-1}, namely, the $3x+1$ conjugacy map, has the same ergodic domain as Q.

Let us point out that every ergodic set in Theorem 5 is closed, hence it is the ω-limit set in $\mathbb{Z}_2$ of any point of the associated cycle C. For convenience, we write $\omega(C)$ to refer to this set.

In order to identify the cycles having an ever-doubling period, it is convenient to use the following criterion (see Theorem 3.1 in [6]) whose original formulation and vocabulary have been significantly modified.

Theorem 6. (Bernstein, Lagarias) *Let $m \geq k \geq 2$ and let C be a cycle of Q_m of length 2^k. If C is part of a cycle of Q_{m+2} of length 2^{k+2}, then C has an ever-doubling period.*

After some straightforward numerical computations, we find, by applying Theorems 5 and 6 above, that Q is ergodic on the ω-limit set of every cycle in Table

1. Though it is not the case for any other odd set of 2-adic measure at least 2^{-6}, there are many smaller odd ergodic sets. In Table 2, we provide their respective numbers when sorted by size (see also Table 2.2 in [6]). We obtain that the total measure of the odd ergodic sets exceeds 0.48, whereas it is trivially upper bounded by $\mu(B(1,1/2)) = 1/2$.

For each odd ergodic set of measure 2^{-k} and each $m \geq 1$, we easily get, by applying the autoconjugacy (17) repeatedly m times, an even ergodic set of measure 2^{-k-m}. It yields that the ergodic domain has a measure greater than 0.96.

Conjecture 3. (*Ergodicity Conjecture*) The ergodic domain of Q has full 2-adic measure.

We expect this conjecture to be closely related to the distribution of periodic orbits, about which little is known.

3.6. Cycles

The search of the periodic points of the function Q is far from trivial due to the fact that it is nowhere differentiable, as proved by Müller in [25] (see also [5] for a short proof).

Hereafter, we call *Q-cycle* (resp. *T-cycle*) a periodic orbit of the function Q (resp. T). Unlike T-cycles, it is not known whether the Q-cycles are all rational. Note that the set of even Q-cycles may be easily deduced from the odd ones by using the functional equation (17), and by adding the fixed point 0.

As a consequence of Theorem 3 (§3.5), the period of any Q-cycle is always a power of 2. In contrast with the cycles having an ever-doubling period introduced in Definition 8 (§3.5), a Q-cycle corresponds to some cycle of Q_n whose period remains unchanged for all sufficiently large n, and that systematically splits into two cycles of Q_{n+1}, one of which splits again, and so on as n increases. From a heuristic point of view, this resembles an infinite branching process that allows one to estimate the number of short cycles of Q_n for large n, as in [6, §6].

One observes in Example 3 (§3.2) that -1 is a fixed point for the function Q, as for T. In fact, there are infinitely many since -2^k and $2^k/3$ are fixed points for all $k \geq 0$, in addition to the trivial fixed point 0. It is conjectured that -1 and $1/3$ are the only odd ones (Fixed Point Conjecture, in [6]). Numerically, it is easy to verify that any such point is necessarily very close, if not equal, to -1 or $1/3$.

In the same paper, Bernstein and Lagarias also mentioned the existence of the odd rational cycle $(-1/3, 1)$ of period 2, and conjectured that there are finitely many odd cycles for any given period 2^j ($3x+1$ Conjugacy Finiteness Conjecture).

Lately, I found that $(-1/5, 5/7)$ is another odd rational cycle of period 2. Indeed,

$$Q\left(-\frac{1}{5}\right) = \overline{0011}_2 = \frac{5}{7} \quad \text{and} \quad Q\left(\frac{5}{7}\right) = \overline{0011}_2 = -\frac{1}{5}.$$

Period	Q-cycle	T-cycles
1	$(-1) = (\overline{1}_2)$ $(\frac{1}{3}) = (\overline{01}_2)$	(-1) $(1, 2)$
2	$(-\frac{1}{3}, 1) = (0\overline{1}_2, \overline{01}_2)$ $(-\frac{1}{5}, \frac{5}{7}) = (\overline{0011}_2, 00\overline{11}_2)$	(0) and $(1, 2)$ $(\frac{1}{5}, \frac{4}{5}, \frac{2}{5})$ and $(\frac{5}{7}, \frac{11}{7}, \frac{20}{7}, \frac{10}{7})$

Table 3: Odd Q-cycles and corresponding T-cycles.

Next, I conducted a numerical verification up to period 16 on the set of rationals of the form p/q where p, q are odd coprime integers lower than 1000 in absolute value. Working modulo 2^{40} was enough to rule out the candidates that are not part of a known Q-cycle.

In Table 3, we list the known odd Q-cycles, and, for each rational element, the T-cycle appearing in its orbit of T iterates. So far, no Q-cycle was found having a prime period strictly greater than 2. This leads one to think that there is none.

Conjecture 4. *(Odd Cycles Conjecture)* The function Q has exactly two odd fixed points, -1 and $\frac{1}{3}$, and two odd cycles of prime period 2, $(-\frac{1}{3}, 1)$ and $(-\frac{1}{5}, \frac{5}{7})$. There exists no other odd cycle, rational or not.

4. The $3x + 1$ Set

4.1. Euclidean Embedding

Overall, the automorphism Q and its inverse remain somewhat mysterious. One may wish to somehow visualize their action on $\mathbb{Z}_2$. Recall that the space $\mathbb{Z}_2$ is not Euclidean and totally disconnected, which makes it difficult to represent graphically [8, 11]. It is known to be homeomorphic to the Cantor ternary set, which has Lebesgue measure 0. We propose to apply a continuous function M that sends $\mathbb{Z}_2$ to the real interval $[0, 2]$. The map M, as defined below, is very similar to the Monna[4] map [24].

Definition 11. Let M denote the continuous 2-Lipschitz map from $\mathbb{Z}_2$ to $[0, 2]$

$$M : \sum_{k=0}^{\infty} r_k 2^k \longmapsto \sum_{k=0}^{\infty} r_k 2^{-k}, \quad \text{where } r_k = 0 \text{ or } 1.$$

The action of the map M may seem counter-intuitive, as it does not preserve the usual order between the rational numbers. For instance, the images of positive and negative rational numbers are deeply intertwined. But, conveniently, M sends

[4] The original Monna map sends $\mathbb{Z}_2$ to $[0, 1]$.

2-adic balls with radius 2^{-k} onto real intervals of length 2^{1-k}, so that the set of odd 2-adic integers is entirely mapped on the interval $[1, 2]$. We call it the "odd" side, whereas $[0, 1]$ may be regarded as the "even" side.

Let us point out that M is not one-to-one, since

$$M(1) = 1 = \sum_{k=1}^{\infty} 2^{-k} = M(-2),$$

and more generally,

$$M\left(n + 2^k\right) = M\left(n - 2^{k+1}\right) = M(n) + 2^{-k} \quad \text{for } 0 \leq n \leq 2^k - 1.$$

As a result, the mapping M is not truly an embedding, although when restricted to $\mathbb{Z}_2 \setminus \mathbb{Z}$ it is one-to-one. Further, it is easily seen that the set $M(\mathbb{Z})$ coincides with the set of dyadic numbers, namely, rationals whose denominator is a power of 2, from the interval $[0, 2]$.

Definition 12. Letting $X = M$ and $Y = M \circ Q$, we call "$3x+1$" set the parametric set of the plane $\mathbb{R}^2$ denoted by $\mathbf{Q}_{3x+1}$ and defined by

$$\mathbf{Q}_{3x+1} = (X, Y)(\mathbb{Z}_2) = \{(M(r), M(Q(r))) : r \in \mathbb{Z}_2\}.$$

As shown below, each point of $\mathbf{Q}_{3x+1}$ corresponds to a unique parity sequence. Somehow, the $3x + 1$ set "fully" encodes the dynamics of the $3x + 1$ map.

Lemma 2. *The function $(X, Y) : \mathbb{Z}_2 \to [0, 2]^2$ is one-to-one and continuous with respect to the 2-adic measure on its domain.*

Proof. Suppose there exist two distinct 2-adic integers a and b such that $X(a) = X(b)$ and $Y(a) = Y(b)$. We infer that a, b, $Q(a)$, and $Q(b)$ are all in $\mathbb{Z}$, and $Q(a)$ or $Q(b)$ is positive. Say $Q(a)$ is a positive integer, so it has a finite binary expansion. From the inverse formula (10), it follows that a is rational with denominator a power of 3 strictly greater than 1, and numerator coprime to 3. Since a is a rational integer, there yields a contradiction. Hence, (X, Y) is one-to-one.

Moreover, it is also continuous as a composition of continuous functions. □

Viewing (X, Y) as a continuous bijection between the compact sets $\mathbb{Z}_2$ and $\mathbf{Q}_{3x+1}$, then its inverse is known to be continuous. Therefore (X, Y) is an embedding from the parameter space $\mathbb{Z}_2$ into the Euclidean space $\mathbb{R}^2$.

The fact that rationality is always preserved by the map M leads to an immediate reformulation of the Periodicity Conjecture.

Conjecture 5. (*Rational Points Conjecture*) All points in $\mathbf{Q}_{3x+1}$ have coordinates that are either both rational or both irrational.

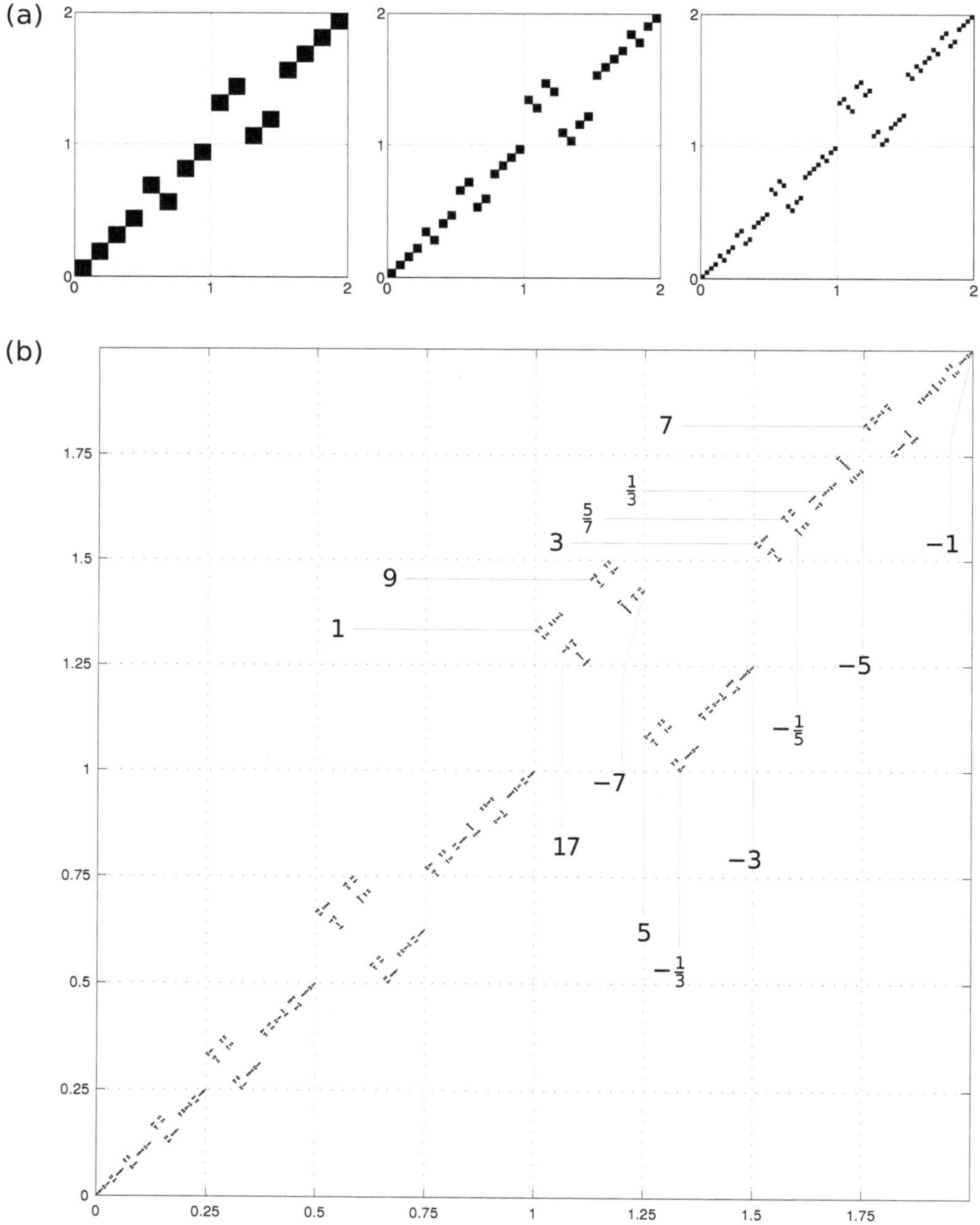

Figure 1: **(a)** Coverings of $\mathbf{Q}_{3x+1}$ made of 2^k squares of side length 2^{1-k} for $k = 4$, 5, and 6, from left to right. **(b)** The set $\mathbf{Q}_{3x+1}$ with (green) line segments indicating the rational points from Table 4, along with their respective parameter value.

r	$Q(r)$	$X(r)$	$Y(r)$	r	$Q(r)$	$X(r)$	$Y(r)$
1	$-\frac{1}{3}$	1	$\frac{4}{3}$	3	$-\frac{23}{3}$	$\frac{3}{2}$	$\frac{37}{24}$
17	$-\frac{401}{3}$	$\frac{17}{16}$	$\frac{493}{384}$	$\frac{5}{7}$	$-\frac{1}{5}$	$\frac{11}{7}$	$\frac{8}{5}$
9	$-\frac{6377}{3}$	$\frac{9}{8}$	$\frac{8941}{6144}$	$-\frac{1}{5}$	$\frac{5}{7}$	$\frac{8}{5}$	$\frac{11}{7}$
-7	$-\frac{5}{7}$	$\frac{5}{4}$	$\frac{10}{7}$	$\frac{1}{3}$	$\frac{1}{3}$	$\frac{5}{3}$	$\frac{5}{3}$
5	$-\frac{13}{3}$	$\frac{5}{4}$	$\frac{13}{12}$	-5	$-\frac{3}{7}$	$\frac{7}{4}$	$\frac{12}{7}$
$-\frac{1}{3}$	1	$\frac{4}{3}$	1	7	$-\frac{1595}{3}$	$\frac{7}{4}$	$\frac{2797}{1536}$
-3	-7	$\frac{3}{2}$	$\frac{5}{4}$	-1	-1	2	2

Table 4: A few rational points in $\mathbf{Q}_{3x+1}$ associated with an odd rational value of the 2-adic parameter r and sorted by increasing abscissa.

In Table 4, we provide the coordinates of various rational points from the set $\mathbf{Q}_{3x+1}$. Among them, six points are associated with one of the odd Q-cycles in Table 3 (§3.6), whose respective parameter values are -1, $-1/3$, $-1/5$, $1/3$, $5/7$, and 1.

Regardless of the 2-adic parametrization of $\mathbf{Q}_{3x+1}$, one can visualize it by taking only natural integers. This is due to the density of $\mathbb{N}$ in $\mathbb{Z}_2$, and to the fact that both functions M and Q are Lipschitz. Practically, it suffices to calculate the parity vectors of length k for every nonnegative integer up to 2^k for some k reasonably large, and apply M on the resulting binary expansions. We took $k = 12$ in Figure 1b.

The same method has been already used, e.g., by Hashimoto in [16], to represent the "graph" of the map[5] T acting on $\mathbb{Z}_2$.

A slightly different construction of the $3x+1$ set can be achieved by considering a sequence of nested sets made of finitely many squares (Figure 1a). Although their respective areas tend to zero as the number of squares increase, we prove in Lemma 3 that they all cover $\mathbf{Q}_{3x+1}$.

Lemma 3. *For all positive integers k, we have*

$$\mathbf{Q}_{3x+1} \subset \bigcup_{n=0}^{2^k-1} \left[X(n), X(n)+2^{1-k}\right] \times \left[Y_k(n), Y_k(n)+2^{1-k}\right]$$

where $Y_k(n) = M\left(Q(n) \mod 2^k\right)$.

[5]Most often, a variant of the map T is considered, leading to a slower or faster dynamics.

Proof. Let r be a 2-adic integer, and let $k \geq 0$. Setting $n = r \mod 2^k$, we have

$$X(r) \in X\left(B(n, 2^{-k})\right) = \left[X(n), X(n) + 2^{1-k}\right]$$

and

$$Y(r) \in Y\left(B(n, 2^{-k})\right) = M\left(B(Q(n), 2^{-k})\right) = \left[Y_k(n), Y_k(n) + 2^{1-k}\right].$$

The inclusion claimed in Lemma 3 follows. □

From this result, we infer that

$$\mathbf{Q}_{3x+1} \subset \bigcap_{k=0}^{\infty} \bigcup_{n=0}^{2^k-1} \left[X(n), X(n) + 2^{1-k}\right] \times \left[Y_k(n), Y_k(n) + 2^{1-k}\right]. \tag{24}$$

In fact, the equality holds, as both sets have exactly the same number of intersections with every line parallel to the y-axis.

Another corollary of Lemmas 2 and 3 is the presence of infinitely many discontinuities in $\mathbf{Q}_{3x+1}$ for the Euclidean metric, at each point whose abscissa is dyadic, except the extremal points (0,0) and (2,2).

Let us observe in Figure 1b that it has a rather symmetric aspect with respect to the diagonal $\Delta = \{(x, x) : 0 \leq x \leq 2\}$. This is mainly due to the congruence (22) in Corollary 3. An underlying question would be to determine how much does the function Q differ from its inverse, about which little is known. From Figure 1a, it is clear that the symmetry is broken only at a rather small scale. It turns out that few points are effectively symmetric. Despite the non-injectivity of the map M, it is most likely that only those points whose parameter values are part of a Q-cycle of period at most 2, are symmetric. We obtain thereby a symmetric subset that is expected to contain exactly six points in the upper right quarter of the set $\mathbf{Q}_{3x+1}$, two of them being on Δ (see Conjecture 4 in §3.6, and Table 4).

One may further notice a number of affine self-similarities, some of which are made explicit in the next section (§4.2).

4.2. Self-similarity

As a result of the functional equations (17) and (18) satisfied by Q, it is possible to delimit regions of $\mathbf{Q}_{3x+1}$ that are identical through an affine transformation. To this aim, we first introduce two infinite families of real intervals, which realize a covering of the half-open interval $[0, 2)$.

Definition 13. For every integer $k \geq 2$, let

$$\alpha_k = -1 - (-2)^{k-2} \mod 2^k, \quad m_k = 2^{k-2} - 1, \quad n_k = 3 \cdot 2^{k-2} - 1,$$

so that $\alpha_k = m_k$ if k is odd, and $\alpha_k = n_k$ otherwise. Then define the real intervals

$$I_k = [M(m_k), M(n_k)] = \left[2 - 2^{3-k}, 2 - 3 \cdot 2^{1-k}\right]$$

and

$$J_k = [M(n_k), M(m_{k+1})] = \left[2 - 3 \cdot 2^{1-k}, 2 - 2^{2-k}\right]$$

of length 2^{1-k}. The mapping M sends the 2-adic ball $B(\alpha_k, 2^{-k})$ onto I_k or J_k alternatively, according to the parity of k.

The next lemma, along with Corollary 5, will prove useful to delimit in the $3x+1$ set all parts corresponding to parametric values in the same congruence class as m_k or n_k modulo 2^k.

Lemma 4. *The integers $(m_k)_{k\geq 2}$ and $(n_k)_{k\geq 2}$ have the properties*

$$Q(m_k) \equiv m_k \pmod{2^k} \quad \text{and} \quad Q(n_k) \equiv n_k \pmod{2^k} \quad \text{for } k \text{ even,} \tag{25}$$

$$Q(m_k) \equiv n_k \pmod{2^k} \quad \text{and} \quad Q(n_k) \equiv m_k \pmod{2^k} \quad \text{for } k \text{ odd.} \tag{26}$$

Proof. The function Q induces a permutation on $\mathbb{Z}/2^k\mathbb{Z}$. Thus, we can reason on Q^{-1} instead of Q.

Let us write, first, the binary representations

$$m_k = \underbrace{0011\ldots1_2}_{k} \quad \text{and} \quad n_k = \underbrace{1011\ldots1_2}_{k}.$$

Applying formula (11) from Corollary 2, we get

$$\begin{aligned} Q^{-1}(m_k) &\equiv -1 - 2^{k-2}3^{2-k} - 2^{k-1}3^{2-k} \pmod{2^k} \\ &\equiv -1 - 2^{k-2}3^{1-k} \pmod{2^k} \\ &\equiv -1 + (-2)^{k-2} \pmod{2^k} \end{aligned}$$

since $\frac{1}{3} \equiv -1 \pmod 4$, and similarly,

$$\begin{aligned} Q^{-1}(n_k) &\equiv -1 - 2^{k-2}3^{2-k} \pmod{2^k} \\ &\equiv -1 - (-2)^{k-2} \pmod{2^k}. \end{aligned}$$

The properties (25) and (26) follow by considering the parity of k. □

Corollary 5. *For every $k \geq 2$,*

$$Y\left(B(\alpha_k, 2^{-k})\right) = J_k \quad \textit{and} \quad Y\left(B(2\alpha_k + 1, 2^{-k-1})\right) = I_{k+1},$$

where $Y = M \circ Q$ and $B(\alpha_k, 2^{-k})$ is the closed ball with center α_k and radius 2^{-k} in $\mathbb{Z}_2$ (see Definitions 9 in §3.5, and 11 in §4.1).

Proof. From Lemma 4, we have

$$Q(\alpha_k) \equiv n_k \pmod{2^k} \quad \text{and} \quad Q(2\alpha_k + 1) \equiv m_{k+1} \pmod{2^{k+1}},$$

whatever the parity of k. Since Q is an isometry, it yields

$$Q(B(\alpha_k, 2^{-k})) = B(n_k, 2^{-k}) \quad \text{and} \quad Q(B(2\alpha_k + 1, 2^{-k-1})) = B(m_{k+1}, 2^{-k-1}).$$

Hence the result. □

We now establish the existence of infinitely many affine relationships within $\mathbf{Q}_{3x+1}$ and give their analytic expressions.

Theorem 7. *The set* $\mathbf{Q}_{3x+1} = (X, Y)(\mathbb{Z}_2)$ *admits the self-affine relationships*

$$(X, Y)(2r) = \frac{1}{2}(X, Y)(r) \tag{27}$$

for $r \in \mathbb{Z}_2$, *and*

$$(X, Y)(2r + 1) = \frac{1}{2}(X, Y)(r) + (1, 1 - 2^{-k}) \tag{28}$$

for $r \equiv \alpha_k \pmod{2^k}$ *and* $k \geq 2$, *where* $X = M$ *and* $Y = M \circ Q$ *as in Definition 11 (§4.1).*

Proof of (27). For all 2-adic integers r,

$$M(2r) = \frac{1}{2}M(r)$$

and, by the functional equation (17),

$$M(Q(2r)) = M(2Q(r)) = \frac{1}{2}M(Q(r)).$$

□

Proof of (28). Let $r \equiv \alpha_k \pmod{2^k}$. It is easily seen that

$$M(2r + 1) = 1 + \frac{1}{2}M(r).$$

Now, recall the functional equation (18):

$$Q(2r + 1) = 2Q(r) + 1 - 2^k.$$

Lemma 4 gives $Q(r) \equiv n_k \pmod{2^k}$, yielding the 2-adic expansion

$$Q(r) = 1 + 2 + 2^2 + \ldots + 2^{k-3} + 2^{k-1} + \ldots.$$

Hence, we obtain

$$M(Q(2r + 1)) = 1 + \frac{1}{2}M(Q(r)) - 2^{-k}.$$

□

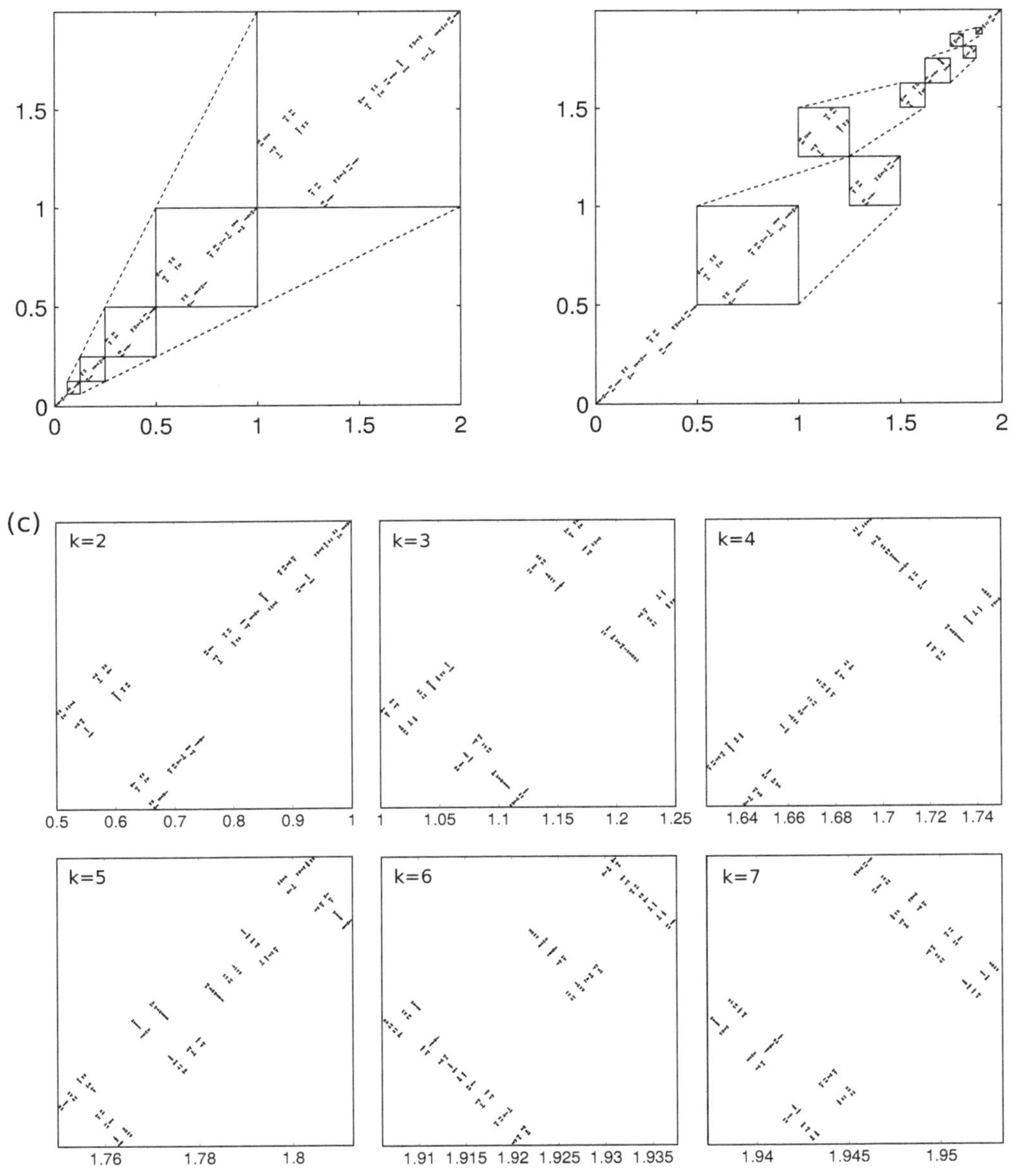

Figure 2: **(a-b)** Identical parts of $\mathbf{Q}_{3x+1}$ through the affine transformations (27) and (28) in (a) and (b) respectively. **(c)** Enlarged parts of $\mathbf{Q}_{3x+1}$ delimited by some of the boxes in (b), namely, J_k^2 for k even and $I_k \times J_k$ for k odd, with $2 \leq k \leq 7$.

Parity	Box 1	Box 2	Affine transformation
-	$[0,2]^2$	$[0,1]^2$	$(x,y) \longmapsto \left(\frac{x}{2}, \frac{y}{2}\right)$
k even	J_k^2	$J_{k+1} \times I_{k+1}$	$(x,y) \longmapsto \left(1+\frac{x}{2}, 1+\frac{y}{2}-2^{-k}\right)$
k odd	$I_k \times J_k$	I_{k+1}^2	$(x,y) \longmapsto \left(1+\frac{x}{2}, 1+\frac{y}{2}-2^{-k}\right)$

Table 5: Pairs of boxes covering two parts of $\mathbf{Q}_{3x+1}$ that coincide modulo an affine transformation with scaling factor $\frac{1}{2}$. Note that $k \geq 2$.

It follows from Theorem 7 together with Corollary 5 that some parts of the $3x+1$ set described in Table 5 are identical through the affine transformations represented on Figures 2a and 2b. The first line of Table 5 is linked to the autoconjugacy (17). The fact that the boxes $[0,2]^2$ and $[0,1]^2$ are nested leads to an infinite descent on the "even" side of $\mathbf{Q}_{3x+1}$ towards the origin. On the other hand, the equation (28) applies only for odd values of the parameter r, except when $k = 2$. From the corresponding pairs of boxes in Table 5, we obtain a covering of the "odd" side of $\mathbf{Q}_{3x+1}$, which takes the form in Figure 2b of an infinite "cascade" along the diagonal Δ.

Other functional equations like (16) and (21) do not imply self-similarity on our plane representation in Figure 1b, due to the value 3 not being a power of 2.

On Figure 2c, we show enlarged parts of $\mathbf{Q}_{3x+1}$, mainly from the "odd" side, that are delimited by squares of side-length 2^{1-k} for $2 \leq k \leq 7$. Putting together all the affine similarities, it yields that the content of the first square J_2^2 is made entirely of small copies of J_k^2 for even $k \geq 4$, and $I_k \times J_k$ for odd $k \geq 3$, plus an extra point at $\left(\frac{1}{2}, \frac{1}{2}\right)$. It is a puzzling question whether every square on Figure 2c is also made of small pieces taken elsewhere within the odd side of $\mathbf{Q}_{3x+1}$ and outside $J_3 \times I_3$.

Nonetheless, one observes some relative diversity of patterns. Unlike for the Cantor ternary set, there seems to be no simple geometric scheme able to reproduce $\mathbf{Q}_{3x+1}$, in the sense that more and more calculations are required for refining the shape of each pattern.

Finally, the unveiling of self-similarity at all scales raises the question of the Hausdorff dimension, to which the following theorem answers without much difficulty.

Theorem 8. *The set* $\mathbf{Q}_{3x+1}$ *has Hausdorff dimension 1.*

Proof. First, observe that the Hausdorff dimension of $\mathbf{Q}_{3x+1}$ is at least 1, as it contains at least one point for each abscissa taken in the interval $[0,2]$.

For all $k \geq 0$, we obtain from Lemma 3 (§4.1) a covering of $\mathbf{Q}_{3x+1}$ made of

$\nu_k = 2^k$ boxes of side-length $l_k = 2^{1-k}$. The number of boxes is minimal because the number of intervals of length l_k required to cover $[0, 2]$ is at least 2^k.

Therefore the "box-counting" dimension of $\mathbf{Q}_{3x+1}$ is equal to

$$\lim_{k\to\infty} \frac{\log \nu_k}{\log\left(\frac{1}{l_k}\right)} = \lim_{k\to\infty} \frac{k}{k-1} = 1,$$

which is an upper bound of its Hausdorff dimension. □

Acknowledgements. I am indebted to the anonymous referee for his valuable comments that helped to improve the exposition of the $3x + 1$ set.

References

[1] E. Akin, Why is the 3x+1 problem hard?, *Contemp. Math.* **356** (2004), 1–20.

[2] V. Anashin, Ergodic transformations in the space of p-adic integers, in *p-Adic Mathematical Physic*, AIP Conference Proceedings, Vol. 826, Belgrade, 2006, 3–24. Available at `https://arxiv.org/abs/math/0602083` .

[3] V. Anashin, A. Khrennikov, E. Yurova, Ergodicity criteria for non-expanding transformations of 2-adic spheres, *Discrete Contin. Dyn. Syst.* **34** (2014), 367–377.

[4] P. Andaloro, On total stopping times under 3x+1 iteration, *Fibonacci Quart.* **38** (2000), 73–78.

[5] D. J. Bernstein, A non-iterative 2-adic statement of the $3N + 1$ conjecture, *Proc. Amer. Math. Soc.* **121** (1994), 405–408.

[6] D. J. Bernstein, J. C. Lagarias, The $3x + 1$ conjugacy map, *Can. J. Math.* **48** (1996), 1154–1169.

[7] C. Böhm, G. Sontacchi, On the existence of cycles of given length in integer sequences like $x_{n+1} = x_n/2$ if x_n even, and $x_{n+1} = 3x_n + 1$ otherwise, *Atti Accad. Naz. Lincei Sci. Fis. Mat. Natur.* **64** (1978), 260–264.

[8] D. V. Chistyakov, Fractal geometry for images of continuous embeddings of p-adic numbers and solenoids into Euclidean spaces, *Theoret. Math. Phys.* **109** (1996), 1495–1507.

[9] J. Coquet, A summation formula related to the binary digits, *Invent. Math.* **73** (1983), 107–115.

[10] R. E. Crandall, On the "$3x + 1$" problem, *Math. Comp.* **32** (1978), 1281–1292.

[11] A. A. Cuoco, Visualizing the p-adic integers, *Amer. Math. Monthly* **98** (1991), 355–364.

[12] A. Edgington, The autoconjugacy of a generalized Collatz map, preprint available at `https://arxiv.org/abs/1206.0553` (2012), 1–6.

[13] C. J. Everett, Iteration of the number theoretic function $f(2n) = n, f(2n+1) = 3n+2$, *Adv. Math.* **25** (1977), 42–45.

[14] L. E. Garner, On heights in the Collatz 3n+ 1 problem, *Discrete Math.* **55** (1985), 57–64.

[15] R. K. Guy, *Unsolved Problems in Number Theory*, Third Edition, Springer, 2004.

[16] Y. Hashimoto, A fractal set associated with the Collatz problem, *Bull. of Aichi Univ. of Education (Natural science)* **56** (2007), 1–6. Available at `https://www.researchgate.net/profile/Yukihiro_Hashimoto` .

[17] J. C. Lagarias, The $3x+1$ problem and its generalizations, *Amer. Math. Monthly* **92** (1985), 3–23.

[18] J. C. Lagarias, The set of rational cycles for the $3x + 1$ problem, *Acta Arith.* **56** (1990), 33–53.

[19] J. C. Lagarias (editor), *The Ultimate Challenge: The 3x+1 Problem*, Amer. Math. Soc., 2010.

[20] J. López, P. Stoll, The 3x+1 conjugacy map over a Sturmian word, *Integers* **9** (2009), 141–162.

[21] K. R. Matthews, A. Watts, A generalization of Hasse's generalization of the Syracuse algorithm, *Acta Arith.* **43** (1984), 167–175.

[22] H. Möller, Über Hasse's Verallgemeinerung des Syracuse-Algorithmus (Kakutani's Problem), *Acta Arith.* **34** (1978), 219–226.

[23] K. G. Monks, J. Yazinski, The autoconjugacy of the $3x + 1$ function, *Discrete Math.* **275** (2004), 219–236.

[24] A. F. Monna, Sur une transformation simple des nombres P-adiques en nombres réels, *Indag. Math.* **14** (1952), 1–9.

[25] H. Müller, Das '3n+1' Problem, *Mitt. Math. Ges. Hamburg* **12** (1991), 231–251.

[26] O. Rozier, The $3x + 1$ problem: a lower bound hypothesis, *Funct. Approx. Comment. Math.* **56** (2017), 7–23.

[27] R. Simonetto, *Conjecture de Syracuse : Avancées Inédites*, 2016. Available at `https://mathsyracuse.wordpress.com` .

[28] R. Terras, A stopping time problem on the positive integers, *Acta Arith.* **30** (1976), 241–252.

4

ESTIMATING THE DENSITY OF THE ABUNDANT NUMBERS

Dominic Klyve
Dept. of Mathematics, Central Washington University, Ellensburg, Washington
Dominic.Klyve@cwu.edu

Melissa Pidde
Dept. of Mathematics, Central Washington University, Ellensburg, Washington

Kathryn E. Temple
Dept. of Mathematics, Central Washington University, Ellensburg, Washington
Kathryn.Temple@cwu.edu

Abstract

Mathematicians have been interested in properties of abundant numbers – those which are smaller than the sum of their proper factors – for over 2,000 years. During the last century, one line of research has focused in particular on determining the density of abundant numbers in the integers. Current estimates have brought the upper and lower bounds on this density to within about 10^{-4}, with a value of $K \approx 0.2476$, but more precise values seem difficult to obtain. In this paper, we employ computational data and tools from inferential statistics to get more insight into this value. We also put a lower bound on the quantity of abundants in any interval of size 10^6. Finally, we consider the "time series" nature of our data, and consider the possibility of employing tools from this branch of statistics to more carefully refine our statistical estimates.

1. Introduction and Background

In the first century CE, Nichomachus defined an *abundant number* to be an integer the sum of whose proper factors is greater than the integer itself. The smallest abundant number is 12, with factors $1+2+3+4+6 = 16$; there are 22 such integers less than or equal to 100. It is easy to see that there are infinitely many abundant numbers – every multiple of a perfect or abundant number is itself abundant. This leads us quickly to the question: what is the density of the abundant numbers?

It is not immediately apparent that abundant numbers even have a density. In what follows, we shall use the concept of *natural density*, which roughly asks for the fraction of integers which are members of a given set. More formally, we use the following:

Definition 1. Given a set S of positive integers, let $S(n)$ be the number of integers $x \leq n$ with $x \in S$. We define the *natural density* of S, $\mathbf{d}S$, as:

$$\mathbf{d}S := \lim_{n\to\infty} \frac{S(n)}{n},$$

supposing the limit exists [10]. If the limit does not exist, the set does not have a natural density.

Let $\mathcal{A}$ be the set of abundant numbers, and let $\mathbf{d}\mathcal{A}$ represent their (natural) density. With this definition of density, Davenport [3] showed in 1933 that the density of the abundant numbers exists. Finding lower bounds for $\mathbf{d}\mathcal{A}$ is easy. Since every multiple of 6 is abundant, for example, we have $\mathbf{d}\mathcal{A} \geq 1/6$. Similarly, since 20 is the first abundant not a multiple of 6, we can put a stronger lower bound on the density of abundant numbers using a standard inclusion-exclusion argument, noting that $\mathbf{d}\mathcal{A} > 1/6 + 1/20 - 1/60 = 1/5$ (since multiples of 60 are double-counted by the first two terms). Similarly, stronger lower bounds can be found by considering more abundant numbers, again using standard ideas of inclusion and exclusion; we shall revisit this idea in Section 9.1.

Finding upper bounds on $\mathbf{d}\mathcal{A}$ is quite a bit harder; indeed it is not completely trivial to show that $\mathbf{d}\mathcal{A} < 1$. In 1932, however, Behrend [1] showed that $\mathbf{d}\mathcal{A} < 0.47$. In fact, he showed the slightly stronger statement that:

$$\frac{\mathcal{A}(n)}{n} < 0.47$$

for all integers n. Since that time, the best known bounds on $\mathbf{d}\mathcal{A}$ have steadily improved. The progress of the state of the art is shown in Table 1.

Year	Author	Lower	Upper
1932	Behrend [1]		0.47
1933	Behrend [2]	0.241	0.314
1955	Salié [13]	0.246	
1971 – 72	Wall, et al. [15, 14]	0.246	0.2909
1998	Deléglise [4]	0.2474	0.2480
2010	Kobayashi [6]	0.2476171	0.2476475

Table 1: History of the best-known bounds on the density of abundant numbers.

Our goal in this paper is to use computational and statistical methods to explore some questions concerning the distribution of abundant numbers over various ranges. We shall start with the distribution over a small range starting from 1, and then "zoom out" to larger ranges, pausing to describe what we can learn about the density of abundant numbers from each of these snapshots. Over larger ranges, we will naturally use larger subintervals of integers as our unit of study. Rather than try to improve the best known bounds on $\mathbf{d}\mathcal{A}$ (currently due to Kobayashi, whose work involved hundreds of CPU hours, several clever custom algorithms, and was the primary work of his Ph.D. dissertation and a paper [6, 7]), we use our computational and statistical methods to observe several features of the distribution of abundant numbers in various ranges. We shall also introduce the idea of using techniques from the theory of Time Series Analysis to provide insight into the behavior of number-theoretic functions.

While the final interval in Table 1 provides the best proven bounds on the density of abundant numbers, there are some theoretical reasons to think that the true density is at the lower end of Kobayashi's range. We shall thus use this lower bound throughout the paper as a possible conjectural value, and shall call this value $K = 0.2476171$.

2. Values of $\mathcal{A}(n)/n$ to 10^4

As a warm-up to our bigger study, we first examine the density of abundant numbers up to 10,000. Since the first abundant number is 12, we have $\mathcal{A}(n)/n = 0$ for $n < 12$, and thus it is perhaps not surprising to see that $\mathcal{A}(n)/n$ grows fairly monotonically at the beginning.

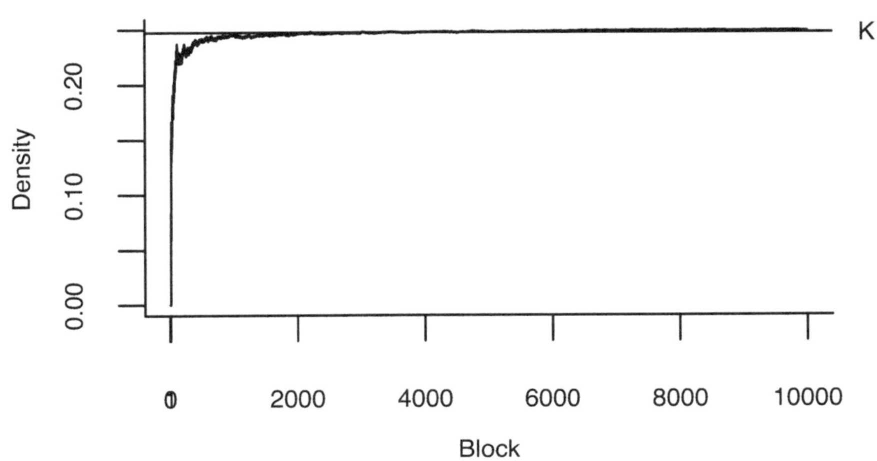

Figure 1: Density of abundant numbers up to 10^4.

Indeed, $\mathcal{A}(n)/n$ is so much smaller for values less than about 10^3 than for larger values this is is difficult to graph all of these on the same axis (see Figure 1). For this reason, we take a closer look at the behavior of $\mathcal{A}(n)/n$ over the region $[10^3, 10^4]$ in Figure 2.

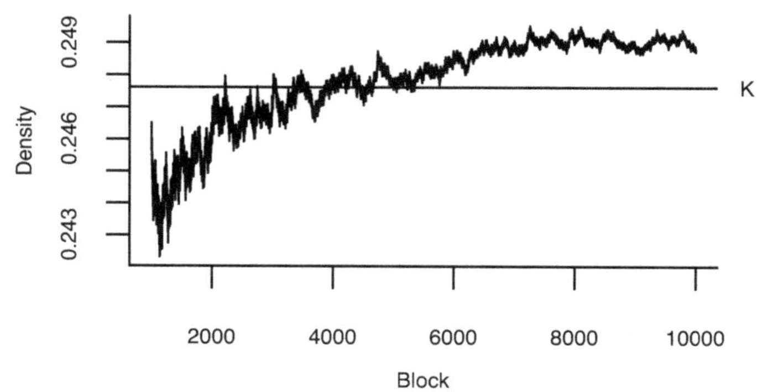

Figure 2: Density of abundant numbers in the range $[10^3, 10^4]$.

We see in this second plot that the density seems to be increasing. Somewhat surprisingly, $\mathcal{A}(n)/n$ passes Kobayashi's upper bound at about $n = 3000$, and does not drop back into his range in the interval $[3000, 10000]$. (Note that Kobayashi's upper and lower bounds on $d\mathcal{A}$ are graphically indistinguishable on the scale of this graph.) Of course, eventually it must do so, and we shall take our first step out in order to try to locate when this might happen.

3. Values of $\mathcal{A}(n)/n$ to 10^7

If we expand our "window" by three orders of magnitude, we find a similar theme on a larger scale. Before we examine the results, however, we establish a few notational conventions we shall follow as we move to larger ranges. In practice, there are computational limits on the number of integers for which we can store the value of $\mathcal{A}(n)$. We therefore collected data about the quantity of abundant numbers in blocks of size 10^6. Expanding on our previous notation, we make the following definition:

Definition 2. We shall use the notation $\mathbf{d}\mathcal{A}_{10^6,n}$ to denote the density of abundant integers in the range $[(n-1)10^6 + 1, n10^6]$.

Thus, $\mathbf{d}\mathcal{A}_{10^6,1}$ is the density of abundant numbers in the first block of one million integers, and $\mathbf{d}\mathcal{A}_{10^6,7}$ is the density of the abundant numbers in the range $[6000001, 7000000]$. A quick computer check gives the values in Table 2.

n	$\mathbf{d}\mathcal{A}_{10^6,n}$
1	0.247545
2	0.247491
3	0.247509
4	0.247766
5	0.247704
6	0.247681
7	0.247713
8	0.247782
9	0.247739
10	0.247807

Table 2: Values of $\mathbf{d}\mathcal{A}_{10^6,n}$ for $1 \leq n \leq 10$. Each gives the proportion of abundant integers in the nth block of size one million.

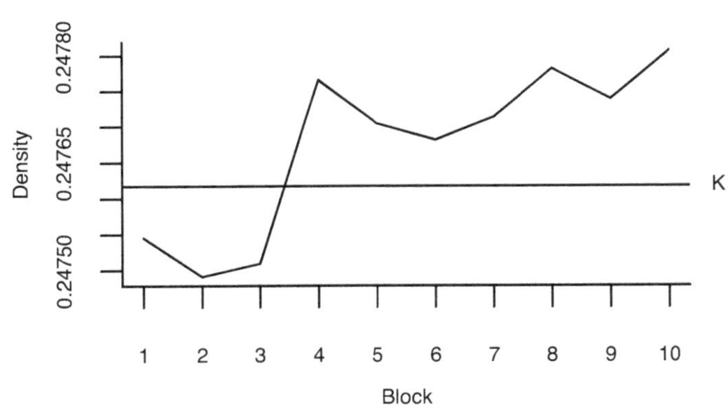

Figure 3: The density of abundant numbers, $\mathbf{d}\mathcal{A}_{10^6,n}$ for $n \in [1, 10]$. The horizontal line is at $K = .2476171$, which is Kobayashi's lower bound. Note that the values are being plotted here for each additional block, and are not cumulative.

As described above, we see in Figure 3 a pattern similar to that seen on the interval $[1, 10^4]$. In the first three blocks of size 10^6, the density of abundant numbers is less than Kobayashi's lower bound, after which they are all greater. Once again, we find that values of $\mathcal{A}(n)/n$ seem to be generally increasing.

It is encouraging that we are able to identify large-scale changes in the density of abundant numbers using these block values. Indeed, from this point on most of the study of abundant numbers in this work will center on these values of $\mathbf{d}\mathcal{A}_{10^6,n}$.

4. Values of $\mathcal{A}(n)/n$ to 10^8

Given that the density of abundant numbers is generally increasing until 10^7, and given that it cannot continue to do so indefinitely, it is reasonable for us to ask when the density levels out or decreases. If we zoom out one more order of magnitude, we find a partial answer to this question; a graph of the block densities $\mathbf{d}\mathcal{A}_{10^6,n}$ for $n \in [1, 100]$ is in Figure 4.

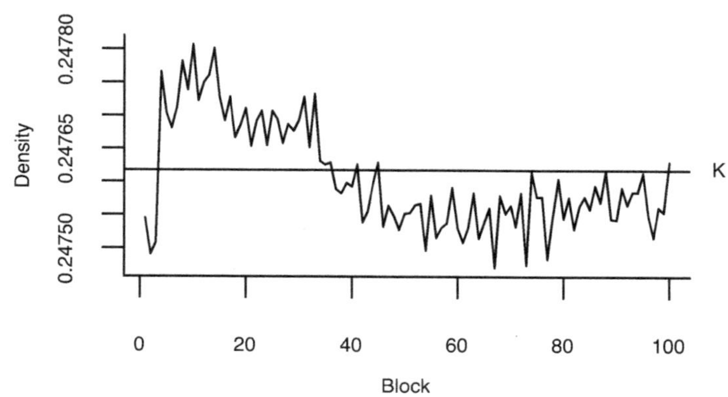

Figure 4: Values of $\mathbf{d}\mathcal{A}_{10^6,n}$ for $n \in [1, 100]$.

The increasing trend, it seems, was short-lived. After about 4×10^7, values of $\mathbf{d}\mathcal{A}_{10^6,n}$ drop below K, and stay below for almost all of these blocks up to 10^8.

A graph of the running averages $\mathcal{A}(n)/n$ is in Figure 5. It shows that, by approximately $80 \cdot 10^6$, $\mathcal{A}(n)/n$ has dropped below K and stays there through the end of this range.

4.1. An Analogy: Counting Primes With $\pi(x)$, li(x) and $\frac{x}{\log x}$

As we consider the relationship between $\mathcal{A}(n)/n$ and $d\mathcal{A}$, we note that the relationship between the asymptotic limit of a number-theoretic ratio and its finite values can take two general forms. Consider, for example, the relationship between the prime counting function $\pi(x)$, the logarithmic integral $\text{li}(x) := \int_2^x \frac{1}{\log t} dt$, and the function $x/\log(x)$. While it is true that asymptotically both $\pi(x) \sim \text{li}(x)$ and $\pi(x) \sim x/\log(x)$, in the latter case the asymptotic belies a subtle inequality. In fact, for all $x \geq 17, \pi(x) > \frac{x}{\log x}$, and thus $\pi(x) - \frac{x}{\log x} > 0$ for all $x \geq 17$ [12]. On the other hand, Littlewood proved that the sign of $\pi(x) - \text{li}(x)$ changes infinitely often [9] .

Based on the evidence so far, it becomes increasingly tempting to conjecture that $\mathcal{A}(n)$ will oscillate above and below $d\mathcal{A}$ infinitely often. More data might help us

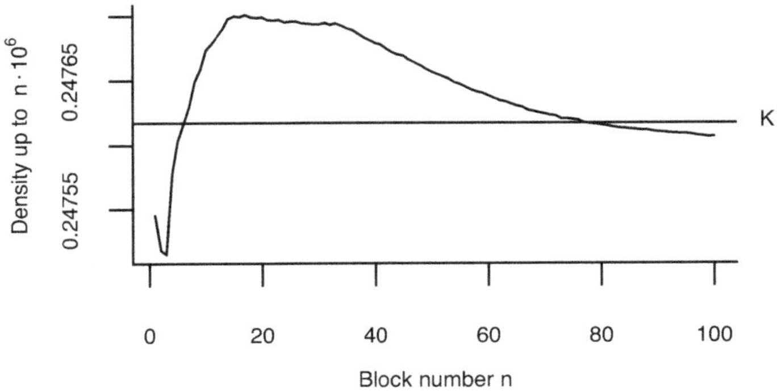

Figure 5: Cumulative density of the abundant numbers up to $n \times 10^6$ for all $n \in [0, 100]$. The horizontal line is at height K.

be more certain, however, and to that end we consider values over a larger scale.

5. Values of $\mathcal{A}(n)/n$ to 10^{11}

We next consider integers up to 10^{11}, giving us 100,000 blocks of size 10^6. At this point, there is enough data that plotting it on a single graph renders the graph unreadable, so instead we include as representative of this data graphs of the first 1000 blocks (integers up to size 10^9), block numbers 9000 - 10,000 (integers between 9×10^9 and 10^{10}) and block numbers 99,000 through 100,000 (integers between 9.9×10^{10} and 10^{11}). These are Figures 6, 7, and 8, respectively.

While there is definitely some initial variability in the densities, this appears to have diminished as the block number increases, and the last two sets of blocks visually appear to be quite similar. There is still substantial block-to-block variation, but overall the pattern does not seem to be changing much, if at all.

Note that because we have an order structure given by the block number (and indirectly by the ordering of the integers), we can think of the data series as being analogous to a time series - a sequence of data points indexed by time. Loosely speaking, statisticians refer to time homogeneity as *stationarity*. We'll discuss stationarity in more detail in Section 7. For now, we simply note that this suggests that it might be useful to apply statistical reasoning and techniques to develop more insight into the behavior of abundant numbers, and specifically into the values $\mathbf{d}\mathcal{A}_{10^6,n}$. In the next section, therefore, we begin to use statistical techniques to study these values.

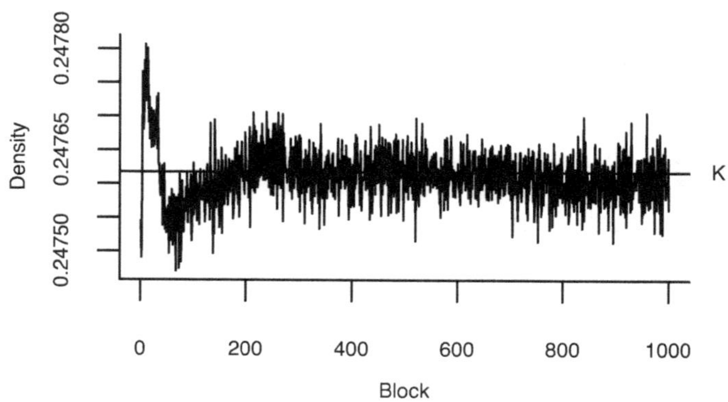

Figure 6: Density of abundant numbers in the first 1000 blocks of size 10^6.

6. A Statistical Approach to Abundant Numbers

Given an integer n, we let $a(n)$ be the function that returns 1 if n is abundant, and 0 otherwise. Intuitively, we might expect $a(n)$ to behave like a random variable taking values from the set $\{0, 1\}$, with $a(n)$ taking the value 1 with probability $\mathbf{d}\mathcal{A}$, and taking the value 0 with probability $1 - \mathbf{d}\mathcal{A}$. To that end, let us define just such a random variable, X, as follows:

$$X := \begin{cases} 0 & \text{with probability } 1 - \mathbf{d}\mathcal{A}; \\ 1 & \text{with probability } \mathbf{d}\mathcal{A}. \end{cases}$$

A random variable X which takes two distinct values with given probabilities is a *Bernoulli* variable. It has a mean $\mu_X = \mathbf{d}\mathcal{A}$ and standard deviation $\sigma_X = \sqrt{\mathbf{d}\mathcal{A}(1 - \mathbf{d}\mathcal{A})}$. Since we are working under the assumption that X behaves like $a(n)$, we use the midpoint of Kobayashi's bounds on $\mathbf{d}\mathcal{A}$ above to estimate $\mu_X \approx 0.247632$ and $\sigma_X \approx 0.431637$.

Because we are interested in the distribution of values of $\mathbf{d}\mathcal{A}_{10^6,n}$, we also define a new random variable Y to model the proportion of abundant integers in blocks of size 10^6 as

$$Y := \frac{1}{10^6} \sum_{i=1}^{10^6} X_i.$$

We can model the distribution of Y by recalling the Central Limit Theorem.

Theorem 3 (Central Limit Theorem). *Given a set of independent, identically distributed variables $X_1, X_2, \ldots X_n$, whose common distribution has mean μ and*

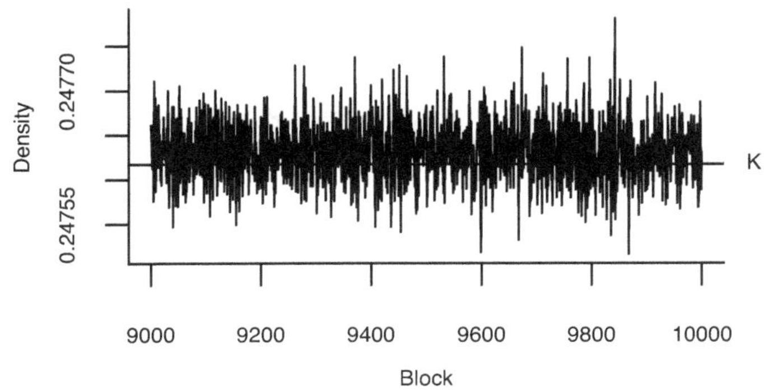

Figure 7: Density of abundant numbers in an interval of 1000 consecutive blocks of size 10^6, starting with block 9000.

finite standard deviation σ, for large n their mean $\bar{X}$ is approximately normally distributed with mean μ and standard deviation $\sigma/\sqrt{n}$; that is, the distribution of $\bar{X}$ can be approximated by the $N(\mu, \sigma/\sqrt{n})$ distribution.

We have previously defined the random variable X, trying to use it as a model for whether a single integer is abundant. Let X_i correspond to the ith integer. Since the variables X_i are indeed independent, identically distributed variables, then Y is approximately normally distributed with mean $\mathbf{d}\mathcal{A}$ and standard deviation $\sqrt{\frac{\mathbf{d}\mathcal{A}(1-\mathbf{d}\mathcal{A})}{10^6}}$.

Using the Central Limit Theorem, we can try to predict the behavior of the values of $\mathbf{d}\mathcal{A}_{10^6,n}$ over different values of n. We have:

$$\mathbf{d}\mathcal{A}_{10^6,n} = \frac{1}{10^6} \sum_{i=(n-1)10^6+1}^{n\cdot 10^6} a(i).$$

If we assume that the $a(i)$ are independent, identically distributed variables, we would expect that Y behaves like $\mathbf{d}\mathcal{A}_{10^6,n}$, or more formally,

Conjecture 4. The values of $\mathbf{d}\mathcal{A}_{10^6,n}$ over all values of n are normally distributed with mean $\mathbf{d}\mathcal{A}$ and standard deviation $\sqrt{\frac{\mathbf{d}\mathcal{A}(1-\mathbf{d}\mathcal{A})}{10^6}}$. Taking the midpoint of Kobayashi's interval as our value for $\mathbf{d}\mathcal{A}$, we expect more precisely that the values of $\mathbf{d}\mathcal{A}_{10^6,n}$ will fall in an $N(0.247632, 0.0004316)$ distribution.

It now remains to check this conjecture against the computational data.

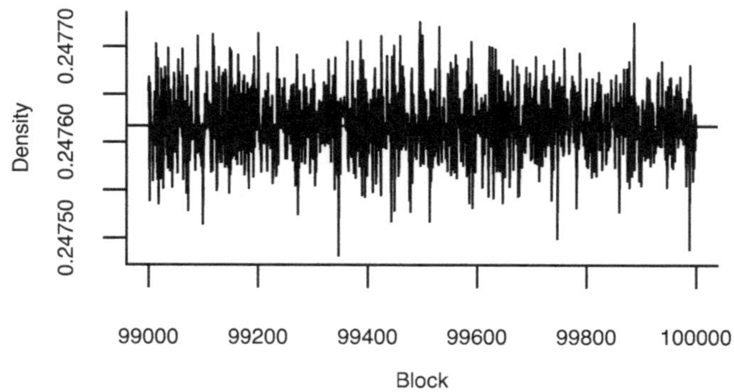

Figure 8: Density of abundant numbers in an interval of 1000 consecutive blocks of size 10^6, starting with block 99000.

6.1. Missing Variability in $\mathbf{d}\mathcal{A}_{10^6,n}$

As expected, the values of $\mathbf{d}\mathcal{A}_{10^6,n}$ are normally distributed (Figure 9). However, the standard deviation is far from what we conjectured above. We expected to find a standard deviation of about 4.32×10^{-4}. Instead, the value is only 3.54×10^{-5} – more than an order of magnitude less. We next need to address the question of why the values of $\mathbf{d}\mathcal{A}_{10^6,n}$ seem to be clustered together so much more than we would expect.

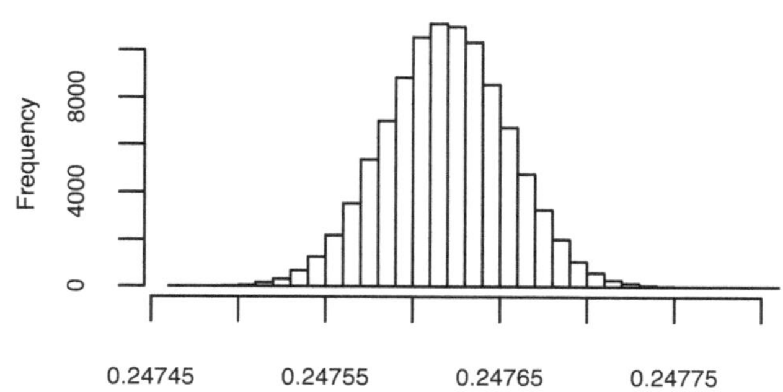

Figure 9: A histogram giving the number of blocks with given density from the values of $\mathbf{d}\mathcal{A}_{10^6,n}$ for all $n \in [1, 100\ 000]$.

In formulating Conjecture 4, we used the hypothesis that the individual values of $a(n)$ were independent. That is, we assumed that whether any particular integer is abundant does not influence whether any other integer is abundant. We first need to examine this assumption more closely.

6.2. Independence of Values of $a(n)$

Recall that our function $a(n)$ is an indicator function on whether an integer n is abundant. Before considering the trickier question of the behavior of the values of $\mathbf{d}\mathcal{A}_{10^6,n}$, let us first look at the independence of $a(n)$. For example, one implication of independence would be that knowing that an integer n is abundant would not change the likelihood that $n+1$ is abundant.

Even without answering the question of independence precisely, we can note several things. The first is that even numbers are considerably more likely to be abundant than odd numbers – this alone suggests that if n is an abundant even integer, $n+1$ is unlikely to be abundant. A related consideration concerns short runs. Since every multiple of 6 is abundant, we see that the abundants occur with more regularity than they would if $a(n)$ truly behaved like a sequence of independent random variables. Therefore, we see that the values of $a(n)$ are not independent, and that we shall therefore need more sophisticated methods to describe the distribution of $\mathbf{d}\mathcal{A}_{10^6,n}$.

6.3. Back to the Distribution of $\mathbf{d}\mathcal{A}_{10^6,n}$

We prove in Section 9 that any block of size 10^6 must necessarily contain at least $237,110$ abundant numbers. This lower bound helps explain the unusually small standard deviation in the values of $\mathbf{d}\mathcal{A}_{10^6,n}$. Since the blocks contain about 247,600 abundant numbers on average, the "random" nature of the distribution will occur entirely in the remaining $10,000$ or so values. That is, we should expect a much smaller standard deviation in the values of $\mathbf{d}\mathcal{A}_{10^6,n}$ than our naïve heuristic suggested.

6.4. Independence of Values of $\mathbf{d}\mathcal{A}_{10^6,n}$

We know that values of $a(n)$ aren't really independent, but what about $\mathbf{d}\mathcal{A}_{10^6,n}$? Because each of these values incorporates information about the abundance of 10^6 different integers, it seems likely that the issues arising from dependence of $a(n)$ will be "smoothed out", letting us make the following conjecture.

Conjecture 5. The values of $\mathbf{d}\mathcal{A}_{10^6,n}$ over all values of n form a sequence of independent, identically distributed variables.

We shall undertake a careful exploration of Conjecture 5 in Section 7.

If Conjecture 5 is true, we can use the first 100,000 blocks described in Section 5 to calculate an approximate 95% confidence interval for the true mean value

$\mathbf{d}\mathcal{A}_{10^6,n}$, and therefore the value of $\mathbf{d}\mathcal{A}$. Over these first 100,000 values, we first find $\bar{x} = 0.24761949309$ and $s = 3.54164 \times 10^{-5}$, and thus our confidence interval becomes

$$0.24761949309 \pm 1.96 * \frac{(3.54164 \times 10^{-5})}{\sqrt{10^5}},$$

or $(0.247619274, 0.247619713)$. Note that this interval lies near the bottom of Kobayashi's interval of $(0.2476171, 0.2476475)$, and thus if our estimate is correct, his lower bound is very near the truth.

However, we must be quite careful here. Some of the graphs presented earlier suggest that there are long-term trends in the values of $\mathbf{d}\mathcal{A}_{10^6,n}$, and that there may be some time series structure to these data.

If the values of $\mathbf{d}\mathcal{A}_{10^6,n}$ are not independent, then the confidence interval calculated above is not valid. We therefore turn next to the field of time series analysis to consider whether the values of $\mathbf{d}\mathcal{A}_{10^6,n}$ are, in fact, independent, and if not to apply some alternate techniques to calculate a confidence interval for $\mathbf{d}\mathcal{A}$.

7. Abundant Densities as Time Series

Since the values $\mathbf{d}\mathcal{A}_{10^6,n}$ have an order structure (they are ordered by the block index n), we can analyze the independence of $\mathbf{d}\mathcal{A}_{10^6,n}$ using tools from time series analysis, where the block index n plays the role of the time variable. A standard tool in time series analysis to look at the order structure of a time series (if any) is the autocorrelation function. The autocorrelation function of a time series Y_t, or ACF, is defined by $\rho(k) = Corr(Y_t, Y_{t+k})$, where $Corr(\cdot, \cdot)$ denotes the correlation between two random variables. The observant reader will note that implicit in this notation is the assumption that the correlation does not depend on the time, t, but only on the distance between the observations, k, which is often referred to as the lag. This will be true provided that the series Y_t is *second-order stationary* or *weakly stationary*, that is, the means of the observations, $E[Y_t]$, variances, $Var(Y_t)$, and covariances, $Cov(Y_t, Y_{t+k})$, are independent of t. For our series $\mathbf{d}\mathcal{A}_{10^6,n}$, the graph suggests that the values from about block number 10,000 to block number 100,000 do, in fact, represent a stationary time series. Since the beginning of the series seems to represent 'burn-in' or initial noisiness not typical of the rest of the series, we will work with the data set starting at block 10,001; this is well past any unusual initial behavior.

Since the data appear to be stationary, we consider the sample ACF, $\hat{\rho}(k)$. If the densities in successive blocks are truly independent, then all of the sample autocorrelations will be near 0. However, $\hat{\rho}(1) = -.169$, so we see that successive blocks are not independent and, in fact, are negatively correlated. Considering edge effects, this seems reasonable: a slightly higher proportion of abundant numbers in

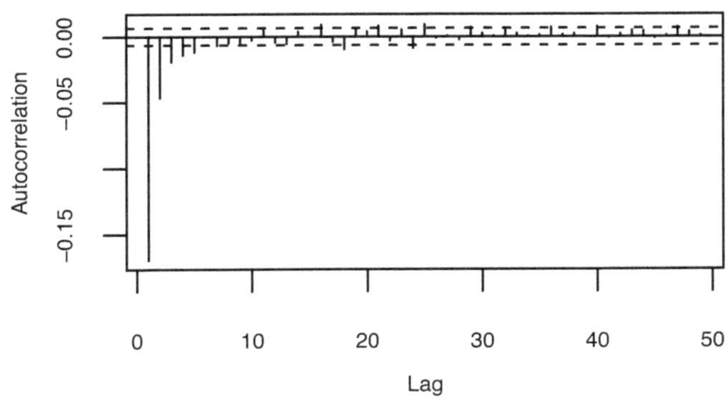

Figure 10: Sample ACF for $\mathbf{d}\mathcal{A}_{10^6,n}$.

one block may mean that it had an extra multiple of 6, 20, etc; the next block then might have a slightly lower proportion of abundant numbers. The second autocorrelation is also negative, but with smaller magnitude, and the magnitude then falls off somewhat quickly (Figure 10).

Unfortunately, what this goes to establish is that the block densities $\mathbf{d}\mathcal{A}_{10^6,n}$ are not independent in n, which means that the conditions for the standard normal-based confidence interval discussed above are not met. However, the problem of giving a confidence interval for the mean of a stationary time series is a standard one in the time series literature. There are various approaches proposed for calculating such a confidence interval; we will implement a bootstrap-based method called the *moving block bootstrap.* This method was chosen because of its relatively mild assumptions and, in particular, the fact that it does not require a particular model for the time series to be specified.

The moving block bootstrap method was originally proposed by Künsch in [8]. (Note that there is no relationship between the general term 'block' in the bootstrap method and our use of it to refer to a set of integers in this paper.) For this method, our original series of 90,000 observations is divided into $(90,000 - b + 1)$ overlapping blocks of block size b. Then $90,000/b$ of these blocks are chosen with replacement and used to form a new series of 90,000 observations. We compute the mean of the new series and then repeat the whole process many times. The resulting means are viewed as a sample from an empirical distribution, and then the bootstrap confidence interval uses the percentiles of this sample to give a confidence interval for the population mean. The table below gives results for several choices of block size from $b = 100$ to $b = 10,000$. In each case, the re-sampling procedure was repeated 10,000 times and the 2.5th and 97.5th percentiles of the resulting empirical distributions were taken as lower and upper confidence limits. Along with the confidence interval itself, the table also gives the width of each confidence interval. Notice that the confidence intervals are consistent with the confidence interval calculated using the standard normal-based techniques, and the narrowest of the intervals is slightly narrower than the classical confidence interval.

Block size	Lower endpoint	Upper endpoint	Width
100	0.24761881	0.24761918	3.7×10^{-7}
500	0.24761867	0.24761930	6.3×10^{-7}
1000	0.24761855	0.24761940	8.5×10^{-7}
5000	0.24761816	0.24761961	1.45×10^{-6}
10,000	0.24761801	0.24761943	1.42×10^{-6}
Classical CI	0.247619274	0.247619713	4.39×10^{-7}

These techniques seem promising, but do require that the time series be stationary (since estimating the mean requires that there be a time-invariant mean to estimate). In our efforts to see whether it is plausible that these conditions hold, we made a final zoom out to integers of size 10^{12}.

8. Values of $\mathcal{A}(n)/n$ to 10^{12}

In an effort to see whether this behavior continues, we then ran an additional 900,000 blocks, giving us 10^6 blocks each of size 10^6 (representing abundant integers out to size 10^{12}). With this much data, however, it turns out that the noisiness between blocks is actually hiding some of the underlying behavior of the data. This is more obvious if we consider somewhat larger blocks. If we take these 10^{12} integers but use blocks of size 10^9 rather than 10^6 (giving us 1000 data points rather than 10^6), we see that our conjecture that the series was a stationary time series is, unfortunately, not the case.

Figure 11 shows some very interesting behavior; the averages are below Kobayashi's lower bound for a sizable range approximately between blocks 200 and 400, corresponding to integers between 2×10^{11} and 4×10^{11}. Beyond this there appears to be a very slow upward trend through the range calculated. We see that the range up to block 100 at this scale (representing the first 100,000 blocks at the previous scale) do appear to be stationary and fairly stable, but that the behavior changes noticeably shortly thereafter.

Unfortunately, this means that further analysis based on looking at the block densities as a stationary time series is not possible, since there is a noticeable slow upward trend. The fact that there seem to be oscillations leads to the following conjecture:

Conjecture 6. The values of $\frac{\mathcal{A}(n)}{n} - d\mathcal{A}$ will change sign infinitely often.

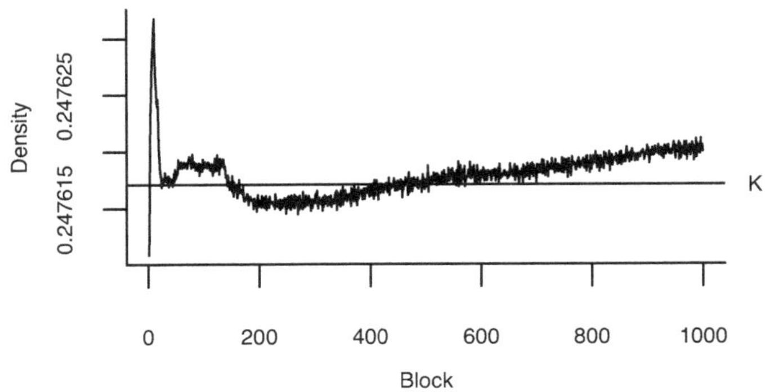

Figure 11: Density of abundant numbers using a block size of 10^9, over the first 1000 blocks. The horizontal line is Kobayashi's lower bound.

Despite the fact that our statistical efforts have not been successful in determining new truths about the behavior of $\frac{\mathcal{A}(n)}{n}$ at infinity, we have found some interesting results concerning bounds on the density of abundant numbers in short intervals, a question which it seems has not been considered before. We shall turn to this in the in final section of this paper.

9. Bounds on the Density of Abundant Numbers in Short Intervals

Having found both some promising leads and some disappointing dead ends in our statistical analysis, we conclude this paper with a pair of theorems which determine upper and lower limits on the number of abundants in any consecutive set of 10^6 integers – that is, we explore what limitations exist on the values of $\mathbf{d}\mathcal{A}_{10^6,n}$.

9.1. A Lower Bound on $\mathbf{d}\mathcal{A}_{10^6,n}$

In Section 1, we included a short inclusion-exclusion argument that the density of abundant numbers is at least $1/5$. We can use a more precise version of this reasoning to put an absolute lower bound on the quantity of abundant numbers in any block of 10^6 consecutive integers. We will use the following notation:

Definition 7. Given a set of integers X, let $f_X(d)$ the number of elements of X divisible by d. If there is no confusion about the set in question, we will shorten this notation simply to $f(d)$.

Now let X be any block of 10^6 consecutive integers. Then $f(6) = f_X(6)$, the number of integers in the set divisible by 6, satisfies $\lfloor 10^6/6 \rfloor \leq f(6) \leq \lceil 10^6/6 \rceil$

(where $\lfloor \cdot \rfloor$ and $\lceil \cdot \rceil$ represent the floor and ceiling functions, respectively), with $f(6)$ taking the higher or lower of these values depending on the value of the smallest element of $X \pmod 6$. Since each multiple of 6 is abundant, we can say that there are at least 166666 abundant numbers in any such block. Furthermore, $f(20)$ satisfies $\lfloor 10^6/20 \rfloor \leq f(20) \leq \lceil 10^6/20 \rceil$ and so on. Using this same idea and our inclusion-exclusion argument from Section 1, we find that $\mathbf{d}\mathcal{A}_X$, the proportion of integers in X which are abundant, can be bounded as

$$\begin{aligned}\mathbf{d}\mathcal{A}_X &\geq \frac{1}{10^6}(f_X(6) + f_X(20) - f_X(60)) \\ &\geq \frac{1}{10^6}(\lfloor 10^6/6 \rfloor + \lfloor 10^6/20 \rfloor - \lceil 10^6/60 \rceil) = .199999.\end{aligned}$$

This lower bound is slightly less than the bound of 1/5 given above. The difference reflects the fact that we are now considering the density of abundant numbers in intervals, and thus each term in the inclusion-exclusion argument has the potential to contribute errors due to rounding. In the following we shall work to balance the benefits of including more terms in the argument with the fact that each additional term carries this potential for rounding error.

If we are going to use more values for our inclusion-exclusion argument (beyond 6 and 20), which values should we choose? The natural choice is the set of *primitive nondeficient* numbers. The set of nondeficient numbers is slightly larger than the set of abundant numbers; it includes both abundant numbers and perfect numbers – those integers precisely equal to the sum of their proper factors. A nondeficient number is *primitive* if it is not a multiple of another nondeficient number.

More generally, we can find explicit bounds on the number of abundant integers in an interval using the Bonferroni inequalities. We shall begin with the general setting for the Bonferroni inequalities as given in [11, Theorem 6.8].

Theorem 8 (Bonferroni inequalities). *Let X be a nonempty, finite set of N objects, and let $P_1, \ldots, P_r$ be properties that elements of X may have. For each subset $I \subset \{1, 2, \ldots, r\}$, let $N(I)$ denote the number of elements of X that have each of the properties indexed by the elements of I. Let N_0 denote the number of elements of X with none of these properties. Then if m is a nonnegative even integer,*

$$N_0 \leq \sum_{k=0}^{m} (-1)^k \sum_{\substack{I \subset \{1,2,\ldots,r\} \\ |I|=k}} N(I),$$

while if m is a nonnegative odd integer,

$$N_0 \geq \sum_{k=0}^{m} (-1)^k \sum_{\substack{I \subset \{1,2,\ldots,r\} \\ |I|=k}} N(I).$$

(Here we take $N(\emptyset)$ to be $|X|$, the cardinality of X.)

We shall take the set X in this theorem to be our set above, that is, a set of 10^6 consecutive integers. We shall let P_j be the property that an integer is divisible by j, so $N(\{n_1, n_2, \ldots, n_k\}) = f(n)$, where n is the least common multiple of $\{n_1, n_2, \ldots, n_k\}$. Then N_0 will be the number of elements of X not divisible by any of $n_1, n_2, \ldots, n_r$.

The value N_0 bears a useful relationship with the number of nondeficient integers (hereafter, nondeficients) in X, which we will denote by $\mathcal{A}_X$. In particular, for any set of nondeficient integers I, the union of the nondeficients in X and those integers counted by N_0 contains all of X, so we have $|X| \leq \mathcal{A}_X + N_0$, or

$$\mathcal{A}_X \geq |X| - N_0. \tag{1}$$

We shall combine these ideas in Theorem 10 to develop a method to find better explicit lower bounds on the density of nondeficients in intervals.

We will first find it useful to have one more definition:

Definition 9. Given a set of integers I, we shall denote by $\operatorname{lcm}(I)$ the least common multiple of all elements of I.

Now applying the Bonferroni inequalities to our setting, we have the following:

Theorem 10. *Let X be a set of n consecutive integers and let d be any positive integer, and recall that $f(k)$ represents the number of elements of X divisible by k. As above, let $\mathcal{A}_X$ denote the number of nondeficient integers in X. Finally, let A_d represent the set of the first d primitive nondeficient integers. Then if m is a nonnegative even integer, we have*

$$\mathcal{A}_X \geq \sum_{\substack{I \subset A_d \\ 0<|I|\leq m}} (-1)^{|I|+1} f(\operatorname{lcm}(I)),$$

Proof. From Theorem 8, we know that for nonnegative even integers m, we have

$$N_0 \leq \sum_{k=0}^{m} (-1)^k \sum_{\substack{I \subset A_d \\ |I|=k}} N(I).$$

Thus, by (1), the quantity of nondeficient integers in X, $\mathcal{A}_X$, satisfies

$$\mathcal{A}_X \geq |X| - \sum_{k=0}^{m} (-1)^k \sum_{\substack{I \subset A_d \\ |I|=k}} N(I).$$

Note that in our notation, we have that $N(I) = f(\operatorname{lcm}(I))$, and thus we can write this as

$$\mathcal{A}_X \geq |X| - \sum_{k=0}^{m} (-1)^k \sum_{\substack{I \subset A_d \\ |I|=k}} f(\operatorname{lcm}(I)).$$

The double sum above is taken over all subsets of A_d of size at most m. We can thus rewrite the double sum directly, and find

$$\mathcal{A}_X \geq |X| - \sum_{\substack{I \subset A_d \\ |I| \leq m}} (-1)^{|I|} f(\mathrm{lcm}(I)).$$

Finally, note that by the definition of $f_X(n)$, when I is the empty set, we have $(-1)^{|I|} \cdot f(\mathrm{lcm}(I)) = |X|$, and so we can rewrite the above as

$$\mathcal{A}_X \geq - \sum_{\substack{I \subset A_d \\ 0<|I| \leq m}} (-1)^{|I|} f(\mathrm{lcm}(I)) = \sum_{\substack{I \subset A_d \\ 0<|I| \leq m}} (-1)^{|I|+1} f(\mathrm{lcm}(I)),$$

proving the theorem. □

We can use Theorem 10 to find lower bounds on $\mathcal{A}_X$, the number of nondeficients in a set X, by choosing values of m and d. In practice, we will find it useful to replace each of the summands in Theorem 10 with an easily computable lower bound. We shall do this using the following definition.

Definition 11. Given a set of integers I, let $g(I)$ be defined as

$$g(I) := \begin{cases} \lfloor \frac{10^6}{\mathrm{lcm}(I)} \rfloor & \text{if } |I| \text{ is odd;} \\ \lceil \frac{10^6}{\mathrm{lcm}(I)} \rceil & \text{if } |I| \text{ is even.} \end{cases}$$

Note that we now have, for each of the summands in Theorem 10,

$$(-1)^{|I|+1} f(\mathrm{lcm}(I)) > (-1)^{|I|+1} g(I).$$

We therefore have the following.

Theorem 12. *Given X, m, and d as in Theorem 10, we can put a lower bound on the quantity of nondeficients in X, $\mathcal{A}_X$ as*

$$\mathcal{A}_X \geq \sum_{\substack{I \subset A_d \\ 0<|I| \leq m}} (-1)^{|I|+1} g(I).$$

Returning to the particular context in which we are working, we apply this theorem to blocks of size 10^6 and write:

Theorem 13. *Let A_d be the set of the first d primitive nondeficient integers. Then for any nonnegative even integer $m \leq d$ and any $n \geq 0$, we have that the proportion of nondeficient integers in any interval of length 10^6 is at least*

$$\frac{1}{10^6} \sum_{\substack{I \subset A_d \\ 0<|I| \leq m}} (-1)^{|I|+1} g(I).$$

Proof. The theorem follows by applying Theorem 12, setting X to be any set of 10^6 consecutive integers. □

The reader should confirm that choosing $d = 2$, $m = 2$ in Proposition 13 would give the result that the proportion of nondeficient integers in any interval of length 10^6 is at least $\frac{1}{10^6}\left(\lfloor 10^6/6\rfloor + \lfloor 10^6/20\rfloor - \lceil 10^6/60\rceil\right) = 0.199999$.

It remains for us to find the best lower bound by finding the best choices of d and m. In fact, for any given d, there will an "optimal" value of m – that is, a choice of m which gives the best lower bound. The basic idea here is that increasing m gives us more terms in our inclusion-exclusion calculation (and thus a better bound), but includes more rounding errors (and thus a worse bound). In practice, finding a general function for predicting the best value of m for a given d in these cases is difficult or impossible, and instead we computationally check all possible values of m to find the best choice.

In Table 3, for each choice of d, the optimal value of $m \leq d$ is shown, together with the resulting bound on the proportion of nondeficient integers in any interval of length 10^6. The bolded row is the optimal lower bound using this technique for those values of d tested (and likely for all d).

d	m	Bound
2	2	0.199999
3	2	0.216665
4	4	0.223803
5	4	0.228882
6	6	0.232969
7	6	0.234347
8	6	0.235469
9	6	0.236277
10	6	0.236765
11	6	0.237024
12	**4**	**0.237111**
13	4	0.236887
14	4	0.236527
15	4	0.236103
16	4	0.235535
17	4	0.235077
18	4	0.234633
19	4	0.234025
20	4	0.233248

Table 3: Best choices of m and corresponding best lower bounds on $\mathbf{d}\mathcal{A}_{10^6,n}$ from Proposition 13 for small values of d.

By choosing $d = 12$ and $m = 4$ in Proposition 13, we can show that 0.237111 is a lower bound of the proportion of nondeficient integers in any interval of length 10^6. (Also note that this seems to be the best possible lower bound using inclusion-exclusion techniques alone – stronger results may be possible, but would require other techniques.) The work and computations above serve to prove the following:

Proposition 14. *Any consecutive sequence of* 10^6 *integers contains at least* 237111 *nondeficient numbers.*

The count of nondeficient numbers, in practice, is very close to the count of abundant numbers (since perfect numbers are rare). It follows from the Euclid-Euler theorem on even perfect numbers that no block of size 10^6 (with smallest value at least 8128) can contain more than such value. (From Table 2, in the first 10^6 integers there are 247545 abundant numbers, so any block with smallest value below 8128 must contain at least 239417 abundant integers.) We can thus also state the following:

Corollary 15. *Assuming the non-existence of odd perfect numbers, any consecutive sequence of* 10^6 *integers contains at least* 237110 *abundant numbers.*

9.2. An Upper Bound on $\mathbf{d}\mathcal{A}_{10^6,n}$

It is not hard to show that the only upper bound on $\mathbf{d}\mathcal{A}_{10^6,n}$ is 10^6, that is, there are 10^6 consecutive abundant numbers. We could, in theory, even find these numbers by using the Chinese Remainder Theorem and solving appropriate modular equations using appropriate "small" abundant numbers.

In fact, Erdős proved the stronger result in 1935 that the longest sequence of consecutive abundant numbers up to n is of the order $\log\log\log n$ [5]. We should thus expect a consecutive sequence of 10^6 abundant numbers somewhere in the range of $\exp(\exp(\exp(10^6)))$!

10. Conclusions

Although the hypotheses required for proper inference using the statistical techniques described in this paper were not satisfied, the fact that all of the estimates made and intervals calculated fell near the bottom of Kobayashi's range suggest that in fact his lower bound may be near the true value of $\mathbf{d}\mathcal{A}$. In fact, we feel comfortable with the conjecture that $0.24761 < \mathbf{d}\mathcal{A} < 0.24762$, which, if true, would fix the value of $\mathbf{d}\mathcal{A}$ to one more decimal digit.

The techniques used in Theorem 13 can be used to put lower bounds on the number of abundant integers in intervals of any length. We note, for example, that taking $d = 22$ and $m = 8$ in the theorem gives that any consecutive sequence of 10^9 integers contains at least $240,770,557$ nondeficient numbers.

We also note that these statistical techniques have the potential to be used to address other number-theoretic questions. In particular, we would like to see Time Series Analysis applied to help gain insight into other open problems.

Acknowledgements. The authors would like to thank the referee for careful reading and valuable suggestions which improved the paper.

References

[1] F. Behrend, Über numeri abundantes, *Sitzungsber. Akad. Berlin* (1932), 322–328.

[2] F. Behrend, Über numeri abundantes, ii, *Sitzungsber. Akad. Berlin* (1933), 280–293.

[3] H. Davenport, Über numeri abundantes, *Sitzungsber. Akad. Berlin* (1933), 830–837.

[4] M. Deléglise, Bounds for the density of abundant integers, *Exp. Math.* **7**, no. 2 (1998), 137–143.

[5] P. Erdős, Note on consecutive abundant numbers, *J. Lond. Math. Soc.* **S1-10**, no. 1 (1935), 128.

[6] M. Kobayashi, On the Density of Abundant Numbers, Ph.D. thesis, Dartmouth College, 2010.

[7] M. Kobayashi, A new series for the density of abundant numbers, *Int. J. Number Theory* **10**, no. 1 (2014), 73–84.

[8] H R. Künsch, The jackknife and the bootstrap for general stationary observations, *Ann. Statist.* **17** no. 3 (1989), 1217–1241.

[9] J. E. Littlewood, Sur la distribution des nombres premiers, *C. R. Math. Acad. Sci. Paris* **158** (1914), 1869–1872.

[10] I. Niven, The asymptotic density of sequences, *Bull. Amer. Math. Soc.* **57**, no. 6 (1951), 420–434.

[11] P. Pollack, *Not Always Buried Deep: A Second Course in Elementary Number Theory*, American Mathematical Society, Providence, RI, 2009.

[12] J. B. Rosser and L. Schoenfeld, Approximate formulas for some functions of prime numbers, *Illinois J. Math.* **6** (1962), 64–94.

[13] Hans Salié, Über die Dichte abundanter Zahlen, *Math. Nachr.* **14**, no. 1 (1955), 39–46.

[14] C. R Wall, P. L. Crews, and D. B. Johnson, Density bounds for the sum of divisors function, *Math. Comp.* **26**, no. 119 (1972), 773–777.

[15] C. R. Wall, Topics Related to the Sum of Unitary Divisors of an Integer, Ph.D. thesis, University of Tennessee, 1970.

5

EXPLICIT EXPRESSIONS FOR THE MOMENTS OF THE SIZE OF AN (N, $DN-$ 1)-CORE PARTITION WITH DISTINCT PARTS

Anthony Zaleski [1]
Department of Mathematics, Rutgers University, Piscataway, New Jersey
anthony.zaleski@rutgers.edu

Abstract

In a previous paper, we used computer-assisted methods to find explicit expressions for the moments of the size of a uniform random $(n, n+1)$-core partition with distinct parts. In particular, we conjectured that the distribution is asymptotically normal. However, our analysis hinged on a characterization of $(n, n+1)$-core partitions given by Straub, which is not readily generalized to other families of simultaneous core partitions. In another paper with Doron Zeilberger, we made use of the characterization in terms of posets to analyze $(2n+1, 2n+3)$-core partitions with distinct parts; here, the distribution was found *not* to be asymptotically normal. Our method involved finding recursive structure in the relevant sequence of posets. We remarked that this method is applicable to other families of core partitions, provided that one can understand the corresponding posets. Here, we use the poset method (and, as before, a computer) to analyze $(n, dn-1)$-core partitions with distinct parts, where d is a natural number. We exhibit formulas for the moments of the size, as functions of d with n fixed, and vice versa. We conjecture that the distribution is asymptotically normal as n approaches infinity. Finally, we find formulas for the first few moments, as functions of *both* n and d.

1. Introduction

1.1. (s, t)-core Partitions

Recall that the *hook length* of a box in the Young diagram of a partition is the number of boxes to the right (the arm) plus the number of boxes below it (the leg) plus one (the head). (We use the English convention for Young diagrams; see Figure 1.) A partition is an *s-core* if its Young diagram avoids hook length s and an (s, t)-*core* if it avoids hook lengths s and t [3]. For example, the partition $9 = 4+3+1+1$ in Figure 1 is a $(6, 8)$-core but not a $(6, 7)$-core.

---
[1] now working in industry

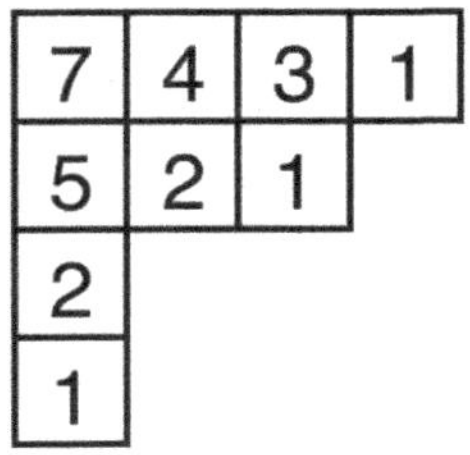

Figure 1: Young diagram of the partition $9 = 4+3+1+1$, showing the hook lengths of each box.

The number of (s,t)-core partitions is finite if and only if s and t are coprime, which we shall assume from now on [2]. Let $X_{s,t}$ be the random variable "size of an (s,t)-core partition," where the sample space is the set of all (s,t)-core partitions, equipped with the uniform distribution. In [5], with the help of Maple, Zeilberger derived explicit expressions (as polynomials in s and t) for the expectation, variance, and numerous higher moments of $X_{s,t}$. The original paper noted that "From the 'religious-fanatical' viewpoint of the current 'mainstream' mathematician, they are 'just' conjectures, but nevertheless, they are *absolutely certain* (well, at least as absolutely certain as most proved theorems)," and a donation to the OEIS was offered for the theory to make the results rigorous. Later, it was found that such a theory did exist and the results are entirely rigorous; see the updates at the paper's site.

Zeilberger also computed some standardized central moments of $X_{s,t}$ and the limit of these expressions as $s, t \to \infty$ with $s - t$ fixed. From this he conjectured the limiting distribution. Perhaps surprisingly, it is abnormal.

1.2. $(n, n+1)$-core Partitions With Distinct Parts

Things are not as easy if we look at the random variable "size of an (s,t)-core partition *with distinct parts*." In this case, there does not seem to a single formula for the moments in terms of s and t. However, we can consider certain indexed families of core partitions and look for explicit formulas for the moments.

For example, Amdeberhan considered $(n, n+1)$-cores with distinct parts and conjectured that they are enumerated by the Fibonacci numbers [1]. He also made conjectures about the maximum and average size of these partitions as functions of n. Xiong proved these conjectures [10]. Independently, Straub proved the enumeration conjecture and generalized it to $(n, dn-1)$-cores (more on this later) [9]. He provides a lemma that a partition into distinct parts is an $(n, n+1)$-core if and only if its perimeter (largest hook length) is less than n. In [12], we used this characterization to expand upon the result about the average size, deriving explicit expressions for the higher moments of the size, as rational functions of Fibonacci numbers. Our results give strong evidence that the limiting distribution as $n \to \infty$ *is* normal in this case.

1.3. The Poset Characterization

Unfortunately, we do not see how to generalize the characterization above to other indexed families of partitions. However, there is another characterization of (s,t)-core partitions due to Anderson [2].

Given coprime s and t, define the poset

$$P_{s,t} := \mathbb{N} \setminus (s\mathbb{N} + t\mathbb{N}), \tag{1}$$

where $\mathbb{N} = \{0, 1, 2, 3, \dots, \}$, where the partial-order relation $c \leq_P d$ holds whenever $d - c$ can be expressed as $\alpha s + \beta t$ for some $\alpha, \beta \in \mathbb{N}$. (It can be shown that $P_{s,t}$ is finite with largest element $st - s - t$.)

Anderson defined a bijection between (s,t)-core partitions and *order ideals* of $P_{s,t}$. (Under our convention, a subset I of a poset P is an order ideal if and only if $x \in I, y \leq_P x$ implies $y \in I$.) Further, the hook lengths of the boxes in the leftmost column in the partition's Young diagram correspond to the labels of the order ideal.

Now, a partition has distinct parts if and only if there are no consecutive hook lengths in the leftmost column. Thus:

(s,t)-core partitions with distinct parts are bijective to order ideals of $P_{s,t}$ with no consecutive labels.

1.4. $(2n+1, 2n+3)$-core Partitions With Distinct Parts

Given a family of core partitions, we have a corresponding sequence of posets which we can hopefully understand. In [13], we plotted some of the posets corresponding to $(2n+1, 2n+3)$-core partitions with distinct parts and were able to see a recursive structure. This led to recursions for the generating functions of the partitions. We were again able to verify the enumeration result and derive explicit expressions for the moments in terms of n. We showed that the distribution of size is not asymptotically normal as $n \to \infty$.

We should note that several other authors have independently approached the $(2n+1, 2n+3)$ case. For example, Baek, Nam, and Yu proved the enumeration result using bijective methods [4]. Yan et al. also derived the enumeration result and proved a conjecture regarding the size of the largest part [6]. Finally, Paramonov approached the problem using the abacus characterization and Dyck paths [8].

2. $(n, dn-1)$-core Partitions With Distinct Parts

In [9], Straub generalizes the problem in Section 1.2 by considering $(n, dn-1)$-core partitions with distinct parts, where n and d are natural numbers. He proves in Theorem 4.1 that the number of such partitions, call it $N_d(n)$, satisfies a generalized Fibonacci recurrence:

$$\begin{aligned} &N_d(1) = 1, \;\; N_d(2) = d, \\ &N_d(n) = N_d(n-1) + dN_d(n-2). \end{aligned} \tag{2}$$

Of course, this reduces to the usual Fibonacci numbers when $d = 1$. Note that we can view $N_d(n)$ as a sequence of polynomials in d. Also, Nath and Sellers [7] and Paramonov [8] independently proved a similar enumeration result for $(n, dn+1)$-cores. They made use of the abacus characterisation.

Here, we shall use the poset characterization in Section 1.3 to easily recover Straub's result and discover new conjectures about the distribution of the sizes of the partitions.

2.1. Understanding the Posets

By Section 1.3, we know that $(n, dn-1)$-core partitions with distinct parts are bijective with order ideals of $P_{n,dn-1}$ containing no consecutive labels. We can use the procedure `PW` in the Maple package `Armin` accompanying [13] to plot $P_{n,dn-1}$ for various n and d.

For example, Figure 2 depicts the poset $P_{4,19}$, i.e., the $n = 4, d = 5$ case. (When plotting $P_{s,t}$, we use the convention to increment s in the $\downarrow$ direction and t in the $\leftarrow$ direction. Thus, the largest label, $st - s - t$, is in the lower left corner.) It is easy to show that this general trapezoidal shape persists for arbitrary values of n and d. We can also see $P_{n,dn-1}$ as a colonnade of $n-1$ vertical pillars with heights $d(n-1)-1, d(n-2)-1, \ldots, d-1$. Further, the tops of the pillars have labels $1, 2, \ldots, n-1$.

2.2. Characterizing the Order Ideals

Next, we recover the recursion (2) by enumerating the order ideals of $P_{n,dn-1}$ with no consecutive labels.

Referring to the $n = 4, d = 5$ example, let I be an order ideal of $P_{4,19}$. Let I_k be the part of I contained in the kth pillar.

- If $I_1 = \emptyset$, then I is isomorphic to an order ideal of $P_{n-1,d(n-1)-1} = P_{3,5\cdot 3-1}$.
- Otherwise, let x be the largest member of I_1. Then $x \in \{1, 1+n, \ldots, 1+(d-1)n\} = \{1, 5, 9, 13, 17\}$. For if $21 \in I_1$, then $1, 2 \in I$, contradicting the assumption that I contains no consecutive labels. So there are $d = 5$ choices

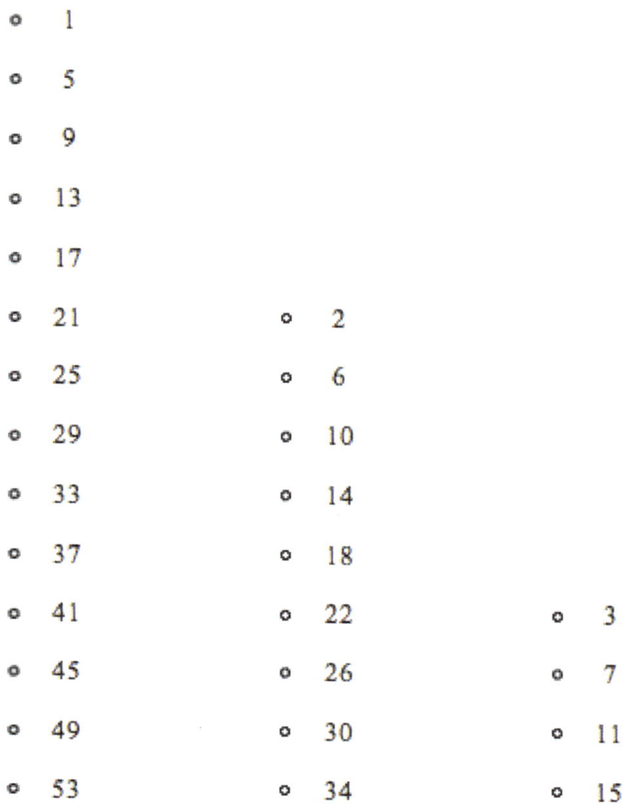

Figure 2: The poset $P_{4,19} = P_{4,4\cdot 5-1}$.

for I_1. Further, I_2 must be empty; otherwise, again, we have $1, 2 \in I$. Thus the remainder of I is isomorphic to an order ideal of $P_{n-2,d(n-2)-1} = P_{2,5\cdot 2-1}$.

To summarize, if I is an order ideal of $P_{n,dn-1}$ with no consecutive labels, then either I is isomorphic to an order ideal of $P_{n-1,d(n-1)-1}$, or I_1 has d options and the rest of I is isomorphic to an order ideal of $P_{n-2,d(n-2)-1}$. This proves (2).

From the above observation, we have the following characterization:

Any order ideal of $P_{n,dn-1}$ with no consecutive labels is of the form $I = I_1 \cup \cdots \cup I_{n-1}$, where

- Each I_k is either empty or of the form $\{k, k+n, \cdots, k+i_k n\}$, where $i_k \leq d-1$ if $k < n-1$, and $i_{n-1} < d-1$.
- If I_k is nonempty, then I_{k+1} is empty.

2.3. Computing the Generating Function

Our ultimate goal is to investigate the distribution of the size of $(n, dn-1)$-core partitions with distinct parts. To this end, we define the generating function

$$G_{d,n}(q) := \sum_{p} q^{|p|}, \tag{3}$$

where p ranges over $(n, dn-1)$-core partitions with distinct parts, and $|p|$ denotes the size of the partition p, i.e., the sum of its parts. We shall give an efficient scheme for computing $G_{d,n}(q)$ for fixed d and n.

Proceeding as in [13], we first compute the auxiliary generating function

$$F_{d,n}(q,t) := \sum_{I} q^{w(I)} t^{|I|}, \tag{4}$$

where I ranges over all order ideals of $P_{n,dn-1}$ with no consecutive labels; $w(I)$ is the sum of the labels in I; and $|I|$ is the number of labels in I. Then, as explained in [13], we can obtain $G_{d,n}(q)$ by replacing occurrences of t^k in $F_{d,n}(q,t)$ with $q^{-k(k-1)/2}$.

To compute $F_{d,n}(q,t)$, we use the reasoning of the previous section, but this time we keep track of the weight of the order ideal.

First, we introduce yet another auxiliary generating function. For $1 \le k \le n-1$, let $P^k_{n,dn-1}$ be the sub-poset of $P_{n,dn-1}$ obtained by chopping off everything to the left of the kth column (note $P^1_{n,dn-1} = P_{n,dn-1}$). Define $F^k_{d,n}(q,t)$ as in (4), except I ranges over order ideals of $P^k_{n,dn-1}$ with no consecutive labels (note $F^1_{d,n}(q,t) = F_{d,n}(q,t)$).

By the reasoning of the previous section, the first column of an order ideal of $P^k_{n,dn-1}$ is either empty or of the form $\{k, k+n, \cdots, k+in\}$, where $0 \le i \le d-1$). Since the latter set has weight

$$q^{\sum_{j=0}^{i}(k+jn)} t^{i+1} = q^{(i+1)(in/2+k)} t^{i+1},$$

we have the recursion

$$\begin{aligned} F^k_{d,n}(q,t) &= F^{k+1}_{d,n}(q,t) + \left(\sum_{i=0}^{d-1} q^{(i+1)(in/2+k)} t^{i+1} \right) F^{k+2}_{d,n}(q,t) \text{ for } 1 \le k \le n-2, \\ F^{n-1}_{d,n}(q,t) &= \sum_{i=0}^{d-2} q^{(i+1)(in/2+k)} t^{i+1}, \\ F^{n}_{d,n}(q,t) &:= 1. \end{aligned} \tag{5}$$

Note that this is a recursion in the auxiliary index k, not in n and d.

Given n and d, we can use (5) to find $F^1_{d,n}(q,t) = F_{d,n}(q,t)$. Finally, we make the substitution $t^k \to q^{-k(k-1)/2}$ to find $G_{d,n}(q)$. All of this is done in the procedure `Gdn` in the Maple package.

3. Distribution of the Size

Given fixed n and d, we can pick a uniform random $(n, dn-1)$-core partition with distinct parts, and consider its size, call it $X_{n,d}$. Then $X_{d,n}$ is a random variable, so it makes sense to inquire about its distribution. Since $G_{d,n}$ is the generating function for $X_{d,n}$, we can easily compute as many moments of the distribution as we please, for fixed n and d.

Using this information, we can investigate how the moments behave as functions of n and d. We consider two cases: n is variable and d is fixed, and vice versa. In each case, we consider the behavior of $X_{d,n}$ as the variable tends to infinity; in particular, we address the question of asymptotic normality. Finally, we derive formulas for the first few moments as functions of *both* n and d.

3.1. Varying d and Fixing n

First, we introduce some terminology. Given a natural number k, let us denote

$$m_k(d,n) := \left[\left(q\frac{d}{dq}\right)^k G_{d,n}(q)\right]_{q=1}$$

the kth "pre-moment" of $X_{d,n}$. Thus,

$$M_k(d,n) := \frac{m_k}{G_{d,n}(1)} = \frac{m_k}{N_d(n)} = \mathbb{E}[X_{d,n}^k]$$

is the kth (straight) moment of $X_{d,n}$. For example, the mean is $\mu_{d,n} = M_1$.

We denote the kth *central moment* by

$$M_k^c(d,n) := \mathbb{E}[(X_{d,n} - \mu)^k].$$

For example, the variance is $\sigma_{d,n}^2 := M_2^c$.

Finally, the kth *standardized moment* is

$$M_k^s(d,n) := \frac{M_k^c}{\sigma^k}.$$

Note that the central, straight, and standardized moments can easily be computed from the pre-moments.

Now, for numeric values of d and n, we can easily compute all the quantities above. Analyzing moment data for many values of d and n confirms the following.

Conjecture 1. For each n, the kth pre-moment $m_k(d,n)$ of $X_{d,n}$ is a polynomial in d. Further, the degree of this polynomial is $2k + \lfloor n/2 \rfloor$.

For example, our experimental evidence indicates that

$$\begin{aligned}
\{\lim_{d\to\infty} M_k^s(d,3)\}_{k=3}^{\infty} &= 2/7\sqrt{5}, \frac{15}{7}, \frac{100}{77}\sqrt{5}, \frac{6625}{1001}, \frac{750}{143}\sqrt{5}, \dots \\
&\approx .641, 2.14, 2.91, 6.62, 11.7, \dots \\
\{\lim_{d\to\infty} M_k^s(d,4)\}_{k=3}^{\infty} &\approx .162, 2.08, 1.19, 6.20, 7.05, \dots \\
\{\lim_{d\to\infty} M_k^s(d,5)\}_{k=3}^{\infty} &\approx .237, 2.22, 1.76, 7.43, 10.8, \dots \\
\{\lim_{d\to\infty} M_k^s(d,6)\}_{k=3}^{\infty} &\approx .052, 2.36, .671, 7.80, 5.15, \dots \\
&\dots \\
\{\lim_{d\to\infty} M_k^s(d,10)\}_{k=3}^{\infty} &\approx -0.001, 2.62, .130, 10.1, 2.17, \dots .
\end{aligned} \tag{6}$$

Recall that the standard normal distribution has moments $0, 1, 0, 3, 0, 15, \dots$. The sequences above seem to approach this as $n \to \infty$, leading us to the following:

Conjecture 2. For each fixed n, the distribution of $X_{d,n}$ is *not* asymptotically normal; in fact, $(X_{d,n} - \mu_{d,n})/\sigma_{d,n}$ tends to some abnormal distribution X_n as $d \to \infty$. However, X_n *is* asymptotically normal; that is, $(X_n - \mu)/\sigma$ tends to the standard normal distribution as $n \to \infty$.

3.2. Varying n and Fixing d

Next, we fix d, and look at $X_{d,n}$ as a sequence of random variables indexed by n. The $d = 1$ case was already addressed in [13], where we found that the pre-moments are given by polynomials in n and the Fibonacci numbers. In light of (2), we might expect the same to be true for arbitrary d, except we use the generalized Fibonacci numbers $N_d(n)$:

Conjecture 3. For each d, the kth pre-moment $m_k(d, n)$ of $X_{d,n}$ is of the form $a(n)N_d(n) + b(n)N_d(n + 1)$, where a and b are polynomials in n.

Again, experimental evidence verifies this claim. The one anomalous case seems to be $d = 2$, for which $N_d(n) = 2^{n-1}$. In this case, our methods do not yield nice formulas for the moments.

Upon computing the limits' standardized moments, we do get the familiar sequence $0, 1, 0, 3, 0, 15, \dots$ in this case, leading to the following conjecture.

Conjecture 4. For fixed d, the distribution of $X_{d,n}$ is asymptotically normal. That is, $(X_{d,n} - \mu_{d,n})/\sigma_{d,n}$ approaches the standard normal distribution as $n \to \infty$.

3.3. Varying n and d

Finally, it is possible (but computationally taxing) to obtain a single formula for the kth moment as a function of *both* k and d. In the previous section, we *fixed* d

and k, we and used `GuessRecPol` to fit the sequence $\{m_k(d,n)\}_{n=2}^{\infty}$ to the ansatz $a(n)N_d(n) + b(n)N_d(n+1)$, where $a(n)$ and $b(n)$ are polynomials. However, using the methods Section 3.1, we can fix *only* k and look at $\{m_k(d,n)\}_{n=2}^{\infty}$ as a sequence of polynomials in d. Further, due to Maple's ability to handle linear systems with symbolic coefficients, `GuessRecPol` can still be applied to this sequence. Of course, now $a(n)$ and $b(n)$ will be polynomials in n whose coefficients are rational functions in d.

Conjecture 5. The kth pre-moment $m_k(d,n)$ of $X_{d,n}$ is of the form $A(n,d)N_d(n)+B(n,d)N_d(n+1)$, where A and B are degree $2k$ polynomials in n whose coefficients are rational functions in d.

Due to the amount of data needed to fit the kth moment to the ansatz, it takes a few minutes even to generate the formula for the 3rd moment.

3.4. Sample Formulas for the Moments

Here, we present a small taste of the conjectures yielded by our Maple package. In the first two conjectures to follow, we could easily have presented formulas for many more moments, but we omit them to save space. See the next section for instructions to replicate these results and many more for yourself.

In general, we conjectured that $M_k(d,n)$ is a rational function in d for n fixed. However, for $n=3$ the straight moments seem to be polynomials:

Conjecture 6. The expectation of $X_{d,3}$ is $d^2/3 + d/4 - 1/12$, and the variance is $4d^4/45 + d^3/12 - d^2/144 + d/24 + 31/720$.

Here is an example where we fix d.

Conjecture 7. The expectation of $X_{3,n}$ is

$$\frac{25}{39}n^2 - \frac{479}{507}n + \frac{406}{507} + \frac{N_3(n+1)}{N_3(n)}\left(-\frac{1}{39}n^2 + \frac{29}{169}n - \frac{158}{507}\right).$$

Finally, here is the expectation once and for all, in terms of both n and d.

Conjecture 8. The expectation of $X_{d,n}$ is

$$\frac{\left(5\,d^3+7\,d^2+d-1\right)n^2}{24(4\,d+1)} - \frac{\left(8\,d^5+21\,d^4+7\,d^3-d^2+3\,d-2\right)n}{24(16\,d^3-24\,d^2-15\,d-2)}$$
$$+\frac{17\,d^4+13\,d^3-9\,d^2-7\,d-2}{12(16\,d^3-24\,d^2-15\,d-2)} + \frac{N_d(n+1)}{N_d(n)}$$
$$\cdot\left(-\frac{\left(d^2-1\right)n^2}{24(4\,d+1)} - \frac{\left(2\,d^4-9\,d^3-16\,d^2-3\,d+2\right)n}{8(16\,d^3-24\,d^2-15\,d-2)} - \frac{d^4+20\,d^3+9\,d^2-20\,d-10}{12\,(d-2)\,(4\,d+1)^2}\right).$$

Note that this formula is singular at $d = 2$, explaining the anomaly mentioned earlier. However, we can still make sense of the $d = 2$ case by first plugging in a numeric value of n (so that the N_d's become polynomials in d), then taking the limit as $d \to 2$. So this formula effectively works for *all* n and d.

4. Recent Work by Xiong and Zang

After the writing of this paper, Xiong and Zang informed us of their recent arXiv preprint [11] proving our conjectures about the degree and polynomality of the moments. Based on their results, our conjectured formulas for the moments are made rigorous. However, the remarks about asymptotic normality still stand as conjectures.

5. Using the Maple Package

The Maple package core2.txt accompanying this paper may be found at the following URL:
http://www.math.rutgers.edu/~az202/Z.

To use the Maple package, place core2.txt in the working directory and execute read('core.txt');.

To see the main procedures, execute Help();. For help on a specific procedure, use Help(<procedure name>);.

Here are some things to try:

- Gdn(q,3,7); gives the generating function (according to size) of $(3, 3 \cdot 7 - 1)$-core partitions with distinct parts.
- MdnK(d,3,1); and McdnK(d,3,2); reproduce Conjecture 3.6.
- MdnK(3,n,1); reproduces Conjecture 3.7.
- MdnK(d,n,1); reproduces Conjecture 3.8.
- map(p->limit(p,d=infinity),MsdnK(d,3,7)); reproduces the first equation in (6).

Acknowledgements. The author thanks Doron Zeilberger for introducing this project to him and guiding his research in the right direction. The author also thanks the journal referees and Neil Sloane for their helpful comments on the draft.

References

[1] T. Amdeberhan, Theorems, problems and conjectures, arXiv:1207.4045 (2015).

[2] J. Anderson, Partitions which are simultaneously t_1- and t_2-core, *Discrete Math.* **51** (2014), 205-220.

[3] D. Armstrong, C. R. H. Hanusa, and B. C. Jones, Results and conjectures on simultaneous core partitions, *European J. Combin.* **41** (2014), 205-220.

[4] J. Baek, H. Nam, and M. Yu, A bijective proof of Amdeberhan's conjecture on the number of $(s, s+2)$-core partitions with distinct parts, arXiv:1705.02691 (2017).

[5] S. B. Ekhad and D. Zeilberger, Explicit expressions for the variance and higher moments of the size of a simultaneous core partition and its limiting distribution, *The Personal Journal of Shalosh B. Ekhad and Doron Zeilberger* (2015), http://www.math.rutgers.edu/~zeilberg/mamarim/mamarimhtml/stcore.html.

[6] Z. Jin, G. Qin, S. H. F. Yan, and R. D. P. Zhou, On $(2k+1, 2k+3)$-core partitions with distinct parts, *Discrete Math.* **340** (2017), 1191-1202.

[7] R. Nath and J. Sellers, Abaci structures of $(s, ms \pm 1)$-core partitions, *Electron. J. Combin.* **24** (2017).

[8] K. Paramonov, Cores with distinct parts and bigraded Fibonacci numbers, *Discrete Math.* **341** (2018), 875-888.

[9] A. Straub, Core partitions into distinct parts and an analog of Euler's theorem, *European J. Combin.* **57** (2016), 40-49.

[10] H. Xiong, Core partitions with distinct parts, *Electron. J. Combin.* **25** (2018).

[11] H. Xiong and W. Zang, On the polynomiality and asymptotics of moments of sizes for random $(n, dn \pm 1)$-core partitions with distinct parts, arXiv:1804.09091 (2018).

[12] A. Zaleski, Explicit expressions for the moments of the size of an $(s, s+1)$-core partition with distinct parts, *Adv. Appl. Math.* **84** (2017), 1-7.

[13] A. Zaleski and D. Zeilberger, Explicit expressions for the expectation, variance, and higher moments of the size of a $(2n+1, 2n+3)$-core partition with distinct parts, *J. Difference Equ. Appl.* **23** (2017).

6

SOME NEW EXAMPLES OF INFINITE IMAGE PARTITION REGULAR MATRICES

Neil Hindman
Department of Mathematics, Howard University, Washington, DC
nhindman@aol.com

Dona Strauss
Department of Pure Mathematics, University of Leeds, Leeds, UK
d.strauss@hull.ac.uk

Abstract

A matrix A, finite or infinite, is *image partition regular* (over the set $\mathbb{N}$ of positive integers) if and only if, whenever $\mathbb{N}$ is finitely colored, there is a vector $\vec{x}$ of the appropriate size with entries in $\mathbb{N}$ such that all entries of $A\vec{x}$ are the same color (or *monochromatic*). A large number of characterizations of finite matrices that are image partition regular are known. There is no known characterization of infinite image partition regular matrices, and the classes of infinite matrices that are known to be image partition regular have been rather limited; we present a list of those classes of which we are aware. Extending an idea of Patra and Ghosh, we produce several new classes of infinite image partition regular matrices.

1. Introduction

We shall be concerned in this paper with finite or infinite matrices whose entries are rational numbers.

Definition 1.1. A matrix (finite or infinite) is *admissible* if and only if it has entries from $\mathbb{Q}$, no row equal to $\vec{0}$, and finitely many nonzero entries in each row.

We will index the rows and columns of a matrix A by ordinals (here always countable). The first infinite ordinal is $\omega = \mathbb{N} \cup \{0\}$. Recall that an ordinal is the set of its predecessors, so for an ordinal α, the statements $x < \alpha$ and $x \in \alpha$ are synonymous.

If A is an $\alpha \times \delta$ matrix, B is a $\gamma \times \tau$ matrix, and $\mathbf{O}$ represents a matrix with all entries equal to 0 of the appropriate size, then $\begin{pmatrix} A & \mathbf{O} \\ \mathbf{O} & B \end{pmatrix}$ is an $(\alpha + \gamma) \times (\delta + \tau)$ matrix.

We will follow the custom of denoting the entries of a matrix by the lower case letter corresponding to the upper case letter which is the name of the matrix.

Definition 1.2. Let α and δ be positive ordinals. An $\alpha \times \delta$ matrix A is *image partition regular* (IPR) if and only if A is admissible and whenever $\mathbb{N}$ is finitely colored, there is some $\vec{x} \in \mathbb{N}^{\delta}$ such that the entries of $A\vec{x}$ are monochromatic.

Many of the classical results of Ramsey Theory can be naturally viewed as the assertion that certain matrices are image partition regular. For example, Schur's Theorem and the length 4 version of van der Waerden's theorem are the assertions that the matrices

$$\begin{pmatrix} 1 & 0 \\ 0 & 1 \\ 1 & 1 \end{pmatrix} \text{ and } \begin{pmatrix} 1 & 0 \\ 1 & 1 \\ 1 & 2 \\ 1 & 3 \end{pmatrix}$$

are IPR.

The first class of matrices shown to be image partition regular were the matrices that produced Deuber's (m, p, c) *sets* [2]. Deuber used these sets to prove Rado's Conjecture. This was the assertion that if a "large" set was finitely colored, then one of the color classes must be "large", where a large set was defined to be one that contains solutions to every (finite) partition regular system of homogeneous linear equations.

The matrices producing the (m, p, c) sets are special cases of *first entries* matrices.

Definition 1.3. Let $\alpha, \delta \in \mathbb{N}$. An $\alpha \times \delta$ matrix A is a *first entries* matrix if and only if A is admissible and for $i, j < \alpha$, if $k = \min\{t < \delta : a_{i,t} \neq 0\} = \min\{t < \delta : a_{j,t} \neq 0\}$, then $a_{i,k} = a_{j,k} > 0$.

Using Deuber's methods, it was easy to see that all first entries matrices are IPR. The first characterizations of image partition regular matrices were found in [7]. One of these was that a finite matrix A is image partition regular if and only if it has among its images all of the images of some first entries matrix B.

There are now at least a dozen known characterizations of IPR matrices – see [13, Theorem 15.24]. Of these, one that is important from the point of view of this paper is that a finite matrix A is IPR if and only if it has images contained in every *central* subset of $\mathbb{N}$. This has the important consequence that if A and B are finite IPR matrices, then $\begin{pmatrix} A & \mathbf{O} \\ \mathbf{O} & B \end{pmatrix}$ is also IPR.

In order to define the notion of *central* that we just mentioned, we need to give a brief introduction to the algebraic structure of $\beta\mathbb{N}$. And we will need to use that structure for most of our proofs in Section 3. In particular, we will use the structures of both $(\beta\mathbb{N}, +)$ and $(\beta\mathbb{N}, \cdot)$, so we will write here about an arbitrary semigroup S.

Given a discrete space S, we take the Stone-Čech compactification of βS of S to be the set of ultrafilters on S, the principal ultrafilters being identified with the points of S. Given $A \subseteq S$, let $\overline{A} = \{p \in \beta S : A \in p\}$. Then $\{\overline{A} : A \subseteq S\}$ is a basis for the open sets and a basis for the closed sets of βS. If $(S, \cdot)$ is a discrete semigroup, the operation extends to βS so that $(\beta S, \cdot)$ becomes a right topological semigroup with S contained in its topological center. That is, for any $p \in \beta S$, the function $q \mapsto q \cdot p$ from βS to itself is continuous and for any $x \in S$, the function $q \mapsto x \cdot q$ from βS to itself is continuous.

Since $(\beta S, \cdot)$ is a compact Hausdorff right topological semigroup, it has a smallest two sided ideal, $K(\beta S, \cdot)$, which is the union of all of the minimal right ideals of $(\beta S, \cdot)$ and is also the union of all of the minimal left ideals of $(\beta S, \cdot)$. The intersection of any minimal left ideal with any minimal right ideal is a group and any two such groups are isomorphic. Given $p, q \in \beta S$ and $A \subseteq S$, $A \in p \cdot q$ if and only if $\{x \in S : x^{-1}A \in q\} \in p$ where $x^{-1}A = \{y \in S : x \cdot y \in A\}$. Any compact Hausdorff right topological semigroup has an idempotent. Given idempotents p and q in $(\beta S, \cdot)$, $p \leq q$ if and only if $p = p \cdot q = q \cdot p$. An idempotent p is minimal with respect to this ordering if and only if $p \in K(\beta S, \cdot)$. See Part I of [13] for much more information about the structure of βS.

One fact about $(\beta\mathbb{Z}, +)$ and $(\beta\mathbb{Z}, \cdot)$ that we will repeatedly use is that if $p, q \in \beta\mathbb{Z}$ and $a \in \mathbb{Z}$, then by [13, Lemma 13.1], $a(p + q) = ap + aq$.

Definition 1.4. Let $(S, \cdot)$ be a discrete semigroup and let $A \subseteq S$. Then A is *central* if and only if A is a member of some minimal idempotent in $(\beta S, \cdot)$.

In Section 2 of this paper we present, in historical order of their discovery, most of the known examples of infinite image partition regular matrices of which we are aware, concluding with a recent class of IPR matrices discovered by Patra and Ghosh [16]. In Section 3 we present several new classes, many of which involve extensions of the class produced by Patra and Ghosh.

2. Known Classes of Infinite IPR Matrices

For each $l \in \mathbb{N} \setminus \{1\}$ let

$$W_l = \begin{pmatrix} 1 & 0 \\ 1 & 1 \\ 1 & 2 \\ \vdots & \vdots \\ 1 & l \end{pmatrix}.$$

Then van der Waerden's Theorem [18] says that each W_l is IPR. (In the context of this paper this is immediate from the fact that each W_l is a first entries matrix.) It is an easy consequence of van der Waerden's Theorem that given any finite coloring of $\mathbb{N}$, there is one color which contains arbitrarily long arithmetic progressions.

Consequently, the infinite matrix

$$W = \begin{pmatrix} W_2 & \mathbf{O}_3 & \mathbf{O}_3 & \dots \\ \mathbf{O}_4 & W_3 & \mathbf{O}_4 & \dots \\ \mathbf{O}_5 & \mathbf{O}_5 & W_4 & \dots \\ \vdots & \vdots & \vdots & \ddots \end{pmatrix}$$

is IPR, where $\mathbf{O}_m$ is the $m \times 2$ matrix with all entries equal to 0.

The *Finite Sums Matrix* $\mathbf{F}$ is a matrix whose rows are all possible rows with entries from $\{0, 1\}$, at least one 1, and only finitely many 1's. (Of course there are many such matrices, differing only by the order of their rows. To be definite, one can specify a particular well ordering of the rows.) By the Finite Sums Theorem [6, Theorem 3.1], $\mathbf{F}$ is image partition regular. Given $\vec{x} \in \mathbb{N}^\omega$, the entries of $\mathbf{F}\vec{x}$ are all sums of the form $\sum_{t \in F} x_t$ for $F \in \mathcal{P}_f(\omega)$. (For a set X, we let $\mathcal{P}_f(X)$ be the set of finite nonempty subsets of X.)

Fix an enumeration $\langle B_n \rangle_{n=0}^{\infty}$ of the finite IPR matrices. For each n, assume that B_n is a $u(n) \times v(n)$ matrix. For each $i \in \mathbb{N}$, let $\vec{0}_i$ be the 0 vector with i entries. Let $\mathbf{D}$ be an $\omega \times \omega$ matrix with all rows of the form $\vec{r}_0{}^\frown\vec{r}_1{}^\frown\vec{r}_2{}^\frown \dots$ where each $\vec{r}_i$ is either $\vec{0}_{v(i)}$ or is a row of B_i, and all but finitely many are $\vec{0}_{v(i)}$. Again, if one wants to be definite, one can specify a well ordering of the rows of $\mathbf{D}$.)

It is a consequence of the theorem of [3] and the fact that every finite IPR matrix has its images contained in the images of some (m, p, c) set that $\mathbf{D}$ is image partition regular. Given $\vec{x} \in \mathbb{N}^\omega$, let $\vec{x} = \begin{pmatrix} \vec{x}_0 \\ \vec{x}_1 \\ \vdots \end{pmatrix}$ where each $\vec{x}_i \in \mathbb{N}^{v(i)}$. Then the entries of $\mathbf{D}\vec{x}$ are all sums of the form $\sum_{t \in F} y_t$ where $F \in \mathcal{P}_f(\omega)$ and for $t \in F$, y_t is an entry of $B_t\vec{x}_t$.

As we mentioned in the introduction, we are proceeding in historical order; [18] was published in 1927, [6] was published in 1974, and [3] was published in 1987. Our next example is a whole class of matrices, whose image partition regularity is a consequence of the Milliken-Taylor Theorem obtained independently by Milliken [15] and Taylor [17] and published in 1975 and 1976. So we might seem to be departing from our historical order. However, it was not until the publication of [4] in 1995 that it was noticed that the Milliken-Taylor Theorem guaranteed the existence of a class of infinite partition regular matrices.

Definition 2.1. Let $k \in \omega$ and let $\vec{a} = \langle a_0, a_1, \dots, a_k \rangle$ be a sequence in $\mathbb{Z} \setminus \{0\}$.

(a) $c(\vec{a})$ is the sequence obtained from $\vec{a}$ by deleting any term equal to its immediate predecessor.

(b) $\vec{a}$ is said to be *compressed* if and only if $\vec{a} = c(\vec{a})$.

Definition 2.2. Let $\vec{r} \in \mathbb{Z}^\omega$. Then $d(\vec{r})$ is the sequence obtained from $\vec{r}$ by deleting all occurrences of $\vec{0}$.

Definition 2.3. Let $k \in \omega$ and let $\vec{a} = \langle a_0, a_1, \ldots, a_k \rangle$ be a compressed sequence in $\mathbb{Z} \setminus \{0\}$ with $a_k > 0$. Then $MT(\vec{a})$ is a matrix consisting of all rows $\vec{r}$ with finitely many nonzero entries such that $c(d(\vec{r})) = \vec{a}$.

Again, if one wishes one can specify a particular well-ordering of the rows of $MT(\vec{a})$. In [4] the terms of $\vec{a}$ in the definition of $MT(\vec{a})$ were required to be positive, but the proof of image partition regularity from the Milliken-Taylor Theorem given in [4] can be taken almost verbatim.

As we noted in the introduction, any finite IPR matrix has images in any given central subset of $\mathbb{N}$. Also it is known that the matrices W, $\mathbf{F}$, and $\mathbf{D}$ all have images in any central subset of $\mathbb{N}$. By way of contrast, we have the following theorem.

Theorem 2.4. *Let $k, m \in \omega$ and let $\vec{a} = \langle a_0, a_1, \ldots, a_k \rangle$ and $\vec{b} = \langle b_0, b_1, \ldots, b_m \rangle$ be compressed sequences in $\mathbb{Z} \setminus \{0\}$ with $a_k > 0$ and $b_m > 0$. If there does not exist a positive rational α such that $\vec{a} = \alpha \cdot \vec{b}$, then there exist subsets A and B of $\mathbb{N}$ such that $\mathbb{N} = A \cup B$, there does not exist $\vec{y} \in \mathbb{N}^\omega$ with $MT(\vec{b})\vec{y} \in A^\omega$, and there does not exist $\vec{x} \in \mathbb{N}^\omega$ with $MT(\vec{a})\vec{x} \in B^\omega$.*

Proof. This follows from [8, Corollary 3.9] and the fact that the assumptions that $a_k > 0$ and $b_m > 0$ force the conclusion that there does not exist $\vec{x} \in \mathbb{N}^\omega$ with all entries of $MT(\vec{a})\vec{x}$ less than or equal to 0 or with all entries of $MT(\vec{b})\vec{x}$ less than or equal to 0. □

Definition 2.5. Let α and δ be positive ordinals. An $\alpha \times \delta$ matrix A is *centrally image partition regular* (centrally IPR) if and only if A is admissible and whenever C is a central subset of $\mathbb{N}$, there is some $\vec{x} \in \mathbb{N}^\delta$ such that $A\vec{x} \in C^\alpha$.

We thus have that if $k \in \omega$ and $\vec{a} = \langle a_0, a_1, \ldots, a_k \rangle$ is a compressed sequence in $\mathbb{Z} \setminus \{0\}$ with $a_k > 0$, then $MT(\vec{a})$ is centrally IPR if and only if $k = 0$.

Definition 2.6. Let A be an $\omega \times \omega$ matrix. Then A is a *restricted triangular* matrix if and only if all entries of A are from $\mathbb{Z}$ and there exist $d \in \mathbb{N}$ and an increasing function $j : \omega \to \omega$ such that for all $i \in \omega$,

(1) $a_{i,j(i)} \in \{1, 2, \ldots, d\}$,

(2) for all $l > j(i)$, $a_{i,l} = 0$, and

(3) for all $k > i$ and all $t \in \{1, 2, \ldots, d\}$, $t | a_{k,j(i)}$.

It was shown in [12] that any restricted triangular matrix is centrally IPR. In particular, if each $a_{i,j(i)} = 1$, then condition (3) is automatically satisfied.

The next class of matrices begins a process of extending given IPR matrices.

Theorem 2.7. *Let A be an $\omega \times \omega$ centrally image partition regular matrix and let $\langle b_n \rangle_{n=0}^{\infty}$ be a sequence in $\mathbb{N}$. Let*

$$B = \begin{pmatrix} b_0 & 0 & 0 & \dots \\ 0 & b_1 & 0 & \dots \\ 0 & 0 & b_2 & \dots \\ \vdots & \vdots & \vdots & \ddots \end{pmatrix}. \text{ Then } \begin{pmatrix} \mathbf{O} & B \\ A & \mathbf{O} \\ A & B \end{pmatrix}$$

is centrally image partition regular.

Proof. [12, Theorem 4.7]. □

Theorem 2.8. *Let A be a finite IPR matrix and let B be an infinite IPR matrix. Then $\begin{pmatrix} A & \mathbf{O} \\ \mathbf{O} & B \end{pmatrix}$ is IPR.*

Proof. A combinatorial proof attributed to V. Rödl is in [9, Lemma 2.3]. A short proof using the algebra of $\beta\mathbb{N}$ can be found in [16, Theorem 2.4]. □

We have seen that Milliken-Taylor matrices determined by compressed sequences with more than one entry are not centrally IPR. However, by choosing rows of such matrices with fixed nonzero sums, one does get centrally IPR matrices. If $m > 0$ the conclusion of Theorem 2.9 is trivial, since any central set will include a multiple of m and one can let $\vec{x}$ be a constant sequence. (Theorem 3.7 in [9] asserts more than just that M is centrally IPR.)

Theorem 2.9. *Let $k \in \mathbb{N}$, let $\langle a_0, a_1, \dots, a_k \rangle$ be a compressed sequence in $\mathbb{Z} \setminus \{0\}$ with $a_k > 0$, and let $m \in \mathbb{Z} \setminus \{0\}$. Let M be an admissible matrix such that*

(1) the compressed form of each row is $\langle a_1, a_2, \dots, a_k \rangle$ and

(2) the sum of each row is m.

Then M is centrally IPR.

Proof. [9, Theorem 3.7]. □

In the following theorem, if $\sum_{i=0}^{m} a_i > 0$ and $\sum_{i=0}^{n} b_i > 0$, then the conclusion of the theorem is trivial (although [14, Theorem 2] asserts more than that the matrix is IPR.) But in fact $\sum_{i=0}^{m} a_i$ is allowed to be 0 or negative and $\sum_{i=0}^{n} b_i$ is allowed to be negative.

Theorem 2.10. *Let $m, n \in \omega$ and let $\vec{a} = \langle a_0, a_1, \dots, a_m \rangle$ and $\vec{b} = \langle b_0, b_1, \dots, b_n \rangle$ be sequences in $\mathbb{Z} \setminus \{0\}$. Assume that $a_m > 0$, $b_n > 0$, and $\sum_{i=1}^{n} b_i \neq 0$. Let A be an $\omega \times \omega$ matrix with all rows whose nonzero entries are $a_0, a_1, \dots, a_m$ in order and let B be an $\omega \times \omega$ matrix with all rows whose nonzero entries are $b_0, b_1, \dots, b_m$ in order. Then $\begin{pmatrix} A & \mathbf{O} \\ \mathbf{O} & B \end{pmatrix}$ is IPR.*

Proof. [14, Theorem 2]. □

The fact that the matrix in the next theorem is image partition regular established that it is possible to have an IPR matrix which does not have the property that whenever $\mathbb{R}$ is finitely colored, there must exist monochromatic images arbitrarily close to zero.

Theorem 2.11. *The following matrix is IPR.*

$$\begin{pmatrix} 1 & 0 & 0 & 0 & 0 & 0 & 0 & 0 & \dots \\ 0 & 1 & 0 & 0 & 0 & 0 & 0 & 0 & \dots \\ 2 & 1 & 0 & 0 & 0 & 0 & 0 & 0 & \dots \\ 0 & 0 & 1 & 0 & 0 & 0 & 0 & 0 & \dots \\ 0 & 0 & 0 & 1 & 0 & 0 & 0 & 0 & \dots \\ 4 & 0 & 1 & 1 & 0 & 0 & 0 & 0 & \dots \\ 0 & 0 & 0 & 0 & 1 & 0 & 0 & 0 & \dots \\ 0 & 0 & 0 & 0 & 0 & 1 & 0 & 0 & \dots \\ 0 & 0 & 0 & 0 & 0 & 0 & 1 & 0 & \dots \\ 8 & 0 & 0 & 0 & 1 & 1 & 1 & 0 & \dots \\ \vdots & \vdots & \vdots & \vdots & \vdots & \vdots & \vdots & \vdots & \ddots \end{pmatrix}$$

Proof. [1, Theorem 16]. □

In fact, the matrix of Theorem 2.11 is centrally IPR. The proofs in [1] used the notion of upper asymptotic density. The properties of upper asymptotic density used are shared by *Banach density* and any central subset of $\mathbb{N}$ has positive Banach density. (See [13, Definition 20.1 and Theorems 20.5 and 20.6].)

Theorem 2.12. *Let $k \in \omega$ and let $\vec{a} = \langle a_0, a_1, \dots, a_k \rangle$ be a sequence in $\mathbb{Z} \setminus \{0\}$. Let I be the $\omega \times \omega$ identity matrix, let A be an $\omega \times \omega$ matrix with all rows whose nonzero entries are $a_0, a_1, \dots, a_m$ in order, and let $M = \begin{pmatrix} I \\ A \end{pmatrix}$. If*

(1) $\sum_{i=0}^{k} a_i = 0$ and $a_k = 1$,

(2) $\sum_{i=0}^{k} a_i = 1$, or

(3) $a_0 = a_1 = \dots = a_k = 1$,

then M is IPR.

Proof. [5, Lemma 2.7]. □

It is conjectured in [5] that the conditions of Theorem 2.12 are also necessary, and this conjecture is proved in special cases.

We have seen that Milliken-Taylor matrices determined by compressed sequences of length greater than 1 are not centrally IPR. However, if the final terms are all equal to 1, a shift suffices, as seen in the following theorem.

Theorem 2.13. *Let $m \in \omega$ and for each $i \in \{0, 1, \ldots, m\}$, let $k(i) \in \mathbb{N}$, let $\vec{a}_i = \langle a_{i,0}, a_{i,1}, \ldots, a_{i,k(i)} \rangle$ be a compressed sequence in $\mathbb{Z} \setminus \{0\}$ with $a_{i,k(i)} = 1$. Let $\overline{0}$ and $\overline{1}$ be the length ω constant column vectors. Then*

$$\begin{pmatrix} \overline{1} & \overline{0} & \ldots & \overline{0} & MT(\vec{a}_0) \\ \overline{0} & \overline{1} & \ldots & \overline{0} & MT(\vec{a}_1) \\ \vdots & \vdots & \ddots & \vdots & \vdots \\ \overline{0} & \overline{0} & \ldots & \overline{1} & MT(\vec{a}_m) \\ \overline{0} & \overline{0} & \ldots & \overline{0} & \mathbf{F} \end{pmatrix}$$

is centrally IPR.

Proof. [10, Corollary 6.4]. □

A generalization of Theorem 2.13 (minus the conclusion that the matrix is centrally IPR) is a consequence of Theorem 3.26 below. In that generalization **F** is replaced by an arbitrary Milliken-Taylor matrix and the requirement that $a_{i,k(i)} = 1$ is replaced by the requirement that all of the final entries are the same positive number.

The most recent class of infinite matrices known by us to be IPR (before the results of the present paper) is given in the following theorem.

Theorem 2.14. *Let $u, v \in \mathbb{N}$ and let D be a $u \times v$ IPR matrix. Let A be an $\omega \times v$ matrix all of whose rows are rows of D, let B be an $\omega \times \omega$ centrally IPR matrix, and let M be a Milliken-Taylor matrix. Then $\begin{pmatrix} A & B & \mathbf{O} \\ \mathbf{O} & \mathbf{O} & M \end{pmatrix}$ is image partition regular.*

Proof. [16, Theorem 2.19]. □

3. New Classes of Infinite IPR Matrices

Our first result replaces the assumption in Theorem 2.14 that B is centrally IPR by the assumption that B is simply IPR. As did the authors of [16], we utilize the following set introduced in [9].

Definition 3.1. Let α and δ be positive ordinals and let A be an $\alpha \times \delta$ admissible matrix. Then $C(A) = \{p \in \beta\mathbb{N} :$ for every $P \in p$, there exists $\vec{x} \in \mathbb{N}^\delta$ such that $A\vec{x} \in P^\alpha\}$.

Lemma 3.2. *Let A be an admissible matrix.*

(1) The set $C(A)$ is compact and $C(A) \neq \emptyset$ if and only if A is IPR.

(2) If A is a finite IPR matrix, then $C(A)$ is a subsemigroup of $(\beta\mathbb{N},+)$.

(3) If A is IPR, then $C(A)$ is a left ideal of $(\beta\mathbb{N},\cdot)$.

(4) If A is a finite IPR matrix, then $C(A)$ is a two sided ideal of $(\beta\mathbb{N},\cdot)$.

(5) If A is a finite IPR matrix, then all minimal idempotents of $(\beta\mathbb{N},+)$ are members of $C(A)$.

Proof. Statements (1) and (2) are [9, Lemma 2.5]. Statements (3) and (4) are [16, Lemma 2.3]. Statement (5) holds because finite IPR matrices are centrally IPR. □

Lemma 3.3. *Let $u,v\in\mathbb{N}$ and let A be a $u\times v$ IPR matrix. Let $r\in\beta\mathbb{Z}$ and let p be a minimal idempotent in $(\beta\mathbb{N},+)$.*

(1) If $r\in C(A)$, then $r+p\in C(A)$.

(2) If $r\in -C(A)$, then $r+p\in C(A)$.

(3) If $r\in C(A)$, then $r+(-p)\in -C(A)$.

(4) If $r\in -C(A)$, then $r+(-p)\in -C(A)$.

Proof. [16, Lemma 2.18]. □

In the next lemmas, the products of the form $\frac{a}{b}p$ are computed in $(\beta\mathbb{Q}_d,\cdot)$, where $\mathbb{Q}_d$ is the set of rationals with the discrete topology.

Lemma 3.4. *Let $u,v\in\mathbb{N}$ and let A be a $u\times v$ IPR matrix. Let $k\in\omega$ and let $\langle a_0,a_1,\ldots,a_k\rangle$ be a sequence in $\mathbb{Z}\setminus\{0\}$ with $a_k>0$. Let p be a minimal idempotent in $(\beta\mathbb{N},+)$. Then $a_0p+a_1p+\ldots+a_kp\in C(A)$ and $\frac{a_0}{a_k}p+\frac{a_1}{a_k}p+\ldots+\frac{a_{k-1}}{a_k}p+p\in C(A)$.*

Proof. By [13, Lemma 5.19.2], each $|a_i|\cdot p$ is a minimal idempotent in $(\beta\mathbb{N},+)$ as is each $|\frac{a_i}{a_k}|\cdot p$. Since A is centrally IPR, each a_ip is in $C(A)$ or $-C(A)$ as is each $\frac{a_i}{a_k}p$ depending on whether a_i is positive or negative. The results then follow by repeated applications of Lemma 3.3. □

Lemma 3.5. *Let $k\in\omega$ and let $\vec{a}=\langle a_0,a_1,\ldots,a_k\rangle$ be a compressed sequence in $\mathbb{Z}\setminus\{0\}$ with $a_k>0$. Let p be an idempotent in $(\beta\mathbb{N},+)$. Then*

(1) $a_0p+a_1p+\ldots+a_kp\in C\big(MT(\vec{a})\big)$ and

(2) $\frac{a_0}{a_k}p+\frac{a_1}{a_k}p+\ldots+\frac{a_{k-1}}{a_k}p+p\in C\big(MT(\vec{a})\big)$.

Proof. Conclusion (1) is an immediate consequence of [9, Corollary 3.6]. To verify conclusion (2), let $q=\frac{a_0}{a_k}p+\frac{a_1}{a_k}p+\ldots+\frac{a_{k-1}}{a_k}p+p$ and let $Q\in q$. By (1), $a_kq\in C\big(MT(\vec{a})\big)$. Since p is an idempotent, $a_k\mathbb{N}\in p$. Also $a_kQ\in a_kq$. Thus by [9, Corollary 3.6], we may pick $\vec{x}\in(a_k\mathbb{N})^\omega$ such that $MT(\vec{a})\vec{x}\in(a_kQ)^\omega$. Then $\frac{\vec{x}}{a_k}\in\mathbb{N}^\omega$ and $MT(\vec{a})\frac{\vec{x}}{a_k}\in Q^\omega$. □

Lemma 3.6. *Let $u, v \in \mathbb{N}$ and let D be a $u \times v$ IPR matrix. Let α and δ be positive ordinals, let A be an $\alpha \times v$ matrix, all of whose rows are rows of D, and let B be an $\alpha \times \delta$ matrix which is IPR. If $r \in C(A)$ and $p \in C(B)$, then $r + p \in C\begin{pmatrix} A & B \end{pmatrix}$.*

Proof. Let $H \in r + p$. Then $X = \{z \in \mathbb{N} : -z + H \in p\} \in r$ so pick $\vec{w} \in \mathbb{N}^v$ such that $\vec{y} = A\vec{w} \in X^\alpha$. Then $\{y_i : i < \alpha\}$ is finite so $P = \bigcap_{i<\alpha}(-y_i + H) \in p$. Pick $\vec{z} \in \mathbb{N}^\delta$ such that $B\vec{z} \in P^\alpha$. Then $\begin{pmatrix} A & B \end{pmatrix}\begin{pmatrix} \vec{w} \\ \vec{z} \end{pmatrix} \in H^\alpha$. □

Theorem 3.7. *Let $u, v \in \mathbb{N}$ and let D be a $u \times v$ IPR matrix. Let α and δ be positive ordinals, let A be an $\alpha \times v$ matrix whose rows are the rows of D, let B be an $\alpha \times \delta$ matrix which is IPR, and let M be a Milliken-Taylor matrix. Then $\begin{pmatrix} A & B & \mathbf{O} \\ \mathbf{O} & \mathbf{O} & M \end{pmatrix}$ is IPR.*

Proof. Note that $C(A) = C(D)$ so $C(A)$ is an ideal of $(\beta\mathbb{N}, \cdot)$ and a compact subsemigroup of $(\beta\mathbb{N}, +)$. Also $C(B)$ is a left ideal of $(\beta\mathbb{N}, \cdot)$ so pick $p \in C(A) \cap C(B)$. Pick a compressed sequence $\langle a_i \rangle_{i=0}^k$ in $\mathbb{Z} \setminus \{0\}$ with $a_k > 0$ such that $M = MT(\vec{a})$. We need to show that $C\begin{pmatrix} A & B \end{pmatrix} \cap C(M) \neq \emptyset$.

Now $p \in C(A)$ so $C(A) + p$ is a left ideal of $C(A)$. Pick a minimal left ideal L of $C(A)$ such that $L \subseteq (C(A) + p)$ and pick an idempotent $q \in L$. Now all minimal idempotents of $(\beta\mathbb{N}, +)$ are in $C(A)$ so $K(\beta\mathbb{N}, +) \cap C(A) \neq \emptyset$ and therefore q is minimal in $(\beta\mathbb{N}, +)$ by [13, Theorem 1.65]. Pick $x \in C(A)$ such that $q = x + p$.

Let $r = a_0 q + a_1 q + \ldots + a_k q$. By Lemma 3.5, $r \in C(M)$, so it suffices to show that $r \in C\begin{pmatrix} A & B \end{pmatrix}$. By Lemma 3.4, $r \in C(A)$.

Note that $a_k q = a_k x + a_k p$. Also $a_k q + a_k q = a_k(q + q) = a_k q$ so $r = r + a_k q = r + a_k x + a_k p$. We have that $r + a_k x \in C(A)$ and $a_k p \in C(B)$. By Lemma 3.6, $r + a_k x + a_k p \in C\begin{pmatrix} A & B \end{pmatrix}$. □

Our next new result involves the notion of *thick* sets. A subset A of $\mathbb{N}$ is thick if and only if it contains arbitrarily long integer intervals. Equivalently, by [13, Theorem 4.48], A is thick if and only if there is a left ideal L of $(\beta\mathbb{N}, +)$ with $L \subseteq \overline{A}$. Since any left ideal contains a minimal idempotent, it is immediate that any thick set is central, and thus contains images of any centrally IPR matrix.

Vitaly Bergelson has asked in a personal communication whether every thick set in $\mathbb{N}$ contains solutions to any partition regular system of linear homogeneous equations. By [11, Theorem 2.4] an affirmative answer to Bergelson's question would follow if one could show that any thick set contains images of any infinite IPR matrix. That in turn would follow from the validity of the following conjecture.

Conjecture 3.8. If α and δ are positive ordinals, A is $\alpha \times \delta$ IPR matrix, B is a thick subset of $\mathbb{N}$, $r \in \mathbb{N}$, and $B = \bigcup_{i=1}^r C_i$, then there exist $\vec{x} \in \mathbb{N}^\delta$ and $i \in \{1, 2, \ldots, r\}$ such that $A\vec{x} \in C_i^\alpha$.

Notice that the assertion in Conjecture 3.8 that whenever B is finitely colored, there is a monochromatic image of A is stronger than the assertion that B contains an image of A. Consider for example the following $\omega \times \omega$ matrix:

$$A = \begin{pmatrix} 1 & 1 & 0 & 0 & \dots \\ 2 & 1 & 0 & 0 & \dots \\ 1 & 0 & 1 & 0 & \dots \\ 2 & 0 & 1 & 0 & \dots \\ 3 & 0 & 1 & 0 & \dots \\ 1 & 0 & 0 & 1 & \dots \\ 2 & 0 & 0 & 1 & \dots \\ 3 & 0 & 0 & 1 & \dots \\ 4 & 0 & 0 & 1 & \dots \\ \vdots & \vdots & \vdots & \vdots & \ddots \end{pmatrix}.$$

The assertion that a set B is thick is precisely the assertion that there exists $\vec{x} \in \mathbb{N}^\omega$ such that $A\begin{pmatrix} 1 \\ \vec{x} \end{pmatrix} \in B^\omega$. But it is easy to color $\mathbb{N}$ in two colors so that neither color class contains arbitrarily long arithmetic progressions with a fixed increment.

Definition 3.9. $\mathcal{M} = \{A :$ there exist positive ordinals α and δ such that A is an $\alpha \times \delta$ admissible matrix and whenever T is a thick subset of $\mathbb{N}$ and T is finitely colored, there exists $\vec{x} \in \mathbb{N}^\delta$ such that the entries of $A\vec{x}$ are monochromatic$\}$.

Conjecture 3.8 is the assertion that $\mathcal{M}$ is exactly the set of IPR matrices.

Lemma 3.10. *Let α and δ be positive ordinals, and let A be an admissible $\alpha \times \delta$ matrix. The following statements are equivalent.*

(a) $A \in \mathcal{M}$.

(b) For every minimal left ideal L of $(\beta\mathbb{N}, +)$, $L \cap C(A) \neq \emptyset$.

(c) For every left ideal L of $(\beta\mathbb{N}, +)$, $L \cap C(A) \neq \emptyset$.

Proof. To see that (a) implies (b), assume that $A \in \mathcal{M}$, let L be a minimal left ideal of $(\beta\mathbb{N}, +)$, and suppose that $L \cap C(A) = \emptyset$. For each $p \in L$, pick $C_p \in p$ such that there does not exist $\vec{x} \in \mathbb{N}^\delta$ such that $A\vec{x} \in C_p^\alpha$. Then $L \subseteq \bigcup_{p \in L} \overline{C_p}$. Minimal left ideals of $(\beta\mathbb{N}, +)$ are closed so pick $F \in \mathcal{P}_f(L)$ such that $L \subseteq \bigcup_{p \in F} \overline{C_p}$. Let $B = \bigcup_{p \in F} C_p$. Since $L \subseteq \overline{B}$, B is thick. Since $A \in \mathcal{M}$, pick $\vec{x} \in \mathbb{N}^\delta$ and $p \in F$ such that $A\vec{x} \in C_p^\alpha$. This is a contradiction.

Since every left ideal of $(\beta\mathbb{N}, +)$ contains a minimal left ideal, it is trivial that (b) implies (c).

To see that (c) implies (a), let B be a thick subset of $\mathbb{N}$ and pick a left ideal L of $(\beta\mathbb{N}, +)$ such that $L \subseteq \overline{B}$. Pick $p \in L \cap C(A)$. Let $r \in \mathbb{N}$ and let $B = \bigcup_{i=1}^r C_i$.

Pick $i \in \{1, 2, \ldots, r\}$ such that $C_i \in p$. Since $p \in C(A)$, pick $\vec{x} \in \mathbb{N}^\delta$ such that $A\vec{x} \in C_i^\alpha$. □

Lemma 3.11. *Let A be an admissible matrix. If A is either centrally IPR or a Milliken-Taylor matrix, then $A \in \mathcal{M}$.*

Proof. Assume first that A is centrally IPR. Let L be a minimal left ideal of $(\beta\mathbb{N}, +)$ and pick a minimal idempotent $p \in L$. Then $p \in L \cap C(A)$ so Lemma 3.10 applies.

Now assume that we have $k \in \omega$ and a compressed sequence $\vec{a} = \langle a_0, a_1, \ldots, a_k \rangle$ in $\mathbb{Z} \setminus \{0\}$ with $a_k > 0$ such that $A = MT(\vec{a})$. Let L be a left ideal of $\beta\mathbb{N}$ and let q be an idempotent in L. By [13, Lemma 5.19.2], there is an idempotent $p \in \beta\mathbb{N}$ such that $a_k p = q$. It follows from Lemma 3.5 that $a_0 p + a_1 p + \ldots + a_k p \in C(A) \cap L$. □

Theorem 3.12. *Let B be an IPR matrix and let $M \in \mathcal{M}$. Then $\begin{pmatrix} \overline{1} & B & \mathbf{O} \\ \overline{0} & \mathbf{O} & M \end{pmatrix}$ is IPR.*

Proof. Assume that α and δ are positive ordinals and that B is an $\alpha \times \delta$ IPR matrix. Pick $p \in C(B)$ and let $L = \beta\mathbb{N} + p$. Pick $q \in L \cap C(M)$ and pick $r \in \beta\mathbb{N}$ such that $q = r + p$. Then $r \in \beta\mathbb{N} = C(\overline{1})$ so by Lemma 3.6, $q \in C\begin{pmatrix} \overline{1} & B \end{pmatrix}$. □

The next theorem has a (possibly) stronger hypothesis and stronger conclusion than Theorem 3.12. (As we have noted, if Conjecture 3.8 holds, then the content of Theorems 3.12 and 3.13 are identical.)

Theorem 3.13. *Let $B, M \in \mathcal{M}$. Then $\begin{pmatrix} \overline{1} & B & \mathbf{O} \\ \overline{0} & \mathbf{O} & M \end{pmatrix} \in \mathcal{M}$.*

Proof. Let L be a minimal left ideal of $(\beta\mathbb{N}, +)$. Pick $q \in L \cap C(M)$. We will show that $q \in C\begin{pmatrix} \overline{1} & B \end{pmatrix}$. Pick $p \in L \cap C(B)$. Then $q \in L = \beta\mathbb{N} + p$ so pick $r \in \beta\mathbb{N}$ such that $q = r + p$. By Lemma 3.6, $q \in C\begin{pmatrix} \overline{1} & B \end{pmatrix}$. □

Since the hypothesis on B and the conclusion of Theorem 3.13 are the same, we can iterate the process.

Corollary 3.14. *Let $m \in \omega$ and for $i \in \{0, 1, \ldots, m+1\}$, let $M_i \in \mathcal{M}$. Then*

$$\begin{pmatrix} \overline{1} & M_0 & \mathbf{O} & \mathbf{O} & \ldots & \mathbf{O} & \mathbf{O} & \mathbf{O} \\ \mathbf{O} & \mathbf{O} & \overline{1} & M_1 & \ldots & \mathbf{O} & \mathbf{O} & \mathbf{O} \\ \vdots & \vdots & \vdots & \vdots & \ddots & \vdots & \vdots & \vdots \\ \mathbf{O} & \mathbf{O} & \mathbf{O} & \mathbf{O} & \ldots & \overline{1} & M_m & \mathbf{O} \\ \mathbf{O} & \mathbf{O} & \mathbf{O} & \mathbf{O} & \ldots & \mathbf{O} & \mathbf{O} & M_{m+1} \end{pmatrix} \in \mathcal{M}.$$

Proof. Apply Theorem 3.13 $m+1$ times. □

Theorem 3.15. *Let $u, v \in \mathbb{N}$, let D be a $u \times v$ IPR matrix, and let A be an $\omega \times v$ matrix, whose rows are the rows of D. Let M_0 and M_1 be Milliken-Taylor matrices. Then* $\begin{pmatrix} A & M_1 & \mathbf{O} \\ \mathbf{O} & \mathbf{O} & M_0 \end{pmatrix} \in \mathcal{M}$.

Proof. Let $k, m \in \omega$ and let $\vec{a} = \langle a_0, a_1, \ldots, a_k \rangle$ and $\vec{b} = \langle b_0, b_1, \ldots, b_m \rangle$ be condensed sequences in $\mathbb{Z} \setminus \{0\}$ with $a_k > 0$ and $b_m > 0$ such that $M_0 = MT(\vec{a})$ and $M_1 = MT(\vec{b})$. Let L be a minimal left ideal of $(\beta\mathbb{N}, +)$ and let p be an idempotent in L. Since D is finite and IPR, $p \in C(D)$. Note that by [13, Exercise 4.3.8], $L = \beta\mathbb{N} + p = \beta\mathbb{Z} + p$.

Let $q = \frac{a_0}{a_k}p + \frac{a_1}{a_k}p + \ldots + \frac{a_{k-1}}{a_k}p + p$ and let $q' = \frac{b_0}{b_m}p + \frac{b_1}{b_m}p + \ldots + \frac{b_{m-1}}{b_m}p + p$. Then q and q' are in L. By Lemma 3.4, $q, q' \in C(D) = C(A)$. By Lemma 3.5, $q \in C\big(MT(\vec{a})\big)$ and $q' \in C\big(MT(\vec{b})\big)$.

Let $L' = L \cap C(A)$. Then $q, q' \in L'$. By [13, Theorem 1.65] L' is a minimal left ideal of $C(A)$ so $L' = C(A) + p$. Pick $r \in C(A)$ such that $q = r + q'$. By Lemma 3.6, $q \in C\begin{pmatrix} A & MT(\vec{b}) \end{pmatrix}$ so $L \cap C\begin{pmatrix} A & MT(\vec{b}) \end{pmatrix} \cap C(MT(\vec{a})) \neq \emptyset$. □

Theorem 3.16. *Let $u, v \in \mathbb{N}$ and let D be a $u \times v$ IPR matrix. Let α and δ be positive ordinals and let A be an $\alpha \times v$ matrix, whose rows are the rows of D. Let B be an $\alpha \times \delta$ centrally IPR matrix, and let M be a Milliken-Taylor matrix. Then* $\begin{pmatrix} A & B & \mathbf{O} \\ \mathbf{O} & \mathbf{O} & M \end{pmatrix} \in \mathcal{M}$.

Proof. Let $k \in \omega$ and let $\vec{a} = \langle a_0, a_1, \ldots, a_k \rangle$ be a condensed sequence in $\mathbb{Z} \setminus \{0\}$ with $a_k > 0$ such that $M = MT(\vec{a})$. Let L be a minimal left ideal of $(\beta\mathbb{N}, +)$ and let p be an idempotent in L. Since D is finite and IPR, $p \in C(D)$. Note that by [13, Exercise 4.3.8], $L = \beta\mathbb{N} + p = \beta\mathbb{Z} + p$. Let $q = \frac{a_0}{a_k}p + \frac{a_1}{a_k}p + \ldots + \frac{a_{k-1}}{a_k}p + p$. Then $q \in L$. By Lemma 3.4, $q \in C(D) = C(A)$. By Lemma 3.5, $q \in C(M)$. Since p is an idempotent in L and L is minimal, $q = q + p$. Since B is centrally IPR, $p \in C(B)$. By Lemma 3.6, $q \in C\begin{pmatrix} A & B \end{pmatrix}$. So $L \cap C\begin{pmatrix} A & B \end{pmatrix} \cap C(M) \neq \emptyset$. □

We now introduce two more classes of matrices. As far as we know, it is possible that $\mathcal{M}$, $\mathcal{N}$ and $\mathcal{R}$ are all equal to the set of all IPR matrices.

Definition 3.17. $\mathcal{N} = \{M :$ M is an admissible matrix and for any finite IPR matrix D, $C(D) \cap C(M) \cap K(\beta\mathbb{N}, +) \neq \emptyset\}$.

Definition 3.18. $\mathcal{R} = \{M :$ M is an admissible matrix and for any finite IPR matrix D and any left ideal L of $C(D)$, $L \cap C(M) \neq \emptyset\}$.

Lemma 3.19. $\mathcal{R} \subseteq \mathcal{N}$.

Proof. Let $M \in \mathcal{R}$ and let D be a finite IPR matrix. Pick a minimal left ideal L of $C(D)$. All minimal idempotents of $(\beta\mathbb{N}, +)$ are in $C(D)$ so by [13, Theorem

1.65] we may pick a minimal left ideal T of $(\beta\mathbb{N},+)$ such that $T \cap C(D) = L$ and in particular $L \subseteq K(\beta\mathbb{N},+)$. Thus $\emptyset \neq L \cap C(M) = C(D) \cap C(M) \cap T \subseteq C(D) \cap C(M) \cap K(\beta\mathbb{N},+)$. □

Lemma 3.20. *All centrally IPR matrices and all Milliken-Taylor matrices are in $\mathcal{R}$ (and thus in $\mathcal{N}$).*

Proof. Let D be a finite IPR matrix and let L be a left ideal of $C(D)$. We may presume that L is minimal in $C(D)$. As above, pick a minimal left ideal T of $(\beta\mathbb{N},+)$ such that $T \cap C(D) = L$. Pick an idempotent $p \in T$. If M is centrally IPR, then we have that $p \in L \cap C(M)$.

Assume we have $k \in \omega$ and a condensed sequence $\vec{a} = \langle a_0, a_1, \ldots, a_k \rangle$ in $\mathbb{Z} \setminus \{0\}$ with $a_k > 0$ such that $M = MT(\vec{a})$. Let $q = \frac{a_0}{a_k}p + \frac{a_1}{a_k}p + \ldots + \frac{a_{k-1}}{a_k}p + p$. Since $T = \beta\mathbb{Z} + p$, $q \in T$ and by Lemma 3.4, $q \in C(D)$. By Lemma 3.5, $q \in C(M)$. Therefore $q \in L \cap C(M)$. □

Theorem 3.21. *Let $u, v \in \mathbb{N}$, and let D be a $u \times v$ IPR matrix. Let α and δ be positive ordinals and let A be an $\alpha \times v$ matrix, whose rows are the rows of D. Let B be an $\alpha \times \delta$ member of $\mathcal{R}$ and let $M \in \mathcal{N}$. Then $\begin{pmatrix} A & B & \mathbf{O} \\ \mathbf{O} & \mathbf{O} & M \end{pmatrix} \in \mathcal{N}$.*

Proof. We show first that $C(A) \cap C(M) \cap K(\beta\mathbb{N},+) \subseteq C\begin{pmatrix} A & B & \mathbf{O} \\ \mathbf{O} & \mathbf{O} & M \end{pmatrix}$. So let $q \in C(A) \cap C(M) \cap K(\beta\mathbb{N},+)$. Let $L = C(A) + q$. Since $B \in \mathcal{R}$, pick $r \in L \cap C(B)$. Since $q \in K(\beta\mathbb{N},+) \cap C(A) = K(C(A))$ (by [13, Theorem 1.65]), L is a minimal left ideal of $C(A)$, so $L = L + r$. Pick $s \in L$ such that $q = s + r$. By Lemma 3.6, $q \in C\begin{pmatrix} A & B \end{pmatrix}$ so $q \in C\begin{pmatrix} A & B & \mathbf{O} \\ \mathbf{O} & \mathbf{O} & M \end{pmatrix}$.

Now let $P = \begin{pmatrix} A & B & \mathbf{O} \\ \mathbf{O} & \mathbf{O} & M \end{pmatrix}$. To see that $P \in \mathcal{N}$, let E be a finite IPR matrix. We need to show that $C(E) \cap C(P) \cap K(\beta\mathbb{N},+) \neq \emptyset$.

The rows of $\begin{pmatrix} A & \mathbf{O} \\ \mathbf{O} & E \end{pmatrix}$ are the rows of a finite IPR matrix so, since $M \in \mathcal{N}$,

$$C\begin{pmatrix} A & \mathbf{O} \\ \mathbf{O} & E \end{pmatrix} \cap C(M) \cap K(\beta\mathbb{N},+) \neq \emptyset\,.$$

Also $C\begin{pmatrix} A & \mathbf{O} \\ \mathbf{O} & E \end{pmatrix} \cap C(M) \cap K(\beta\mathbb{N},+) \subseteq C(A) \cap C(M) \cap K(\beta\mathbb{N},+) \subseteq C(P)$.

Therefore $\emptyset \neq C\begin{pmatrix} A & \mathbf{O} \\ \mathbf{O} & E \end{pmatrix} \cap C(M) \cap K(\beta\mathbb{N},+) \subseteq C(E) \cap C(P) \cap K(\beta\mathbb{N},+)$. □

As with Corollary 3.14, we immediately get the following corollary.

Corollary 3.22. *Let $m \in \omega$ and for $i \in \{0,1,\ldots,m\}$, let $M_i \in \mathcal{R}$, let A_i be a matrix with rows indexed by the same positive ordinal as the rows of M_i, and assume that the rows of A_i are all rows of a finite IPR matrix. Let $M_{m+1} \in \mathcal{N}$. Then*

$$\begin{pmatrix} A_0 & M_0 & \mathbf{O} & \mathbf{O} & \ldots & \mathbf{O} & \mathbf{O} & \mathbf{O} \\ \mathbf{O} & \mathbf{O} & A_1 & M_1 & \ldots & \mathbf{O} & \mathbf{O} & \mathbf{O} \\ \vdots & \vdots & \vdots & \vdots & \ddots & \vdots & \vdots & \vdots \\ \mathbf{O} & \mathbf{O} & \mathbf{O} & \mathbf{O} & \ldots & A_m & M_m & \mathbf{O} \\ \mathbf{O} & \mathbf{O} & \mathbf{O} & \mathbf{O} & \ldots & \mathbf{O} & \mathbf{O} & M_{m+1} \end{pmatrix} \in \mathcal{N}.$$

In Corollary 3.22 we could let each M_i be a Milliken-Taylor matrix. If each one is determined by a compressed sequence with the same last term, we will get a stronger result in Theorem 3.26 (except that we only conclude that the resulting matrix is IPR rather than a member of $\mathcal{N}$).

If $F, H \in \mathcal{P}_f(\mathbb{N})$ we write $F < H$ to mean that $\max F < \min H$.

Definition 3.23. Let $k \in \omega$ and let $\vec{a} = \langle a_0, a_1, \ldots, a_k\rangle$ be a compressed sequence in $\mathbb{Z} \setminus \{0\}$. Let $i \in \omega$ and let $\langle x_n\rangle_{n=i}^{\infty}$ be a sequence in $\mathbb{N}$. Then $MT(\vec{a}, \langle x_n\rangle_{n=i}^{\infty}) = \{\sum_{t=0}^{k} a_t \sum_{n\in F_t} x_n$: for each $i \in \{0,1,\ldots,k\}$, $F_i \in \mathcal{P}_f(\mathbb{N})$, $F_0 < F_1 < \ldots < F_k$, and $\min F_0 \geq i\}$.

Note that if $\vec{x} = \langle x_n\rangle_{n=i}^{\infty}$, then $MT(\vec{a}, \langle x_n\rangle_{n=i}^{\infty})$ is the set of entries of $MT(\vec{a})\vec{x}$.

Lemma 3.24. *Let $k \in \mathbb{N}$ and let $\langle a_0, a_1, \ldots a_k\rangle$ be a compressed sequence in $\mathbb{Z} \setminus \{0\}$ with $a_k > 0$. Let p be an idempotent in $\beta\mathbb{N}$, let $q = a_0 p + a_1 p + \ldots + a_n p$, and let $Q \in q$. There exist $D \in p$ and a function $P : \bigcup_{m=1}^{\infty} \mathbb{N}^m \to \mathcal{P}(\mathbb{N})$ such that, if $\langle x_t\rangle_{t=0}^{\infty}$ is a sequence in $\mathbb{N}$ and*

(1) $x_0 \in D$ and

(2) for each $n \in \omega$, $x_{n+1} \in P(x_0, x_1, \ldots, x_n)$,

then $MT(\vec{a}, \langle x_t\rangle_{t=1}^{\infty}) \subseteq D$.

Proof. For $m \in \{0,1,\ldots,k\}$ and $n \in \omega$, let

$$\mathcal{F}_{n,m} = \{\langle F_0, F_1, \ldots, F_m\rangle : \begin{array}{l} \text{for } i \in \{0,1,\ldots,m\},\ F_i \in \mathcal{P}_f(\{0,1,\ldots,n\}) \\ \text{and if } i < m, \text{ then } F_i < F_{i+1}\}, \end{array}$$

noting that if $m > n$, then $\mathcal{F}_{n,m} = \emptyset$.

Let C_0 and for $m \in \{0,1,\ldots,k-1\}$ and $x_0, x_1, \ldots, x_m$ in $\mathbb{N}$, let $B(x_0, x_1, \ldots, x_m)$ be as in the proof of [13, Theorem 15.33].

Let $D = C_0^\star$. Given $n < \omega$ and $\langle x_i\rangle_{i=0}^{n}$ in $\mathbb{N}$, let

$$\begin{aligned} &P(x_0, x_1, \ldots, x_n) = C_0^\star \cap \bigcap\{-\textstyle\sum_{t\in F} x_t + C_0^\star : F \in \mathcal{P}_f(\{0,1,\ldots,n\})\} \\ &\cap \textstyle\bigcap_{m=0}^{k-1} \bigcap\{B(\sum_{t\in F_0} x_t, \ldots, \sum_{t\in F_m} x_t) : \langle F_0, \ldots, F_m\rangle \in \mathcal{F}_{n,m}\} \\ &\cap \textstyle\bigcap_{m=1}^{k} \bigcap\{-\sum_{t\in F_m} x_t + B(\sum_{t\in F_0} x_t, \ldots, \sum_{t\in F_{m-1}} x_t) : \langle F_0, \ldots, F_m\rangle \in \mathcal{F}_{n,m}\}. \end{aligned}$$

(Here for notational convenience we are taking $\bigcap \emptyset = \mathbb{N}$.)

The proof of [13, Theorem 15.33] establishes that D and P are as required. □

Lemma 3.25. *For each $i < \omega$, let $k(i) \in \mathbb{N}$ and let $\vec{a}_i = \langle a_{i,j} \rangle_{j=0}^{k(i)}$ be a compressed sequence in $\mathbb{Z} \setminus \{0\}$ with $a_{i,k(i)} > 0$. Let p be an idempotent in $(\beta\mathbb{N}, +)$, for each $i < \omega$, let $q_i = a_{i,0}p + a_{i,1}p + \ldots + a_{i,k(i)}p$, and let $Q_i \in q_i$. There exists an increasing sequence $\langle x_n \rangle_{n=0}^{\infty}$ in $\mathbb{N}$ such that for each $i < \omega$, $MT(\vec{a}_i, \langle x_n \rangle_{n=i}^{\infty}) \subseteq Q_i$.*

Proof. For $i < \omega$ pick $D_i \in p$ and $P_i : \bigcup_{m=1}^{\infty} \mathbb{N}^m \to \mathcal{P}(\mathbb{N})$ as guaranteed by Lemma 3.24. Pick $x_0 \in D_0$, pick $x_1 \in D_1 \cap P_0(x_0)$, pick $x_2 \in D_2 \cap P_0(x_0, x_1) \cap P_1(x_1)$, and in general, pick $x_{n+1} \in D_{n+1} \cap \bigcap_{i=0}^{n} P_i(x_i, x_{i+1}, \ldots, x_n)$. □

Theorem 3.26. *Let $d \in \mathbb{N}$ and let $m \in \omega$. For $i \in \{0, 1, \ldots, m+1\}$, let $k(i) \in \omega$, let $\vec{a}_i = \langle a_{i,j} \rangle_{j=0}^{k(i)}$ be a compressed sequence in $\mathbb{Z} \setminus \{0\}$ with $a_{i,k(i)} = d$, and let $M_i = MT(\vec{a}_i)$. For $i \in \{0, 1, \ldots, m\}$, let $v_i \in \mathbb{N}$ and let A_i be an $\omega \times v_i$ matrix with all rows in some finite IPR matrix. Then*

$$N = \begin{pmatrix} A_0 & \mathbf{O} & \mathbf{O} & \ldots & \mathbf{O} & M_0 \\ \mathbf{O} & A_1 & \mathbf{O} & \ldots & \mathbf{O} & M_1 \\ \mathbf{O} & \mathbf{O} & A_2 & \ldots & \mathbf{O} & M_2 \\ \vdots & \vdots & \vdots & \ddots & \vdots & \vdots \\ \mathbf{O} & \mathbf{O} & \mathbf{O} & \ldots & A_m & M_m \\ \mathbf{O} & \mathbf{O} & \mathbf{O} & \ldots & \mathbf{O} & M_{m+1} \end{pmatrix} \quad \textit{is IPR.}$$

Proof. Pick a minimal idempotent p in $(\beta\mathbb{N}, +)$. By [13, Lemma 5.19.2] dp is a minimal idempotent so by [13, Exercise 4.3.8] $\beta\mathbb{Z} + dp = \beta\mathbb{N} + dp$. For $i \in \{0, 1, \ldots, m+1\}$, let $q_i = a_{i,0}p + a_{i,1}p + \ldots + a_{i,k(i)}p$. By Lemma 3.4, $q_i \in C(A_j)$ for each $j \in \{0, 1, \ldots, m+1\}$. Given $i \in \{0, 1, \ldots, m\}$, q_i and q_{m+1} are in $(\beta\mathbb{N} + dp) \cap C(A_i)$ which is a minimal left ideal of $C(A_i)$ by [13, Theorem 1.65], so pick $r_i \in C(A_i)$ such that $q_{m+1} = r_i + q_i$. We will show that if $B \in q_{m+1}$ there exists $\vec{w}$ with entries in $\mathbb{N}$ such that all entries of $N\vec{w}$ are in B. So let $B \in q_{m+1}$ be given.

Let $i \in \{0, 1, \ldots, m\}$ and let $D_i = \{x \in \mathbb{N} : -x + B \in q_i\}$. Then $D_i \in r_i$ and $r_i \in C(A_i)$ so pick $\vec{z}_i \in \mathbb{N}^{v_i}$ such that $\vec{y}_i = A_i\vec{z}_i \in D_i^{\omega}$. Now $\{y_{i,j} : j < \omega\}$ is finite so $Q_i = \bigcap_{j<\omega}(-y_{i,j} + B) \in q_i$. Let $Q_{m+1} = B$.

By Lemma 3.25, pick a sequence $\langle x_n \rangle_{n=0}^{\infty}$ in $\mathbb{N}$ such that for each $i \in \{0, 1, \ldots, m+1\}$, $MT(\vec{a}_i, \langle x_n \rangle_{n=i}^{\infty}) \subseteq Q_i$. For $n \in \omega$, let $c_n = x_{m+n+1}$.

$$\vec{w} = \begin{pmatrix} \vec{z}_0 \\ \vec{z}_1 \\ \vdots \\ \vec{z}_m \\ \vec{c} \end{pmatrix}.$$

Then, given $i \in \{0, 1, \ldots, m\}$, $\begin{pmatrix} \mathbf{O} & \ldots & A_i & \ldots & \mathbf{O} & M_i \end{pmatrix} \vec{w} = A_i \vec{z}_i + M_i \vec{c} \in B^\omega$ and $\begin{pmatrix} \mathbf{O} & \mathbf{O} & \ldots & \mathbf{O} & M_{m+1} \end{pmatrix} \vec{w} = M_{m+1} \vec{c} \in B^\omega$ so $\vec{w}$ is as required. □

References

[1] B. Barber, N. Hindman and I. Leader, Partition regularity in the rationals, *J. Comb. Theory Ser. A* **120** (2013), 1590-1599.

[2] W. Deuber, Partitionen und lineare Gleichungssysteme, *Math. Z.* **133** (1973), 109-123.

[3] W. Deuber and N. Hindman, Partitions and sums of (m,p,c)-sets, *J. Combin. Theory Ser. A* **45** (1987), 300-302.

[4] W. Deuber, N. Hindman, I. Leader, and H. Lefmann, Infinite partition regular matrices, *Combinatorica* **15** (1995), 333-355.

[5] D. Gunderson, N. Hindman, and H. Lefmann, Some partition theorems for infinite and finite matrices, *Integers* **14** (2014), Article A12.

[6] N. Hindman, Finite sums from sequences within cells of a partition of $\mathbb{N}$, *J. Combin. Theory Ser. A* **17** (1974), 1-11.

[7] N. Hindman and I. Leader, Image partition regularity of matrices, *Combin. Probab. Comput.* **2** (1993), 437-463.

[8] N. Hindman, I. Leader, and D. Strauss, Separating Milliken-Taylor systems with negative entries, *Proc. Edinb. Math. Soc.* **46** (2003), 45-61.

[9] N. Hindman, I. Leader, and D. Strauss, Infinite partition regular matrices – solutions in central sets, *Trans. Amer. Math. Soc.* **355** (2003), 1213-1235.

[10] N. Hindman, I. Leader, and D. Strauss, Extensions of infinite partition regular systems, *Electron. J. Combin.* **22(2)** (2015), #P2.29.

[11] N. Hindman, I. Leader, and D. Strauss, Duality for image and kernel partition regularity of infinite matrices, *J. Comb.* **8** (2017), 653-672.

[12] N. Hindman and D. Strauss, Infinite partition regular matrices, II – extending the finite results, *Topology Proc.* **25** (2000), 217-255.

[13] N. Hindman and D. Strauss, *Algebra in the Stone-Čech Compactification: Theory and Applications, 2nd edition*, Walter de Gruyter & Co., Berlin, 2012.

[14] I. Leader and P. Russell, Consistency for partition regular equations, *Discrete Math.* **306** (2006), 847-850.

[15] K. Milliken, Ramsey's Theorem with sums or unions, *J. Combin. Theory Ser. A* **18** (1975), 276-290.

[16] S. Patra and S. Ghosh, Concerning partition regular matrices, *Integers* **17** (2017), Article A63.

[17] A. Taylor, A canonical partition relation for finite subsets of ω, J. Combin. Theory Ser. A) **21** (1976), 137-146.

[18] B. van der Waerden, Beweis einer Baudetschen Vermutung, *Nieuw Arch. Wiskd.* **19** (1927), 212-216.

7

PRIMES $p \equiv 1 \bmod d$ AND $a^{(p-1)/d} \equiv 1 \bmod p$

Peng Gao
School of Mathematics and Systems Science, Beihang University, Beijing, China
penggao@buaa.edu.cn

Liangyi Zhao
School of Mathematics and Statistics, University of New South Wales, Sydney, Australia
l.zhao@unsw.edu.au

Abstract

Suppose that $d \in \{2,3,4,6\}$ and $a \in \mathbb{Z}$ with $a \neq -1$ and a is not square. Let $P_{(a,d)}$ be the number of primes p not exceeding x such that $p \equiv 1 \pmod d$ and $a^{(p-1)/d} \equiv 1 \pmod p$. In this paper, we study the mean value of $P_{(a,d)}$.

1. Introduction

Let $\mathbb{P}$ denote the set of prime numbers and $d > 0$ be a squarefree integer. Moreover, suppose that $a \in \mathbb{Z}$ with $a \neq -1$ and a is not a square. Set

$$\mathcal{P}_{(a,d)} = \{p \in \mathbb{P} : p \equiv 1 \pmod d \quad \text{and} \quad a^{(p-1)/d} \equiv 1 \pmod p\}$$

and

$$P_{a,d}(x) = \#\{p \in \mathcal{P}_{(a,d)} : p \leq x\}.$$

It has long been known and studied [9, 5] that the estimation of $P_{(a,d)}$ is connected with Artin's conjecture on primitive roots. This conjecture asserts that the number of primes not exceeding x, and for which a is a primitive root, is asymptotically $c\pi(x)$ where c is some explicitly defined constant whose expression involves a.

Assuming that truth of the Riemann hypothesis for the Dedekind zeta function for the number field $\mathbb{Q}[\sqrt[d]{a}, \sqrt[d]{1}]$, C. Hooley [5] proved that

$$P_{(a,d)}(x) = \frac{\varepsilon(d)}{d\varphi(d)}\mathrm{li}(x) + O\left(\sqrt{x}\log(dx)\right), \tag{1}$$

where $\varepsilon(d)$ is defined in the following way. Writing $a = lm^2$ with l square free, $\varepsilon(d) = 2$ if $l \equiv 1 \pmod 4$ and $2l|d$ and $\varepsilon(d) = 1$ otherwise. Using (1), Hooley was able to prove Artin's conjecture conditionally. For a survey of this conjecture, we refer the reader to [9].

S. Li [8] studied $P_{(a,d)}(x)$ on average and proved that

$$\frac{1}{y}\sum_{2\le a\le y} P_{(a,d)}(x) = \frac{\pi(x;1,d)}{d} + O\left(\frac{xE(x,y)}{\varphi(d)\log x}\right), \tag{2}$$

where

$$E(x,y) = y^{-1/21} \text{ if } y \le x^{2/3} \quad \text{and} \quad E(x,y) = x^{-1/6}\log x \text{ if } y > x^{2/3}.$$

In this paper, we improve the bound (2) for $d = 2$, 3, 4 and 6.

In the case $d = 2$, we prove the following result.

Theorem 1. *We have*

$$\frac{1}{y}\sum_{2\le a\le y} P_{(a,2)}(x) - \frac{\pi(x)}{2} \ll \begin{cases} x\sqrt{\frac{\log x}{y}}, & \text{if } \frac{x}{\log x} \ge y, \\ \sqrt{x}\log x, & \text{if } \frac{x}{\log x} < y. \end{cases}$$

Additionally, for $d = 3$. 4 and 6, we have the following.

Theorem 2. *Let $\varepsilon > 0$ be given. If $d \in \{3,4,6\}$, then*

$$\frac{1}{y}\sum_{2\le a\le y} P_{(a,d)}(x) - \frac{\pi(x;1,d)}{d} \ll (xy)^{\varepsilon}E(x,y),$$

where $E(x,y)$ is defined piece-wise in the following way:

$$E(x,y) = \begin{cases} x^{1/2}, & \text{if } x \le y^{3/5}, \\ x^{4/3}y^{-1/2}, & \text{if } y^{3/5} < x \le y^{6/7}, \\ x^{3/4}, & \text{if } y^{6/7} < x \le y^{6/5}, \\ x^{7/6}y^{-1/2}, & \text{if } y^{6/5} < x \le y^{3/2}, \\ x^{5/6}, & \text{if } y^{3/2} < x \le y^{9/5}, \\ x^{10/9}y^{-1/2}, & \text{if } y^{9/5} < x \le y^{108/55}, \\ x^{1/2}y^{7/10}, & \text{if } y^{108/55} < x \le y^{11/5}, \\ x^{2/3}y^{1/3}, & \text{if } y^{11/5} < x \le y^{5/2}, \\ xy^{-1/2}, & \text{if } y^{5/2} < x. \end{cases}$$

Note that the O-terms in Theorems 1 and 2 are smaller than that in (2) in all xy-ranges. Finally, we shall prove the following smoothed version of Theorem 1, assuming the truth of the generalized Riemann hypothesis.

Theorem 3. *Suppose that* x, $Y > 1$, $U = x^{1/8}Y^{1/4}$ *and set*

$$D(Y) = \{d : Y \leq d \leq 2Y, d \text{ odd and square-free}\}.$$

Let $\Phi_Y(t)$ *be a non-negative smooth function supported on the interval* $(1,2)$ *satisfying the condition* $\Phi_Y(t) = 1$ *for* $t \in (1 + 1/U, 2 - 1/U)$ *and*

$$\Phi_Y^{(j)}(t) \ll_j U^j \tag{3}$$

for all integers $j \geq 0$. *Assuming the truth of the generalized Riemann hypothesis for the Dirichlet L-functions, we have*

$$\begin{aligned} &\frac{1}{\#D(Y)} \sum_{(a,2)=1} \mu^2(a) P_{(8a,2)}(x) \Phi_Y\left(\frac{a}{Y}\right) \\ &\qquad = \frac{\pi(x)}{2} + O\left(\log\log x + \left(\frac{x^{7/8}}{Y^{1/4}} + \frac{x}{Y^{1/2}}\right)\log(xY)\right). \end{aligned} \tag{4}$$

Note that the strength of Theorem 3 is most prominent when Y is large compared to x.

2. Preliminaries

We shall need the following lemma due to Pólya and Vinogradov.

Lemma 1. *Let* $\chi(n)$ *be a non-principal character modulo* q *and* $M \in \mathbb{Z}$ *and* $N \in \mathbb{N}$. *Then we have*

$$\left|\sum_{M<n\leq M+N} \chi(n)\right| \leq 6\sqrt{q}\log q.$$

Proof. This is a standard result and a proof can be found in [6, Theorem 12.5]. □

The following mean-value theorem for quadratic character sums will be useful for us as well.

Lemma 2. *Let* χ *run over all real non-principal characters* $\chi(n) = \left(\frac{D}{n}\right)$ *with* $|D| \leq X$. *We have*

$$\sum_{\chi}\left|\sum_{n\leq Y} \chi(n)\right|^2 \ll XY \log^2 X.$$

Proof. We refer the reader to [7, Lemma 5] for a proof of this lemma. □

Lemma 3. *Assume the truth of the generalized Riemann hypothesis for the Dirichlet L-functions. Then for any non-principal Dirichlet character* $\chi \pmod q$, *we have*

$$\sum_{p\le X}\chi(p) \ll X^{1/2}\log(qX).$$

Proof. This follows readily from [6, Proposition 5.25]. □

Lemma 4 (Large sieve inequality for certain fixed order characters). *Let* $(a_m)_{m\in\mathbb{N}}$ *be an arbitrary sequence of complex number and* $k \in \{3,4,6\}$. *Then, for any* $\varepsilon > 0$,

$$\sum_{Q<q\le 2Q}\ \sideset{}{^\star}\sum_{\substack{\chi \bmod q\\ \chi^k=\chi_0,\chi\neq\chi_0}}\left|\ \sideset{}{^*}\sum_{M<m\le 2M} a_m\chi(m)\right|^2 \ll (QM)^\varepsilon\Delta\sum_{M<m\le 2M}|a_m|^2, \tag{5}$$

where

$$\Delta = \min\left\{Q^{5/3}+M, Q^{4/3}+Q^{1/2}M, Q^{11/9}+Q^{2/3}M, Q+Q^{1/3}M^{5/3}+M^{12/5}\right\}.$$

The implied constant in (5) *depends on* ε, *the star on the sum over* χ *restricts the sum to primitive characters and the asterisks attached to the sum over* m *indicates that* m *ranges over square-free integers.*

Proof. For $k = 4$ and $k = 6$, the outer sums on the left-hand side of (5) include quadratic characters. We use [4, Theorem 1] to bound the contribution of those characters. For the other characters, we use the results in [3, 1]. For $k = 4$, this lemma is an improvement of [3, Theorem 1.2] and the proof goes along the same line as Section 6 of [3]. The only difference is, instead of using [3, Theorem 1.1], one uses [2, Theorem 1.3] at the appropriate places. For $k = 3$ and $k = 6$, this is [1, Theorem 1.4] and [1, Theorem 1.5], respectively. □

3. Proof of Theorem 1

By considering the group structure of $(\mathbb{Z}/p\mathbb{Z})^*$, we easily conclude that $a^{(p-1)/2} \equiv 1 \pmod p$ if and only if $p \nmid a$ and a is a square modulo p. So

$$P_{(a,2)}(x) = \frac{1}{2}\sum_{2<p\le x}\left(\chi_0(a) + \left(\frac{a}{p}\right)\right),$$

where χ_0 is the principal character modulo p. Therefore, we get

$$S := \frac{1}{y}\sum_{2\le a\le y} P_{(a,2)}(x) = \frac{1}{2y}\sum_{2<p\le x}\sum_{a\le y}\chi_0(a) + \frac{1}{2y}\sum_{2<p\le x}\sum_{a\le y}\left(\frac{a}{p}\right) = S_1 + S_2, \tag{6}$$

say. We easily have

$$S_1 = \frac{1}{2y}\sum_{2<p\le x}\left([y]-\left[\frac{y}{p}\right]\right) = \frac{1}{2y}\sum_{2<p\le x}\left(y-\frac{y}{p}+O(1)\right)$$
$$= \frac{\pi(x)}{2}+O\left(\log\log x+\frac{\pi(x)}{y}\right), \tag{7}$$

where we have used the well-known bound

$$\sum_{p\le x}\frac{1}{p} \ll \log\log x. \tag{8}$$

Now

$$S_2 \ll \frac{1}{2y}\sum_{2<p\le x}\sum_{a\le y}\left(\frac{a}{p}\right) \ll \frac{1}{2y}\sum_{2<p\le x}\left|\sum_{a\le y}\left(\frac{a}{p}\right)\right|.$$

Applying Lemma 1, we get that

$$S_2 \ll \frac{1}{2y}\sum_{p\le x}\sqrt{p}\log p \ll \frac{x^{3/2}}{y}, \tag{9}$$

by the prime number theorem. On the other hand, using Cauchy-Schwarz inequality, we conclude that

$$\sum_{p\le x}\left|\sum_{a\le y}\left(\frac{a}{p}\right)\right| \le \pi(x)^{1/2}\left(\sum_{p\le x}\left|\sum_{a\le y}\left(\frac{a}{p}\right)\right|^2\right)^{1/2} \ll xy^{1/2}\log^{1/2} x,$$

where the last inequality comes from Lemma 2 and the prime number theorem. Hence,

$$S_2 \ll \frac{1}{y}xy^{1/2}\log^{1/2} x = \frac{x\log^{1/2} x}{y^{1/2}}. \tag{10}$$

Finally, combining (6), (7), (9) and (10), we arrive at

$$S = \frac{\pi(x)}{2}+O\left(\log\log x+\frac{\pi(x)}{y}+\min\left(\frac{x^{3/2}}{y},\frac{x\log^{1/2} x}{y^{1/2}}\right)\right).$$

The two terms in the minimum above are equal when $x/\log x = y$. The theorem follows by an easy analysis.

4. Proof of Theorem 2

For $d = 3$, 4 and 6, an argument similar to that in the previous section gives that

$$\frac{1}{y}\sum_{a\le y}P_{(a,d)} = \frac{1}{dy}\sum_{\substack{p\le x\\ p\equiv 1 \bmod d}}\sum_{a\le y}\chi_0(a)+\frac{1}{dy}\sum_{\substack{p\le x\\ p\equiv 1 \bmod d}}\sum_{a\le y}\sum_{\substack{\chi \bmod p\\ \chi^d=\chi_0,\ \chi\ne\chi_0}}\chi(a) =: S_1+S_2,$$

say. As before,

$$S_1 = \frac{\pi(x;1,d)}{d} + O\left(\log\log x + \frac{\pi(x;1,d)}{y}\right). \tag{11}$$

To estimate S_2, we note that for all $a \in \mathbb{N}$, a can be written uniquely as l^2m with $l, m \in \mathbb{N}$ and m square-free. Hence, using Cauchy's inequality,

$$S_2^2 \ll \frac{\pi(x;1,d)}{y^2} \sum_{\substack{p\le x \\ p\equiv 1 \bmod d}} \sum_{\substack{\chi \bmod p \\ \chi^d=\chi_0,\ \chi\neq\chi_0}} \left| \sum_{m\le y}^{*} \left(\sum_{l\le\sqrt{y/m}} \chi(l^2) \right) \chi(m) \right|^2,$$

where the asterisks on the sum over m indicates that m runs over square-free integers. Now we will apply Lemma 5 to S_2. Aided by table in [1, p. 897], we get that

$$S_2 \ll (xy)^{\varepsilon}\sqrt{\pi(x;1,d)} \times \begin{cases} 1 & \text{if } x \le y^{3/5} \\ x^{5/6}/y^{1/2} & \text{if } y^{3/5} < x \le y^{6/7} \\ x^{1/4} & \text{if } y^{6/7} < x \le y^{6/5} \\ x^{2/3}/y^{1/2} & \text{if } y^{6/5} < x \le y^{3/2} \\ x^{1/3} & \text{if } y^{3/2} < x \le y^{9/5} \\ x^{11/18}/y^{1/2} & \text{if } y^{9/5} < x \le y^{108/55} \\ y^{7/10} & \text{if } y^{108/55} < x < y^{11/5} \\ x^{1/6}y^{1/3} & \text{if } y^{11/5} < x \le y^{5/2} \\ x^{1/2}/y^{1/2} & \text{if } y^{5/2} < x \end{cases}. \tag{12}$$

Now a quick analysis will give that the O-terms in (11) can be absorbed into the bounds in (12) in all ranges. The result thus follows.

5. Proof of Theorem 3

Set

$$D(Y) = \{d : Y \le d \le 2Y, d \text{ odd and square-free}\}.$$

It is easy to see that

$$\#D(Y) \sim \frac{4Y}{\pi^2}. \tag{13}$$

Now let $\Phi_Y(t)$ be a non-negative smooth function supported on the interval $(1,2)$ satisfying the conditions $\Phi_Y(t) = 1$ for $t \in (1+1/U, 2-1/U)$ and such that

$$\Phi_Y^{(j)}(t) \ll_j U^j \tag{14}$$

for all integers $j \geq 0$. We need to find an asymptotic formula for

$$\begin{aligned} S &= \frac{1}{\#D(Y)} \sum_{(a,2)=1} \mu^2(a) P_{(8a,2)}(x) \Phi_Y\left(\frac{a}{Y}\right) \\ &= \frac{1}{2\#D(Y)} \sum_{2<p\leq x} \sum_{(a,2)=1} \mu^2(a)\chi_0(8a)\Phi_Y\left(\frac{a}{Y}\right) \\ &\quad + \frac{1}{2\#D(Y)} \sum_{2<p\leq x} \sum_{(a,2)=1} \mu^2(a) \left(\frac{8a}{p}\right) \Phi_Y\left(\frac{a}{Y}\right). \end{aligned} \tag{15}$$

Let S_1 and S_2, respectively, denote the two sums above. We shall study them separately.

Recalling that the support of Φ_Y lies inside the interval $(1,2)$ and the bound (3), we have

$$\begin{aligned} S_1 &= \frac{1}{2\#D(Y)} \sum_{2<p\leq x} \left(\sum_{\substack{(a,2)=1 \\ Y\leq a\leq 2Y}} \mu^2(a)\chi_0(8a) + O\left(\frac{Y}{U}\right)\right) \\ &= \frac{1}{2\#D(Y)} \sum_{2<p\leq x} \sum_{\substack{(a,2p)=1 \\ Y\leq a\leq 2Y}} \mu^2(a) + O\left(\frac{\pi(x)}{U}\right). \end{aligned} \tag{16}$$

Using Möbius inversion, we get

$$\begin{aligned} \sum_{2<p\leq x} \sum_{\substack{(a,2p)=1 \\ Y\leq a\leq 2Y}} \mu^2(a) &= \sum_{2<p\leq x} \sum_{\substack{(a,2)=1 \\ Y\leq a\leq 2Y}} \mu^2(a) \sum_{d|(a,p)} \mu(d) \\ &= \sum_{2<p\leq x} \sum_{d|p} \mu(d) \sum_{\substack{(a,2)=1,\ d|a \\ Y\leq a\leq 2Y}} \mu^2(a). \end{aligned} \tag{17}$$

Now it is easy to see that the above can be recast as

$$\sum_{2<p\leq x} \sum_{\substack{Y\leq a\leq 2Y \\ (a,2)=1}} \mu^2(a) + O\left(\sum_{p\leq x} \sum_{\substack{Y\leq a\leq 2Y \\ p|a}} 1 \right) = \pi(x)\#D(Y) + O\left(Y \sum_{p\leq x} \frac{1}{p}\right). \tag{18}$$

Now combining (16), (17), (18) and (8), we arrive at

$$S_1 = \frac{\pi(x)}{2} + O\left(\frac{Y}{\#D(Y)} \log\log x + \frac{\pi(x)}{U}\right). \tag{19}$$

The treatment of S_2 will be more complicated. Let z be a positive number to be chosen later. We start with

$$\mu^2(a) = M_z(a) + R_z(a), \text{ where } M_z(a) = \sum_{\substack{l^2|a \\ l\leq z}} \mu(l) \text{ and } R_z(a) = \sum_{\substack{l^2|a \\ l>z}} \mu(l).$$

We split S_2 as

$$S_2 = S_{2,1} + S_{2,2} \tag{20}$$

where

$$S_{2,1} = \frac{1}{2\#D(Y)} \sum_{2<p\le x} \sum_{(a,2)=1} M_z(a) \left(\frac{8a}{p}\right) \Phi_Y\left(\frac{a}{Y}\right)$$

and

$$S_{2,2} = \frac{1}{2\#D(Y)} \sum_{2<p\le x} \sum_{(a,2)=1} R_z(a) \left(\frac{8a}{p}\right) \Phi_Y\left(\frac{a}{Y}\right).$$

We first deal with $S_{2,2}$. Interchanging summations gives us

$$S_{2,2} \ll \frac{1}{\#D(Y)} \sum_{l>z} \sum_{\substack{m\ge 1\\(m,2)=1}} \Phi_Y\left(\frac{l^2m}{Y}\right) \left| \sum_{2<p\le x} \left(\frac{8l^2m}{p}\right) \right|.$$

Using Lemma 3, we get that the above is

$$S_{2,2} \ll \frac{1}{\#D(Y)} \sum_{l>z} \sum_{m\ge 1} \Phi_Y\left(\frac{l^2m}{Y}\right) x^{1/2} \log(l^2mx) \ll \frac{\#D(Y)}{Y} \frac{x^{1/2}\log(xYz)}{z}. \tag{21}$$

We now evaluate the inner sum of $S_{2,1}$ following a method of K. Soundararajan in [10] by applying the Poisson summation formula to the sum over d. For all odd integers k and all integers m, we introduce the Gauss-type sums

$$\tau_m(k) := \sum_{a \pmod k} \left(\frac{a}{k}\right) e\left(\frac{am}{k}\right) =: \left(\frac{1+i}{2} + \left(\frac{-1}{k}\right)\frac{1-i}{2}\right) G_m(k).$$

We quote [10, Lemma 2.3] which determines $G_m(k)$.

Lemma 5. *If $(k_1,k_2)=1$ then $G_m(k_1k_2) = G_m(k_1)G_m(k_2)$. Suppose that p^a is the largest power of p dividing m (put $a=\infty$ if $m=0$). Then for $b\ge 1$ we have*

$$G_m(p^b) = \begin{cases} 0 & \text{if } b\le a \text{ is odd,}\\ \varphi(p^b) & \text{if } b\le a \text{ is even,}\\ -p^a & \text{if } b=a+1 \text{ is even,}\\ (\frac{m/p^a}{p})p^a\sqrt{p} & \text{if } b=a+1 \text{ is odd,}\\ 0 & \text{if } b\ge a+2. \end{cases}$$

For a Schwartz function F, let

$$\tilde{F}(\xi) = \frac{1+i}{2}\hat{F}(\xi) + \frac{1-i}{2}\hat{F}(-\xi) = \int_{-\infty}^{\infty} \left(\cos(2\pi\xi x) + \sin(2\pi\xi x)\right) F(x)\mathrm{d}x. \tag{22}$$

Moreover, we shall need Lemma 2.6 of [10], quoted below, which determines the inner sum of $S_{2,1}$.

Lemma 6. *Let Φ be a non-negative, smooth function supported in $(1, 2)$. For any odd integer k,*

$$\sum_{\gcd(d,2)=1} M_z(d)\left(\frac{d}{k}\right)\Phi\left(\frac{d}{X}\right) = \frac{X}{2k}\left(\frac{2}{k}\right)\sum_{\substack{\alpha\leq z\\(\alpha,2k)=1}}\frac{\mu(\alpha)}{\alpha^2}\sum_m(-1)^m G_m(k)\tilde{\Phi}\left(\frac{mX}{2\alpha^2 k}\right),$$

where $\tilde{\Phi}$ is as defined in (22).

Applying Lemmas 5 and 6, we transform $S_{2,1}$ into

$$\begin{aligned} S_{2,1} &= \frac{Y}{4\#D(Y)}\sum_{2<p\leq x}\frac{1}{p}\sum_{\substack{\alpha\leq z\\(\alpha,2p)=1}}\frac{\mu(\alpha)}{\alpha^2}\sum_m(-1)^m G_m(p)\tilde{\Phi}\left(\frac{mY}{2\alpha^2 p}\right)\\ &= \frac{Y}{4\#D(Y)}\sum_{2<p\leq x}\frac{1}{\sqrt{p}}\sum_{\substack{\alpha\leq z\\(\alpha,2p)=1}}\frac{\mu(\alpha)}{\alpha^2}\sum_{m\neq 0}(-1)^m\left(\frac{m}{p}\right)\tilde{\Phi}\left(\frac{mY}{2\alpha^2 p}\right) \qquad (23)\\ &= S_{m=\square} + S_{m\neq\square}, \end{aligned}$$

where $S_{m=\square}$ is the part of $S_{2,1}$ with m being a square and $S_{m\neq\square}$ the complementary sum. We shall also use the notations $m = \square$ and $m \neq \square$ to mean, respectively, m is a square and m is not a square.

Now

$$S_{m=\square} = \frac{Y}{2\#D(Y)}\sum_{2<p\leq x}\frac{1}{\sqrt{p}}\sum_{\substack{\alpha\leq z\\(\alpha,2p)=1}}\frac{\mu(\alpha)}{\alpha^2}\sum_{\substack{m=1\\(m,p)=1}}^{\infty}(-1)^m\tilde{\Phi}\left(\frac{m^2Y}{2\alpha^2 p}\right). \qquad (24)$$

Note that trivially estimating the integral in (22) gives

$$\tilde{\Phi}(\xi) \ll 1,$$

and integration by parts yields

$$\tilde{\Phi}(\xi) = \frac{-1}{2\pi\xi}\left(\int_1^{1+1/U} + \int_{2-1/U}^{2}\right)\Phi'(x)\Big(\sin(2\pi\xi x) - \cos(2\pi\xi x)\Big)\mathrm{d}x \ll \frac{1}{|\xi|}.$$

These bounds for $\tilde{\Phi}$ give

$$\sum_{m=1}^{\infty}\left|\tilde{\Phi}\left(\frac{m^2Y}{2\alpha p}\right)\right| \ll \sum_{m\leq(2\alpha^2p/Y)^{1/2}} 1 + \sum_{m>(2\alpha^2p/Y)^{1/2}}\frac{2\alpha^2 p}{m^2 Y} \ll \alpha\sqrt{\frac{p}{Y}}.$$

So inserting the above estimate into (24), we get

$$S_{m=\square} \ll \frac{Y^{1/2}x}{\#D(Y)}\log z. \qquad (25)$$

The estimation of $S_{m\neq\Box}$ is more complicated. Re-arranging the orders of summation and using $\left(\frac{4\alpha^2}{p}\right)$ to detect the condition $(2\alpha,p)=1$, we get

$$S_{m\neq\Box} = \frac{Y}{4\#D(Y)} \sum_{\substack{\alpha\leq z\\(\alpha,2)=1}} \frac{\mu(\alpha)}{\alpha^2} \sum_{m\neq 0,\Box} (-1)^m \sum_{p\leq x} \left(\frac{4\alpha^2 m}{p}\right) \frac{1}{\sqrt{p}} \tilde{\Phi}\left(\frac{mY}{2\alpha^2 p}\right).$$

Using integration by parts and Lemma 3, we get that the inner-most sum in the above displayed equation is

$$\ll \log(\alpha(|m|+2)x) \left|\tilde{\Phi}\left(\frac{mY}{2\alpha^2 x}\right)\right| + \int_1^x \frac{\log(\alpha(|m|+2)V)}{V} \left|\tilde{\Phi}\left(\frac{mY}{2\alpha^2 V}\right)\right| \mathrm{d}V$$
$$+ \int_1^x \frac{Y}{\alpha^2} \frac{\log(\alpha(|m|+2)V)}{V^2} \left|m\tilde{\Phi}'\left(\frac{mY}{2\alpha^2 V}\right)\right| \mathrm{d}V.$$

Thus

$$S_{m\neq\Box} \ll R_1 + R_2 + R_3, \tag{26}$$

where

$$R_1 = \frac{Y}{\#D(Y)} \sum_{\alpha\leq z} \frac{1}{\alpha^2} \sum_m \log(\alpha(|m|+2)X) \left|\tilde{\Phi}\left(\frac{mY}{2\alpha^2 x}\right)\right|,$$

$$R_2 = \frac{Y}{\#D(Y)} \sum_{\alpha\leq z} \frac{1}{\alpha^2} \int_1^x \sum_m \frac{\log(\alpha(|m|+2)V)}{V} \left|\tilde{\Phi}\left(\frac{mY}{2\alpha^2 V}\right)\right| \mathrm{d}V$$

and

$$R_3 = \frac{Y}{\#D(Y)} \sum_{\alpha\leq z} \frac{Y}{\alpha^4} \int_1^x \sum_m \frac{\log(\alpha(|m|+2)V)}{V^2} \left|m\tilde{\Phi}'\left(\frac{mY}{2\alpha^2 V}\right)\right| \mathrm{d}V.$$

Now observe that

$$\begin{aligned}
&\sum_m \log(\alpha(|m|+2)x) \left|\tilde{\Phi}\left(\frac{mY}{2\alpha^2 x}\right)\right| \\
&\quad\ll \sum_{m\leq 2\alpha^2 x/Y} \log(\alpha(|m|+2)x) + \sum_{m>2\alpha^2 x/Y} \log(\alpha(|m|+2)x) \frac{\alpha^4 x^2 U}{m^2 Y^2} \\
&\quad\ll \frac{\alpha^2 x}{Y}\log(\alpha x) + \frac{\log(\alpha x)\alpha^2 x}{Y} U \ll \frac{\log(\alpha x)\alpha^2 x}{Y} U
\end{aligned} \tag{27}$$

and similarly, using (14),

$$\sum_m \log(\alpha(|m|+2)V) \left|m\tilde{\Phi}'\left(\frac{mY}{2\alpha^2 V}\right)\right| \ll \log(\alpha V) \frac{\alpha^4 V^2 U^2}{Y^2}. \tag{28}$$

From (27), it follows that

$$R_1 \ll \frac{xU}{\#D(Y)} \sum_{\alpha \leq z} \log(\alpha x) \ll \frac{xU}{\#D(Y)} z \log(zx) \tag{29}$$

and

$$R_2 \ll \frac{U}{\#D(Y)} \sum_{\alpha \leq z} \int_1^x \log(\alpha V) \mathrm{d}V \ll \frac{U}{\#D(Y)} xz \log(zx). \tag{30}$$

Using (28)

$$R_3 \ll \frac{U^2}{\#D(Y)} \sum_{\alpha \leq z} \int_1^x \log(\alpha V) \mathrm{d}V \ll \frac{U^2}{\#D(Y)} xz \log(zx). \tag{31}$$

Combining (26), (29), (30) and (31), we arrive at

$$S_{m \neq \square} \ll \frac{xz}{\#D(Y)} U^2 \log(zx).$$

From this, together with (25) and (23), it follows that

$$S_{2,1} \ll \frac{Y^{1/2} x}{\#D(Y)} \log z + \frac{xz}{\#D(Y)} U^2 \log(zx).$$

So from the above, together with (21), (20), (19), (15) and (13), we get

$$S = \frac{\pi(x)}{2} + O\left(\log\log x + \frac{\pi(x)}{U} + \frac{x \log z}{Y^{1/2}} + \frac{xz}{Y} U^2 \log(zx) + \frac{x^{1/2} \log(xYz)}{z} \right). \tag{32}$$

Now it still remains to optimize z. To this end, we set

$$z = \frac{Y^{1/2}}{Ux^{1/4}}.$$

So (32) becomes

$$S = \frac{\pi(x)}{2} + O\left(\log\log x + \frac{\pi(x)}{U} + \frac{x \log Y}{Y^{1/2}} + \frac{Ux^{3/4}}{Y^{1/2}} \log(xY) \right).$$

Now recall $U = x^{1/8} Y^{1/4}$ and $\pi(x) \sim x / \log x$. The proof of Theorem 3 is complete.

Acknowledgments. P. G. is supported in part by NSFC grants 11371043 and 11871082 and L. Z. by the FRG grant PS43707 and the Faculty Silverstar Fund PS49334. Parts of this work were done when P. G. visited the University of New South Wales (UNSW) in June 2017. He wishes to thank UNSW for the invitation, financial support and warm hospitality during his pleasant stay. Finally, both authors would like to thank the anonymous referee for his/her careful reading of the manuscript and many helpful comments.

References

[1] S. Baier and M. P. Young, Mean values with cubic characters, *J. Number Theory* **130** (2010), 879-903.

[2] V. Blomer, L. Goldmakher and B. Louvel, L-functions with n-th order twists, *Int. Math. Res. Not. IMRN* **2014** (2014), 1925-1955.

[3] S. Baier and M. P. Young, Large sieve inequalities for quartic character sums, *Q. J. Math.* **63** (2012), 891-917.

[4] D. R. Heath-Brown, A mean value estimate for real character sums, *Acta Arith.* **72** (1995), 235-275.

[5] C. Hooley, On Artin's conjecture, *J. Reine Angew. Math.* **225** (1995), 209-220.

[6] H. Iwaniec and E. Kowalski, *Analytic Number Theory*, American Mathematical Society, Providence, 2004.

[7] M. Jutila, On the mean value of $L(1/2,\chi)$ for real characters, *Analysis* **1** (1981), 149-161.

[8] S. Li, Notes on primes $P \equiv 1 \bmod D$ and $A^{P-1/D} \equiv 1 \bmod P$, *Integers* **15** (2015), A25, 8pp.

[9] M. R. Murty, Artin's conjecture for primitive roots, *Math. Intelligencer* **10** (1988), 59-67.

[10] K. Soundararajan, Nonvanishing of quadratic Dirichlet L-functions at $s=\frac{1}{2}$, *Ann. of Math. (2)* **152** (2000), 447-488.

8

LOWER BOUNDS FOR NUMBERS WITH THREE PRIME FACTORS

Paul Kinlaw
Department of Mathematics, Husson University, Bangor, Maine
KinlawP@husson.edu

Abstract
A *k-almost prime number* is a product of k prime numbers, some of which may be repeated. By a 1900 theorem of Landau, the number of k-almost prime numbers not exceeding x is asymptotic to $x(\log\log x)^{k-1}/((k-1)!\log x)$. We prove a numerically explicit lower bound for 3-almost prime numbers which is asymptotic to Landau's formula, and hence to the actual count. It exceeds Landau's formula for all $x \geq 500194$. We prove an analogous lower bound for products of three distinct prime numbers. This expands on previously known results for $k \leq 2$.

1. Introduction

For a natural number n, the arithmetic functions $\omega(n)$ and $\Omega(n)$ denote the number of prime factors of n, counted without (respectively with) repeated prime factors. Thus for n with prime factorization $n = p_1^{a_1} \cdot \ldots \cdot p_k^{a_k}$, we have $\omega(n) = k$ and $\Omega(n) = a_1 + \ldots + a_k$.

Numbers n such that $\Omega(n) = k$ are called *k-almost primes*. Let $\pi_k(x) = |\{n \leq x : \omega(n) = \Omega(n) = k\}|$ denote the counting function of squarefree k-almost primes, and let $\tau_k(x) = |\{n \leq x : \Omega(n) = k\}|$ denote the counting function of k-almost primes.

For $k = 1$ we have $\pi_1(x) = \tau_1(x) = \pi(x)$, the prime counting function. The prime number theorem asserts that $\pi(x)$ is asymptotic to $x/\log x$. In 1900, Landau [6] proved that for each $k \in \mathbb{N}$, the estimate

$$\pi_k(x) = \frac{x(\log\log x)^{k-1}}{(k-1)!\log x}\left(1 + O\left(\frac{1}{\log\log x}\right)\right)$$

holds, and that the same estimate also holds for $\tau_k(x)$.

Selberg [11] proved that for any $0 < \delta < 1$, uniformly for all $x \geq 3$ and $1 \leq k \leq (2-\delta)\log\log x$, we have

$$\tau_k(x) = G\left(\frac{k-1}{\log\log x}\right)\frac{x(\log\log x)^{k-1}}{(k-1)!\log x}\left(1 + O\left(\frac{k}{(\log\log x)^2}\right)\right),$$

where

$$G(z) = F(1,z)/\Gamma(z+1) \quad \text{and} \quad F(s,z) = \prod_p \left(1 - \frac{z}{p^s}\right)^{-1} \left(1 - \frac{1}{p^s}\right)^z.$$

Classic references [5, 7] provide details on these and similar results. A 1962 paper [10] of Rosser and Schoenfeld and contemporary work [3, 4] of Dusart give explicit bounds for $\pi(x)$, see Lemma 1. A 2018 paper [1] of Bayless et al. established explicit upper bounds for $\pi_k(x)$, $\tau_2(x)$, and $\tau_3(x)$, as well as explicit lower bounds for $\pi_2(x)$ and $\tau_2(x)$. In particular, it is shown [1, Thm. 3.5] that for all $k \geq 2$ and $x \geq 3$, we have

$$\pi_k(x) \leq \frac{1.028x(\log\log x + 0.26153)^{k-1}}{(k-1)!\log x},$$

and [1, Thm. 5.2] that for all $x \geq 10^{12}$, we have

$$\tau_2(x) \geq \pi_2(x) \geq \frac{x(\log\log x + 0.1769)}{\log x}\left(1 + \frac{0.4232}{\log x}\right).$$

Furthermore [1, Thm. 5.3], for all $x \geq 10^{12}$ we have

$$\tau_3(x) \leq \frac{1.028x((\log\log x + 0.26153)^2 + 1.055852)}{2\log x}.$$

A relatively sharp explicit lower bound for $\pi_3(x)$ requires more work, and aside from the explicit results above, the literature consists of implied constants. We prove explicit lower bounds for $\pi_3(x)$ and $\tau_3(x)$ which are asymptotic to Landau's formula, and hence to the actual count. In particular, we prove the following three theorems.

Theorem 1. *For all* $x \geq 500194$,

$$\tau_3(x) > \frac{x(\log\log x)^2}{2\log x}.$$

The constant 500194 is optimal, since the inequality is violated at $x = 500194 - \epsilon$ for all sufficiently small $\epsilon > 0$. We also have an analogue of Theorem 1 for squarefree 3-almost primes, also called *sphenic numbers.*

Theorem 2. *For all* $x \geq 10203553$,

$$\pi_3(x) > \frac{x((\log\log x)^2 - 1)}{2\log x}.$$

The constant 10203553 is also optimal. We also obtain the following.

Theorem 3. *For all sufficiently large* x,

$$\pi_3(x) > \frac{x(\log\log x)^2}{2\log x}.$$

Theorem 3 follows readily from the proof of Theorem 2 by comparing secondary terms, however the reader will see that x must be very large. Without more work, it is unclear where the optimal cutoff c_0 is for Theorem 3 to apply. Lifchitz and Renner [8] computed the values of $\pi_3(10^k)$ for $1 \le k \le 19$, and this data shows that $c_0 > 10^{19}$.

2. Notation and Preliminary Lemmas

Throughout the paper we let p, q, and r denote prime numbers and we let $\log x$ denote the natural logarithm. We define $L = \log\log x$, $x_0 = 10^{12}$, and $y_0 = x_0^{1/3} = 10^4$. Furthermore, $\pi(t)$ denotes the prime counting function and $T(t) = \sum_{p\le t} 1/p$ denotes the sum of reciprocals of prime numbers $p \le t$. We let $B = 0.2614972128\ldots$ denote the *Mertens constant* and $c = 0.26146521$ (see Lemmas 2 and 3).

In bounding expressions, we will make use of manipulations such as

$$\log\log\frac{x}{a} = \log\log x + \log\left(1 - \frac{\log a}{\log x}\right),$$

the Maclaurin series expansion of the righthand term, and the bounds

$$\frac{1}{\log x}\left(1 + \frac{\log a}{\log x}\right) < \frac{1}{\log\frac{x}{a}} \le \frac{1}{\log x} + \frac{1}{\log^2 x}\cdot\frac{\log a\log x_0}{\log\frac{x_0}{a}}$$

for $a > 1$ and $x \ge x_0$. We will use the following bounds [10, Thm. 2] of Rosser and Schoenfeld and [4, Cor. 5.2] of Dusart on the prime counting function.

Lemma 1 (Rosser, Schoenfeld, Dusart). *We have*

$$\frac{x}{\log x} < \pi(x)$$

for all $x \ge 17$. Additionally, we have

$$\frac{x}{\log x}\left(1 + \frac{1}{\log x}\right) \le \pi(x) \le \frac{x}{\log x}\left(1 + \frac{1.2762}{\log x}\right),$$

where the lower bound holds for all $x \ge 599$ and the upper bound holds for all $x > 1$. Furthermore, for all $x > 1$ we have

$$\pi(x) \le \frac{x}{\log x}\left(1 + \frac{1}{\log x} + \frac{2.53816}{\log^2 x}\right).$$

We also use the following bounds [10, Theorem 5] of Rosser and Schoenfeld and [3, Thm. 6.10], [4, Thm. 5.6] of Dusart which give numerically explicit versions of Mertens' second theorem.

Lemma 2 (Rosser, Schoenfeld, Dusart). *Let* $T(t) = \sum_{p \le t} 1/p$ *denote the reciprocal sum of prime numbers up to* t. *We have*

$$T(t) = \log\log t + B + E(t),$$

where $B = 0.2614972128\ldots$ *denotes the Mertens constant, and*

$$-\frac{1}{2\log^2 t} < E(t) < \frac{1}{\log^2 t}$$

for all $t > 1$,

$$|E(t)| < \frac{1}{2\log^2 t}$$

for all $t \ge 286$,

$$|E(t)| \le \frac{1}{10\log^2 t} + \frac{4}{15\log^3 t}$$

for all $t \ge 10372$, *and*

$$|E(t)| \le \frac{1}{5\log^3 t}$$

for all $t \ge 2278383$.

Combining bounds on the prime number reciprocal sum in [10, 4] readily yields the following lower bound.

Lemma 3. *For* $x > 1$,

$$\sum_{p \le x} \frac{1}{p} > \log\log x + 0.26146521.$$

Proof. By Lemma 2, for all $x \ge 2278383$,

$$\sum_{p \le x} \frac{1}{p} \ge \log\log x + B - \frac{0.2}{\log^3 x}, \tag{1}$$

where B is the Mertens constant. By [10, Thm 20],

$$\sum_{p \le x} \frac{1}{p} > \log\log x + B$$

for all $x \le 10^8$. Substituting 10^8 in (1) gives the lemma. □

3. The Proof of Theorem 2

We have

$$\pi_3(x) = |\{n = pqr \leq x : p < q < r\}|$$

where p, q, and r denote prime numbers. To determine the count, we note the possible values for each of p, q, r. We have $p < x^{1/3}$, and $q^2 < qr \leq x/p$, so that $p < q < \sqrt{x/p}$. Since $pqr \leq x$, we also have $q < r \leq x/(pq)$. Therefore, $\pi_3(x)$ is equal to the sum of 1 over all possible values of p, q, r, so that

$$\pi_3(x) = \sum_{p<x^{1/3}} \sum_{p<q<\sqrt{\frac{x}{p}}} \sum_{q<r\leq\frac{x}{pq}} 1 = \sum_{p\leq x^{1/3}} \sum_{p<q\leq\sqrt{\frac{x}{p}}} \left(\pi\left(\frac{x}{pq}\right) - \pi(q)\right).$$

Here we have used the fact that the strict inequalities $p < x^{1/3}$ and $q < \sqrt{x/p}$ can be written to include the case of equality, since the quantity vanishes when $q = \sqrt{x/p}$ or when $p = x^{1/3}$. We checked the inequality using the computer program Pari/GP for the interval $10203553 \leq x \leq 10^{12}$. Thus we may assume $x \geq x_0$. By Lemma 1, since $x/(pq) \geq 599$ we have

$$\pi\left(\frac{x}{pq}\right) \geq \frac{x}{pq\log\frac{x}{pq}}\left(1 + \frac{1}{\log\frac{x}{pq}}\right) \geq \frac{x}{pq\log\frac{x}{pq}}\left(1 + \frac{1}{\log x}\right).$$

We thus have

$$\pi_3(x) \geq \left(1 + \frac{1}{\log x}\right) \sum_{p\leq x^{1/3}} S_1 - \sum_{p\leq x^{1/3}} S_2,$$

where

$$S_1 = \sum_{p<q\leq\sqrt{\frac{x}{p}}} \frac{x}{pq\log\frac{x}{pq}}$$

and

$$S_2 = \sum_{p<q\leq\sqrt{\frac{x}{p}}} \pi(q).$$

We determine a lower bound for S_1 and an upper bound for S_2. By Lemma 7 below, we have

$$\sum_{p\leq x^{1/3}} S_2 \leq x\left(\frac{2L + 0.1436}{\log^2 x} + \frac{10.9113L + 3.1227}{\log^3 x}\right).$$

We next consider S_1. We have

$$S_1 = \frac{x}{p} \sum_{p<q\leq\sqrt{y}} \frac{1}{q\log\frac{y}{q}}$$

where $y = x/p > x^{2/3} \geq x_0^{2/3}$. Recall that the function $T(t) = \sum_{p \leq t} 1/p$ denotes the prime reciprocal sum up to t. We bound S_1 below by applying partial summation to obtain

$$\sum_{p<q\leq\sqrt{y}} \frac{1}{q \log \frac{y}{q}} = \frac{T(\sqrt{y})}{\log \sqrt{y}} - \frac{T(p)}{\log \frac{y}{p}} - \int_p^{\sqrt{y}} \frac{T(t)\, dt}{t \log^2 \frac{y}{t}}. \tag{2}$$

By Lemma 2 this is bounded below by the expression

$$\frac{2(\log\log\sqrt{y} + B - \frac{1}{2\log^2\sqrt{y}})}{\log y} - \frac{\log\log p + B + \frac{1}{\log^2 p}}{\log \frac{y}{p}} \\ - \int_p^{\sqrt{y}} \frac{\log\log t + B + \frac{1}{\log^2 t}}{t \log^2 \frac{y}{t}}\, dt.$$

Substituting $u = \log t$ and integrating, this is equal to

$$\frac{-4}{\log^3 y} - \frac{\log\log p}{\log \frac{y}{p}} - \frac{1}{\log^2 p \log \frac{y}{p}} + \frac{\log\log \frac{y}{p}}{\log y} + \frac{\log p \log\log p}{\log y \log \frac{y}{p}} \\ + \frac{1}{\log^2 y \log \frac{y}{p}} - \frac{1}{\log^2 y \log p} - \frac{2\log\log \frac{y}{p}}{\log^3 y} + \frac{2\log\log p}{\log^3 y}.$$

Thus a lower bound for S_1 is

$$\frac{x}{p\log \frac{x}{p}} \left(\log\log \frac{x}{p^2} + \frac{\log p \log\log p}{\log \frac{x}{p^2}} + \frac{1}{\log \frac{x}{p} \log \frac{x}{p^2}} + \frac{2\log\log p}{\log^2 \frac{x}{p}} \right) \\ - \frac{x}{p\log \frac{x}{p}} \left(\frac{4}{\log^2 \frac{x}{p}} + \log\log p \cdot \frac{\log \frac{x}{p}}{\log \frac{x}{p^2}} + \frac{1}{\log^2 p} \cdot \frac{\log \frac{x}{p}}{\log \frac{x}{p^2}} + \frac{1}{\log \frac{x}{p} \log p} + \frac{2\log\log \frac{x}{p^2}}{\log^2 \frac{x}{p}} \right). \tag{3}$$

To simplify this expression, we compare the second (respectively, third) term of the first line above to the second (respectively, first) term of the second line. We have

$$\frac{\log p \log\log p}{\log \frac{x}{p^2}} - \log\log p \cdot \frac{\log \frac{x}{p}}{\log \frac{x}{p^2}} = -\log\log p.$$

Also, $1/(\log(x/p)\log(x/p^2)) > 1/\log^2(x/p)$. Therefore, a lower bound for expression (3) is

$$\frac{x}{p\log \frac{x}{p}} \left(\log\log \frac{x}{p^2} + \frac{2\log\log p}{\log^2 \frac{x}{p}} \right) \\ - \frac{x}{p\log \frac{x}{p}} \left(\log\log p + \frac{3}{\log^2 \frac{x}{p}} + \frac{1}{\log^2 p} \cdot \frac{\log \frac{x}{p}}{\log \frac{x}{p^2}} + \frac{1}{\log \frac{x}{p} \log p} + \frac{2\log\log \frac{x}{p^2}}{\log^2 \frac{x}{p}} \right). \tag{4}$$

Write the sum of this expression over $p \le x^{1/3}$ as

$$x\left(S_5 + S_6 - (S_7 + S_8 + S_9 + S_{10} + S_{11})\right).$$

Since $2\log\log p - 3$ is negative for $p < 89$ and positive for $p \ge 89$,

$$\begin{aligned} S_6 - S_8 &= \sum_{p \le x^{1/3}} \frac{2\log\log p - 3}{p\log^3 \frac{x}{p}} \\ &\ge \frac{1}{\log^3 x}\sum_{p<89} \frac{2\log\log p - 3}{p}\frac{\log^3 x_0}{\log^3 \frac{x_0}{p}} + \frac{1}{\log^3 x}\sum_{89 \le p \le y_0} \frac{2\log\log p - 3}{p} \\ &\ge \frac{-3.9322}{\log^3 x}. \end{aligned}$$

Note that

$$\begin{aligned} S_{10} = \sum_{p \le x^{1/3}} \frac{1}{p\log p\log^2 \frac{x}{p}} &\le \frac{1}{\log^2 x}\sum_{p \le y_0} \frac{1}{p\log p}\frac{\log^2 x_0}{\log^2 \frac{x_0}{p}} + \frac{2.25}{\log^2 x}\sum_{y_0 < p} \frac{1}{p\log p} \\ &\le \frac{1.7496}{\log^2 x} + \frac{2.25(0.1085)}{\log^2 x} \\ &\le \frac{1.9938}{\log^2 x}. \end{aligned}$$

Here we verified the value $\sum_p 1/(p\log p) = 1.63661\ldots$ found to much higher precision [2, p. 6] by H. Cohen. Similarly,

$$\begin{aligned} S_9 &= \sum_{p \le x^{1/3}} \frac{1}{p\log^2 p\log \frac{x}{p^2}} \\ &\le \sum_{p \le y_0} \frac{1}{p\log^2 p}\left(\frac{1}{\log x} + \frac{1}{\log^2 x}\frac{2\log p\log x_0}{\log \frac{x_0}{p^2}}\right) + \frac{3}{\log x}\sum_{y_0 < p \le x^{1/3}} \frac{1}{p\log^2 p} \\ &\le \frac{1.5151}{\log x} + \frac{3.5585}{\log^2 x} + \frac{3}{\log x}\sum_{y_0 < p} \frac{1}{p\log^2 p} \le \frac{1.5328}{\log x} + \frac{3.5585}{\log^2 x}. \end{aligned}$$

Here we computed the sum over $y_0 < p \le 10^9$ using Pari/GP and then applied partial summation together with Lemmas 2 and 3 to bound the sum over $p > 10^9$. By Lemma 8,

$$\begin{aligned} S_5 - S_7 \ge{} & \frac{0.5L^2 + 0.121434L - 0.22}{\log x} \\ & + \frac{-1.5244L + 1.112}{\log^2 x} + \frac{-2L^2 + 2.9067L + 2.5389}{\log^3 x}, \end{aligned}$$

and by Lemma 5,

$$S_{11} \le \frac{2L^2 + 1.3972L}{\log^3 x}.$$

Finally, note that the application of the bounds

$$L + c < T(x) < L + B + \frac{1}{\log^2 x},$$

valid for all $x > 1$, incur error terms which are particularly large for small values of x. We may improve our bounds as follows. In equation (2) we replaced the quantity $-T(p)/(p\log(x/p^2))$ with

$$-\frac{\log\log p + B + \frac{1}{\log^2 p}}{p\log\frac{x}{p^2}}.$$

Thus we may add the following expression to the lower bound:

$$\sum_{p\le x^{1/3}} \frac{\log\log p + B + \frac{1}{\log^2 p} - T(p)}{p\log\frac{x}{p^2}}.$$

A computation gives a lower bound of

$$\frac{1}{\log\frac{x}{4}} \sum_{p\le y_0} \frac{\log\log p + B + \frac{1}{\log^2 p} - T(p)}{p} \ge \frac{0.9}{\log\frac{x}{4}} \ge \frac{0.9}{\log x}\left(1 + \frac{\log 4}{\log x} + \frac{\log^2 4}{\log^2 x}\right).$$

Again considering equation (2) and the following displayed expression, we may add to the lower bound

$$\begin{aligned}
&\sum_{p\le x^{1/3}} \int_p^{\sqrt{\frac{x}{p}}} \frac{\log\log t + B + \frac{1}{\log^2 t} - T(t)}{t\log^2\frac{x}{pt}}\,dt \\
&\ge \frac{1}{\log^2 x}\sum_{k\le 1228} k\int_{\log p_k}^{\log p_{k+1}} \left(\log u + B + \frac{1}{u^2} - T(p_k)\right) du \\
&\ge \frac{20.4395}{\log^2 x}.
\end{aligned}$$

Here p_k denotes the k-th prime number. Combining all bounds, we obtain

$$\begin{aligned}
\pi_3(x) &\ge \frac{x(0.5L^2 + 0.121434L - 0.8528)}{\log x} + \frac{x(0.5L^2 - 4L + 15)}{\log^2 x} \\
&\quad - \frac{x(4L^2 + 10.9262L)}{\log^3 x} \\
&\ge \frac{x(0.5L^2 + 0.121434L - 0.8528)}{\log x} + \frac{x(0.5L^2 - 4L + 12)}{\log^2 x} \qquad (5) \\
&\ge \frac{x(0.5L^2 + 0.121434L - 0.8528)}{\log x} + \frac{4x}{\log^2 x} \\
&> \frac{x((\log\log x)^2 - 1)}{2\log x} + \frac{4x}{\log^2 x}
\end{aligned}$$

for all $x \ge x_0$. This completes the proof of Theorem 2.

4. Additional Lemmas

We prove several lemmas used in the proof of Theorem 2.

Lemma 4. *For $x \geq x_0$, we have*

$$\sum_{p \leq x^{1/3}} \frac{1}{p \log^2 \frac{x}{p}} \leq \frac{L + 0.071781}{\log^2 x}.$$

Proof. For $x_0 \leq x \leq 10372^3$, we have

$$\sum_{p \leq x^{1/3}} \frac{1}{p \log^2 \frac{x}{p}} \leq \frac{1}{\log^2 x} \sum_{p \leq 10372} \frac{\log^2 x_0}{p \log^2 \frac{x_0}{p}} \leq \frac{\log \log x_0 - 0.0279}{\log^2 x} \leq \frac{L}{\log^2 x},$$

by directly computing the sum over $p \leq 10372$. Suppose next that $x > 10372^3$. By Lemmas 2 and 3 and partial summation,

$$\begin{aligned}\sum_{p \leq x^{1/3}} \frac{1}{p \log^2 \frac{x}{p}} &= \frac{T(x^{1/3})}{\log^2(x^{2/3})} - \int_2^{x^{1/3}} \frac{2T(t)\, dt}{t \log^3 \frac{x}{t}} \\ &\leq \frac{\frac{9}{4}(\log \log x^{1/3} + 0.26300402)}{\log^2 x} - \int_2^{x^{1/3}} \frac{2(\log \log t + c)\, dt}{t \log^3 \frac{x}{t}},\end{aligned}$$

recalling that $c = 0.26146521$. The integral is equal to

$$\left[\frac{\log \log t + c}{\log^2 \frac{x}{t}} + \frac{\log \log \frac{x}{t} - \log \log t}{\log^2 x} - \frac{1}{\log x \log \frac{x}{t}} \right]_2^{x^{1/3}}.$$

Subtracting and bounding the resulting expression, we obtain

$$S_3 \leq \frac{\frac{9}{4}(0.26300402 - c) - \log 2 + \frac{1}{2} + c + \log \log \frac{x}{2}}{\log^2 x} \leq \frac{L + 0.071781}{\log^2 x}$$

for all $x \geq 10372^3$, and therefore for all $x \geq x_0$. □

Lemma 5. *For $x \geq x_0$, we have*

$$\sum_{p \leq x^{1/3}} \frac{1}{p \log^3 \frac{x}{p}} \leq \frac{L + 0.6986}{\log^3 x}.$$

Proof. Note that for $x_0 \leq x \leq 10372^3$, we have

$$\sum_{p \leq x^{1/3}} \frac{1}{p \log^3 \frac{x}{p}} \leq \sum_{p \leq 10372} \frac{1}{p \log^3 x} \frac{\log^3 x_0}{\log^3 \frac{x_0}{p}} \leq \frac{L + 0.5411}{\log^3 x},$$

by directly computing the sum over $p \le 10372$. Suppose next that $x \ge 10372^3$. By partial summation,

$$\sum_{p \le x^{1/3}} \frac{1}{p \log^3 \frac{x}{p}} = \frac{T(x^{1/3})}{\log^3 x^{2/3}} - \int_2^{x^{1/3}} \frac{3T(t)\,dt}{t \log^4 \frac{x}{t}}$$

$$\le \frac{\log\log x^{1/3} + 0.26300402}{\log^3 x^{2/3}} - \int_2^{x^{1/3}} \frac{3(\log\log t + c)\,dt}{t \log^4 \frac{x}{t}},$$

using Lemmas 2 and 3. The integral is equal to

$$\left[\frac{\log\log t + c}{\log^3 \frac{x}{t}} + \frac{\log\log \frac{x}{t} - \log\log t}{\log^3 x} - \frac{1}{\log^2 x \log \frac{x}{t}} - \frac{1}{2\log x \log^2 \frac{x}{t}}\right]_2^{x^{1/3}}.$$

Subtracting and bounding the resulting expression, we find that

$$\sum_{p \le x^{1/3}} \frac{1}{p \log^3 \frac{x}{p}} \le \frac{\frac{27}{8}(0.26300402 - c) - \log 2 + \frac{9}{8} + \log\log \frac{x}{2} + c}{\log^3 x},$$

from which we obtain the lemma. □

Lemma 6. *For $x \ge x_0$ we have*

$$\sum_{p \le x^{1/3}} \frac{1}{\sqrt{p} \log \frac{x}{p}} \le \frac{0.4383 x^{1/6}}{\log x}.$$

Proof. Let $x \ge x_0$. By partial summation,

$$\sum_{p \le x^{1/3}} \frac{1}{\sqrt{p} \log \frac{x}{p}} \le \frac{1.5}{\log x} \sum_{p \le x^{1/3}} \frac{1}{\sqrt{p}}$$

$$= \frac{1.5}{\log x}\left(\frac{\pi(x^{1/3})}{x^{1/6}} + \frac{1}{2}\int_2^{x^{1/3}} \frac{\pi(t)\,dt}{t^{3/2}}\right)$$

$$\le \frac{1.5}{\log x}\left(\frac{\pi(x^{1/3})}{x^{1/6}} + 16.85461 + \frac{1}{2}\int_{y_0}^{x^{1/3}} \frac{\pi(t)\,dt}{t^{3/2}}\right).$$

By Lemma 1, an upper bound for this expression is therefore

$$\frac{1.5}{\log x}\left(\frac{3.4154822 x^{1/6}}{\log x} + 16.85461 + 0.06180521 \int_{y_0}^{x^{1/3}} \frac{dt}{t^{1/2}}\right),$$

from which the lemma readily follows. □

Lemma 7. *For $x \geq x_0$ we have*

$$\sum_{p \leq x^{1/3}} S_2 \leq x \left(\frac{2L + 0.1436}{\log^2 x} + \frac{10.9113L + 3.1227}{\log^3 x} \right).$$

Proof. We have the general formula

$$\sum_{q \leq t} \pi(q) = 1 + 2 + \ldots + \pi(t) = \frac{1}{2}(\pi(t)^2 + \pi(t)).$$

Therefore,

$$S_2 = \frac{1}{2} \left(\pi \left(\sqrt{\frac{x}{p}} \right)^2 + \pi \left(\sqrt{\frac{x}{p}} \right) - \pi(p)^2 - \pi(p) \right).$$

We will address the third term below and drop the fourth term. Let $x \geq x_0$. By Lemma 1, we have

$$\frac{1}{2}\pi \left(\sqrt{\frac{x}{p}} \right) \leq \frac{1.1385\sqrt{x}}{\sqrt{p} \log \frac{x}{p}}$$

and

$$\frac{1}{2}\pi \left(\sqrt{\frac{x}{p}} \right)^2 \leq \frac{2x}{p \log^2 \frac{x}{p}} \left(1 + \frac{2.551155}{\log \frac{x}{p}} \right)^2 \leq \frac{2x}{p \log^2 \frac{x}{p}} + \frac{10.9113x}{p \log^3 \frac{x}{p}}.$$

We thus have

$$\sum_{p \leq x^{1/3}} S_2 \leq x \sum_{p \leq x^{1/3}} \left(\frac{2}{p \log^2 \frac{x}{p}} + \frac{10.9113}{p \log^3 \frac{x}{p}} + \frac{1.1385}{\sqrt{x}} \frac{1}{\sqrt{p} \log \frac{x}{p}} \right).$$

Combining Lemmas 4, 5, and 6, we obtain

$$\sum_{p \leq x^{1/3}} S_2 \leq x \left(\frac{2L + 0.1436}{\log^2 x} + \frac{10.9113L + 7.6227}{\log^3 x} + \frac{1.1}{\log^4 x} \right).$$

We now return to consider the remaining term:

$$\begin{aligned}
\frac{1}{2} \sum_{p \leq x^{1/3}} \pi(p)^2 =& \frac{1}{2} \cdot \frac{\pi(x^{1/3})(\pi(x^{1/3}) + 1)(2\pi(x^{1/3}) + 1)}{6} \\
&\geq \frac{\pi(x^{1/3})^3}{6} \\
&\geq \frac{1}{6} \left(\frac{x^{1/3}}{\log x^{1/3}} \left(1 + \frac{1}{\log x^{1/3}} \right) \right)^3 \\
&\geq \frac{4.5x}{\log^3 x} \left(1 + \frac{9}{\log x} \right) \\
&= \frac{4.5x}{\log^3 x} + \frac{40.5x}{\log^4 x}.
\end{aligned}$$

Here we have used Lemma 1. Combining these bounds, we obtain the lemma. □

Lemma 8. *For all $x \geq x_0$, we have*

$$S_5 - S_7 \geq \frac{0.5L^2 + 0.121434L - 0.22}{\log x} + \frac{-1.5244L + 1.112}{\log^2 x} + \frac{-2L^2 + 2.9067L + 2.5389}{\log^3 x}.$$

Proof. Recall the definitions

$$S_5 = \sum_{p \leq x^{1/3}} \frac{\log\log \frac{x}{p^2}}{p \log \frac{x}{p}}, \qquad S_7 = \sum_{p \leq x^{1/3}} \frac{\log\log p}{p \log \frac{x}{p}}.$$

We apply partial summation with $f(t) = \log\log \dfrac{x}{t^2} / \log \dfrac{x}{t}$ to obtain

$$S_5 = \frac{T(x^{1/3}) \log\log x^{1/3}}{\log x^{2/3}} - \int_2^{x^{1/3}} \frac{T(t) \log\log \frac{x}{t^2}\, dt}{t \log^2 \frac{x}{t}} + \int_2^{x^{1/3}} \frac{2T(t)\, dt}{t \log \frac{x}{t} \log \frac{x}{t^2}}.$$

We next apply partial summation with $g(t) = \log\log t / \log \dfrac{x}{t}$ to obtain

$$S_7 = \frac{T(x^{1/3}) \log\log x^{1/3}}{\log x^{2/3}} - \int_2^{x^{1/3}} \frac{T(t) \log\log t\, dt}{t \log^2 \frac{x}{t}} - \int_2^{x^{1/3}} \frac{T(t)\, dt}{t \log t \log \frac{x}{t}}.$$

Subtracting,

$$S_5 - S_7 = -I_1 + I_2 + I_3 + I_4, \tag{6}$$

where

$$I_1 = \int_2^{x^{1/3}} \frac{T(t) \log\log \frac{x}{t^2}\, dt}{t \log^2 \frac{x}{t}},$$

$$I_2 = \int_2^{x^{1/3}} \frac{2T(t)\, dt}{t \log \frac{x}{t} \log \frac{x}{t^2}},$$

$$I_3 = \int_2^{x^{1/3}} \frac{T(t) \log\log t\, dt}{t \log^2 \frac{x}{t}},$$

and

$$I_4 = \int_2^{x^{1/3}} \frac{T(t)\, dt}{t \log t \log \frac{x}{t}}.$$

Let $x \geq x_0$. Splitting the interval at $x^{1/6}$,

$$\begin{aligned}
I_1 &\leq \left(L - \log\frac{3}{2}\right)\int_{x^{1/6}}^{x^{1/3}} \frac{T(t)}{t\log^2\frac{x}{t}}dt + L\int_2^{x^{1/6}} \frac{T(t)}{t\log^2\frac{x}{t}}dt \\
&\leq \left(L - \log\frac{3}{2}\right)\int_{x^{1/6}}^{x^{1/3}} \frac{\log\log t + B + \frac{1}{\log^2 t}}{t\log^2\frac{x}{t}}dt + L\int_2^{x^{1/6}} \frac{\log\log t + B + \frac{1}{\log^2 t}}{t\log^2\frac{x}{t}}dt \\
&= \left(L - \log\frac{3}{2}\right)\left(F(x^{1/3}) - F(x^{1/6})\right) + L\left(F(x^{1/6}) - F(2)\right) \\
&= L\left(F(x^{1/3}) - F(2)\right) - \left(\log\frac{3}{2}\right)\left(F(x^{1/3}) - F(x^{1/6})\right),
\end{aligned}$$

where the antiderivative F is given by

$$\begin{aligned}
&\frac{\log\log\frac{x}{t}}{\log x} + \frac{\log t\log\log t}{\log x\log\frac{x}{t}} + \frac{B}{\log\frac{x}{t}} + \frac{1}{\log^2 x\log\frac{x}{t}} \\
&\qquad - \frac{1}{\log^2 x\log t} - \frac{2\log\log\frac{x}{t}}{\log^3 x} + \frac{2\log\log t}{\log^3 x}.
\end{aligned}$$

Bounding the resulting expression gives

$$I_1 \leq \frac{0.5L^2 - 0.94565L + 0.1361}{\log x} + \frac{2.2153L}{\log^2 x} + \frac{2L^2 - 2.9067L - 2.0810}{\log^3 x},$$

where we made use of the identity

$$\log\log\frac{x}{2} = \log\log x + \log\left(1 - \frac{\log 2}{\log x}\right)$$

and bounded the Maclaurin series of the right term by comparison to a geometric series. We next bound

$$I_2 \geq \int_2^{x^{1/3}} \frac{2(\log\log t + c)\ dt}{t\log\frac{x}{t}\log\frac{x}{t^2}} = \int_{\log 2}^{\log x^{1/3}} \frac{2(\log u + c)\ du}{(\log x - u)(\log x - 2u)}.$$

Integrating, a lower bound for I_2 is

$$\begin{aligned}
&\frac{2}{\log x}\left[\mathrm{Li}_2\left(\frac{\log t}{\log x}\right) - \mathrm{Li}_2\left(\frac{\log t^2}{\log x}\right) + (\log\log t + c)\left(\log\log\frac{x}{t} - \log\log\frac{x}{t^2}\right)\right]_2^{x^{1/3}} \\
&= \frac{2}{\log x}\left(\mathrm{Li}_2\left(\frac{1}{3}\right) - \mathrm{Li}_2\left(\frac{2}{3}\right) + (\log 2)(L - \log 3 + c)\right) \\
&+ \frac{2}{\log x}\left(-\mathrm{Li}_2\left(\frac{\log 2}{\log x}\right) + \mathrm{Li}_2\left(\frac{\log 4}{\log x}\right) + (\log\log 2 + c)\log\left(1 - \frac{\log 2}{\log\frac{x}{2}}\right)\right) \\
&\geq \frac{(\log 4)L - 2.0947}{\log x} + \frac{(\log 4)(1 - \log\log 2 - c)}{\log^2 x} + \frac{\log^2 2}{2\log^3 x}.
\end{aligned}$$

Here Li_2 denotes the dilogarithm, defined by the improper integral

$$\mathrm{Li}_2(z) = -\int_0^z \frac{\log(1-t)}{t}\,dt.$$

For the terms $(\log 4)/\log^2 x + (\log^2 2)/(2\log^3 x)$ above, we have used the fact that the dilogarithm is concave up, together with the expansion

$$\mathrm{Li}_2(z) = \sum_{k\geq 1} \frac{z^k}{k^2}$$

for $|z| < 1$, putting $z = \log 2/\log x$, to bound

$$\begin{aligned}\mathrm{Li}_2\left(\frac{\log 4}{\log x}\right) - \mathrm{Li}_2\left(\frac{\log 2}{\log x}\right) &\geq \mathrm{Li}_2\left(\frac{\log 4 - \log 2}{\log x}\right) - \mathrm{Li}_2(0)\\ &= \mathrm{Li}_2\left(\frac{\log 2}{\log x}\right) \geq \frac{\log 2}{\log x} + \frac{\log^2 2}{4\log^2 x}.\end{aligned}$$

We have also used the bound $-\log(1-z) \geq z$, here putting $z = \log 2/\log(x/2)$ and noting that $\log\log 2 + c < 0$.

We now turn to I_3. In order to bound I_3 in the right direction, we write

$$I_3 = \int_2^e \frac{T(t)\log\log t\; dt}{t\log^2 \frac{x}{t}} + \int_e^{x^{1/3}} \frac{T(t)\log\log t\; dt}{t\log^2 \frac{x}{t}}.$$

The first integral is equal to

$$\frac{1}{2\log x}\left(\log\left(1 - \frac{1-\log 2}{\log \frac{x}{2}}\right) - \frac{\log 2\log\log 2}{\log\frac{x}{2}}\right) \geq \frac{-0.0304}{\log^2 x}.$$

A lower bound for the second integral is

$$\begin{aligned}\int_e^{x^{1/3}} &\frac{(\log\log t + c)\log\log t\; dt}{t\log^2\frac{x}{t}} = \int_1^{\log x^{1/3}} \frac{(\log u + c)\log u\; du}{(\log x - u)^2}\\ &= \frac{1}{\log x}\left[c\log\left(1 - \frac{u}{\log x}\right) + \frac{cu\log u}{\log x - u} + 2\mathrm{Li}_2\left(\frac{u}{\log x}\right)\right]_1^{\log x^{1/3}}\\ &\quad + \frac{1}{\log x}\left[2(\log u)\log\left(1 - \frac{u}{\log x}\right) + \frac{u\log^2 u}{\log x - u}\right]_1^{\log x^{1/3}}\\ &\geq \frac{0.5L^2 - 1.77881L + 1.9771}{\log x} + \frac{c-2}{\log^2 x} - \frac{0.5188}{\log^3 x}.\end{aligned}$$

Next, note that the use of the inequality $T(t) > \log\log t + c$ allows us to add to the lower bound

$$\int_3^{x^{1/3}} \frac{(T(t)-(\log\log t+c))\log\log t\ dt}{t\log^2\frac{x}{t}}$$
$$\geq \frac{1}{\log^2\frac{x}{3}} \sum_{2\leq k\leq 1228} \int_{\log p_k}^{\log p_{k+1}} (T(p_k)-c-\log u)\log u\ du$$
$$\geq \frac{0.3099}{\log^2 x}\left(1+\frac{\log 9}{\log x}\right).$$

We therefore have

$$I_3 \geq \frac{0.5L^2-1.77881L+1.9771}{\log x} - \frac{1.4591}{\log^2 x} + \frac{0.1621}{\log^3 x}.$$

We now consider

$$I_4 = \int_2^{x^{1/3}} \frac{T(t)\ dt}{t\log t\log\frac{x}{t}} \geq \int_2^{x^{1/3}} \frac{(\log\log t+c)\ dt}{t\log t\log\frac{x}{t}} = \int_{\log 2}^{\log x^{1/3}} \frac{(\log u+c)\ du}{u(\log x-u)}.$$

Integrating, we obtain the expression

$$\frac{1}{\log x}\left[c\log u-(\log u)\log\left(1-\frac{u}{\log x}\right)-c\log(\log x-u)-\mathrm{Li}_2\left(\frac{u}{\log x}\right)+\frac{\log^2 u}{2}\right]_{\log 2}^{\log x^{1/3}}.$$

We bound this expression to obtain

$$I_4 \geq \frac{0.5L^2-0.4317L-0.3608}{\log x} + \frac{0.7658}{\log^2 x} + \frac{0.0556}{\log^3 x}.$$

Note that the inequality $T(t) > \log\log t + c$ was used to bound I_4. We may therefore add the following expression to the lower bound:

$$\int_2^{x^{1/3}} \frac{T(t)-(\log\log t+c)}{t\log t\log\frac{x}{t}}dt \geq \frac{1}{\log\frac{x}{2}} \int_2^{x^{1/3}} \frac{T(t)-(\log\log t+c)}{t\log t}dt.$$

Splitting the interval $[2, y_0]$ into subintervals between primes $[p_k, p_{k+1}]$, $k = 1, 2, \ldots$ 1228, and using the numerical value of $T(p_k)$ in each subinterval, we find by computation in Pari/GP that

$$\frac{1}{\log\frac{x}{2}} \int_2^{x^{1/3}} \frac{T(t)-(\log\log t+c)}{t\log t}dt \geq \frac{0.3945}{\log\frac{x}{2}} \geq \frac{0.3945}{\log x}\left(1+\frac{\log 2}{\log x}\right).$$

A similar argument allows us to improve our upper bound on the integral I_1 in equation (6). Here we used the upper bound

$$L\int_2^{x^{1/6}} \frac{T(t)}{t\log^2\frac{x}{t}}dt \leq L\int_2^{x^{1/6}} \frac{\log\log t+B+\frac{1}{\log^2 t}}{t\log^2\frac{x}{t}}dt.$$

Bounding the difference in the interval $[2, 100]$, we add $0.6909L/\log^2 x$ to the lower bound. Combining the bounds for I_1, I_2, I_3, and I_4 completes the proof of Lemma 8. □

5. The Proof of Theorem 1

We now prove Theorem 1. Note that $\tau_3(x) = \pi_3(x) + N(x)$, where $N(x) = |\{n \le x : n = p^2 q\}|$, and p and q denote (possibly equal) primes. For each such p, the number of possibilities for q is $\pi(x/p^2)$, and we thus have

$$N(x) = \sum_{p \le \sqrt{\frac{x}{2}}} \pi\left(\frac{x}{p^2}\right).$$

The range $500194 \le x \le 10^{12}$ was checked using the computer program Pari/GP. Thus we may assume $x \ge x_0$. Now, a lower bound for $\pi_3(x)$ is given in Theorem 2. The contribution from $N(x)$ is given by the following lemma, from which Theorem 1 follows by an argument analogous to that of inequality (5) in the proof of Theorem 2.

Lemma 9. *For all $x \ge x_0$ we have*

$$\sum_{p \le \sqrt{\frac{x}{2}}} \pi\left(\frac{x}{p^2}\right) \ge x\left(\frac{\alpha}{\log x} + \frac{0.9861}{\log^2 x} + \frac{1.2861}{\log^3 x}\right),$$

where $\alpha = \sum_p 1/p^2 = 0.4522474\ldots$ denotes the reciprocal sum of squares of primes.

Proof. Let $x \ge x_0$. We wish to apply the lower bound $\pi(t) > t/\log t$, valid for $t \ge 17$, in Lemma 1. Thus we write

$$\sum_{p \le \sqrt{\frac{x}{2}}} \pi\left(\frac{x}{p^2}\right) = \sum_{p \le \sqrt{\frac{x}{17}}} \pi\left(\frac{x}{p^2}\right) + \sum_{\sqrt{\frac{x}{17}} < p \le \sqrt{\frac{x}{2}}} \pi\left(\frac{x}{p^2}\right).$$

The right sum is

$$\sum_{\sqrt{\frac{x}{17}} < p \le \sqrt{\frac{x}{2}}} \pi\left(\frac{x}{p^2}\right) = -6 \cdot \pi\left(\sqrt{\frac{x}{17}}\right) + \pi\left(\sqrt{\frac{x}{13}}\right) + \ldots + \pi\left(\sqrt{\frac{x}{2}}\right).$$

Also by Lemma 1, this expression is bounded below by

$$2\sqrt{x}\left(-6 \cdot \frac{1.0972}{\sqrt{17}\log\frac{x}{17}} + \sum_{k=1}^{6} \frac{1}{\sqrt{p_k}\log\frac{x}{p_k}}\right) \ge \frac{1.8188\sqrt{x}}{\log x},$$

where p_k denotes the k-th prime number. For the remaining term, let $S(t) = \sum_{p\le t}(1/p^2)$. By partial summation,

$$\sum_{p\le\sqrt{\frac{x}{17}}} \frac{1}{p^2\log\frac{x}{p^2}} = \frac{S(\sqrt{\frac{x}{17}})}{\log 17} - \int_2^{\sqrt{\frac{x}{17}}} \frac{2S(t)\,dt}{t\log^2\frac{x}{t^2}}$$

$$= \frac{S(\sqrt{\frac{x}{17}})}{\log 17} - \int_2^{\sqrt{\frac{x}{17}}} \frac{2\alpha\,dt}{t\log^2\frac{x}{t^2}} + \int_2^{\sqrt{\frac{x}{17}}} \frac{2(\alpha - S(t))\,dt}{t\log^2\frac{x}{t^2}}.$$

Now,

$$\int_2^{\sqrt{\frac{x}{17}}} \frac{2\alpha\,dt}{t\log^2\frac{x}{t^2}} = \frac{\alpha}{\log 17} - \frac{\alpha}{\log\frac{x}{4}} \le \frac{\alpha}{\log 17} - \frac{\alpha}{\log x}\left(1 + \frac{\log 4}{\log x}\right).$$

Also,

$$\int_2^{\sqrt{\frac{x}{17}}} \frac{2(\alpha - S(t))\,dt}{t\log^2\frac{x}{t^2}} \ge \sum_{k\le 21433} (\alpha - S(p_k))\left[\frac{1}{\log\frac{x}{t^2}}\right]_{p_k}^{p_{k+1}}$$

$$= \sum_{k\le 21433} (\alpha - S(p_k))\frac{\log\frac{p_{k+1}^2}{p_k^2}}{\log\frac{x}{p_{k+1}^2}\log\frac{x}{p_k^2}}$$

$$\ge \frac{0.3592}{\log\frac{x}{9}\log\frac{x}{4}}$$

$$\ge \frac{0.3592}{\log^2 x}\left(1 + \frac{\log 9}{\log x}\right)\left(1 + \frac{\log 4}{\log x}\right)$$

$$\ge \frac{0.3592}{\log^2 x} + \frac{1.2872}{\log^3 x}.$$

Thus,

$$\sum_{p\le\sqrt{\frac{x}{17}}} \frac{1}{p^2\log(\frac{x}{p^2})} \ge \frac{\alpha}{\log x}\left(1 + \frac{\log 4}{\log x}\right) + \frac{0.3592}{\log^2 x} + \frac{1.2872}{\log^3 x} - \frac{1}{\log 17}\sum_{p>\sqrt{\frac{x}{17}}} \frac{1}{p^2}.$$

By [9, Lem. 2.7], we have

$$\frac{1}{\log 17}\sum_{p>\sqrt{\frac{x}{17}}} \frac{1}{p^2} < \frac{1/\log 17}{\sqrt{\frac{x}{17}}\log\sqrt{\frac{x}{17}}} = \frac{2\sqrt{17}/\log 17}{\sqrt{x}\log\frac{x}{17}} \le \frac{3.2431}{\sqrt{x}\log x}.$$

Combining these bounds, we obtain Lemma 9. □

Acknowledgements. The author is very grateful to Jonathan Bayless, Dominic Klyve, and Larry Wilson for helping to improve the mathematics and computer programming, and to the anonymous referee for helpful suggestions improving the quality of the paper.

References

[1] J. Bayless, P. Kinlaw and D. Klyve, Sums over primitive sets with a fixed number of prime factors, *Math. Comp.*, to appear.

[2] H. Cohen, High precision computation of Hardy-Littlewood constants, preprint, `https://www.math.u-bordeaux.fr/~hecohen/` (1999).

[3] P. Dusart, Estimates of some functions over primes without R.H., Arxiv preprint, arXiv:1002.0442 (2010).

[4] P. Dusart, Explicit estimates of some functions over primes, *Ramanujan J.* **45** no. 1 (2018), 227-251.

[5] A. Hildebrand and G. Tenenbaum, On the number of prime factors of an integer, *Duke Mathematical Journal*, **56** no. 3 (1988), 471-501.

[6] E. Landau, Sur quelques problèmes relatifs à la distribution des nombres premiers, *Bull. Soc. Math France* no. 28 (1900), 25–38.

[7] H. L. Montgomery and R. C. Vaughan. *Multiplicative number theory I: Classical theory,* Vol. 97, Cambridge University Press, 2006.

[8] H. Lifchitz and M. Renner, Number of sphenic numbers up to 10^n, *Online Encyclopedia of Integer Sequences*, `http://oeis.org/A215218` (2012).

[9] H. M. Nguyen and C. Pomerance, The reciprocal sum of the amicable numbers, *Math. Comp.*, to appear.

[10] J. B. Rosser and L. Schoenfeld, Approximate formulas for some functions of prime numbers, *Illinois J. Math.* **6** no. 1 (1962), 64–94.

[11] A. Selberg, Note on a paper by L. G. Sathe, *J. Indian Math. Soc.* **18** no. 1 (1954), 83–87.

9

QUASIPERFECT NUMBERS WITH THE SAME EXPONENT

Tomohiro Yamada
Center for Japanese Language and Culture
Osaka University, Aomatanihigashi, Minoo, Osaka, Japan
tyamada1093@gmail.com

Abstract

We study some divisibility properties of quasiperfect numbers. We show that if $N = (p_1 p_2 \cdots p_t)^{2a} = m^2$ is quasiperfect, then $2a+1$ is divisible by 3 and N has at least one prime factor smaller than $e^{721.85}$. Moreover, we find some lower bounds concerning quasiperfect numbers of the form $N = m^2$ with m squarefree.

1. Introduction

A positive integer N is called perfect if $\sigma(N) = 2N$, where $\sigma(N)$ denotes the sum of divisors of N. As is well known, an even integer N is perfect if and only if $N = 2^{k-1}(2^k - 1)$ with $2^k - 1$ prime. In contrast, one of the oldest unsolved problems is whether there exists an odd perfect number or not. Moreover, it is also unknown whether there exists an odd multiperfect number, i.e., an integer N with $\sigma(N) = kN$ for some integer $k \geq 2$ or not.

Cattaneo [1] called a positive integer N quasiperfect if $\sigma(N) = 2N + 1$, and showed that such an integer must be an odd square, that any proper divisor m of N satisfies $\sigma(m) < 2m$, and that any divisor of $\sigma(N)$ is congruent to 1 or 3 modulo 8. Hagis and Cohen [9] showed that if N is quasiperfect, then $N > 10^{35}$ and N has at least 7 distinct prime factors.

It is well known that an odd perfect number must be of the form $q^b p_1^{2a_1} p_2^{2a_2} \cdots p_t^{2a_t}$ for some integers $a_1, a_2, \ldots, a_t, b$ and distinct primes $p_1, p_2, \ldots, p_t, q$ with $q \equiv b \equiv 1 \pmod 4$. In the special case $a_1 = a_2 = \cdots = a_t = a$, several results are known as follows:

(A) If $a_1 = a_2 = \cdots = a_t = a$, we know that $a \geq 9$ and $a \neq 10, 11, 12, 13, 14, 16, 17, 18, 19, 24, 62$, combining results in [3], [10], [13], [15], [16] and [20];

(B) If d divides $2a_i + 1$ for all i, then d cannot be one of $3, 35, 65$, combining results in [6], [15] and [16];

(C) In [21], the author showed that, for any given a, there exist only finitely many odd perfect numbers of the form $q^b(p_1p_2\cdots p_t)^{2a}$;

(D) Fletcher, Nielsen and Ochem [7] showed that an odd perfect number $q^b p_1^{2a_1}p_2^{2a_2}\cdots p_t^{2a_t}$ for which there exists a finite set S of primes such that each $2a_i+1$ is divisible by a prime in S, must have a prime factor below an effectively computable constant C depending on S. In [23], the author gave an explicit upper bound for C.

Some results similar to (A), (B) above are known for quasiperfect numbers of the form $p_1^{2a_1}p_2^{2a_2}\cdots p_t^{2a_t}$ with $a_1=a_2=\cdots=a$ or $d\mid(2a_i+1)(i=1,2,\ldots,t)$ for some integer $d>1$. Cohen [2] showed that if an integer of the form $(p_1p_2\cdots p_t)^{2a}$ is quasiperfect, then a must be congruent to $1,3,5,9$ or $11 \pmod{12}$, and if an integer of the form $p_1^{6a_1+2}p_2^{6a_2+2}\cdots p_t^{6a_t+2}$ is quasiperfect, then $t\geq 230876$.

We would like to begin by giving an extension of this result, which follows from an elementary argument.

Theorem 1. *If $N=p_1^{2a_1}p_2^{2a_2}\cdots p_t^{2a_t}=m^2$ is quasiperfect, then there exists a prime factor $p_j\equiv 1 \pmod 4$ for which $2a_j+1$ does not have a divisor congruent to $5 \pmod 8$. Moreover, if $N=(p_1p_2\cdots p_t)^{2a}=m^2$ is quasiperfect, then:*

(a) Any prime factor of N must be congruent to 1 or $7 \pmod 8$.

(b) $2a+1\equiv 3 \pmod 8$ and, if q is a prime factor of $2a+1$, then q must be congruent to 1 or $3 \pmod 8$.

(c) $2a+1$ must be divisible by 3.

Proof. Since $\sigma(N)=2N+1=2m^2+1\equiv 3 \pmod 8$, we have $\sigma(p_j^{2a_j})\not\equiv 1 \pmod 8$ for some j, for which, by Cattaneo's result mentioned above, $\sigma(p_j^{2a_j})\equiv 3 \pmod 8$ cannot have a prime factor congruent to 5 or 7 modulo 8. If d divides $2a_j+1$, then $\sigma(p_j^{d-1})$ divides $\sigma(p_j^{2a_j})\mid\sigma(N)=2m^2+1$ and therefore has only prime factors congruent to 1 or 3 modulo 8. Since $\sigma(p_j^{2a_j})\equiv 3 \pmod 8$, we must have $p_j\equiv 1 \pmod 4$. If $p_j\equiv 1 \pmod 8$, then $d\equiv 1$ or $3 \pmod 8$, and if $p_j\equiv 5 \pmod 8$, then $d\equiv 1$ or $7 \pmod 8$. Hence, d cannot be congruent to 5 modulo 8. This proves the former part of the theorem.

Next, assume that $N=(p_1p_2\cdots p_t)^{2a}=m^2$ is quasiperfect. By Cohen's result mentioned above, we must have $2a+1\equiv 3 \pmod 4$. Hence, N has no prime factor congruent to $3 \pmod 8$. In particular, $N=m^2$ is not divisible by 3 and therefore $\sigma(N)=2m^2+1$ must be divisible by 3. Hence, there exists a prime factor p_k for which $\sigma(p_k^{2a})$ is divisible by 3. This implies $p_k\equiv a\equiv 1 \pmod 3$. So that $2a+1$ must be divisible by 3 as stated in (c). Since $p_i^2+p_i+1$ divides $\sigma(N)$, $p_i^2+p_i+1$ must be congruent to 1 or 3 modulo 8 and therefore $p_i\equiv\pm 1 \pmod 8$ for any i, proving (a). By the former part of the theorem, $p_s\equiv 1 \pmod 8$ for some s. If

d divides $2a_s + 1$, then $\sigma(p_s^{d-1}) \mid \sigma(p_s^{2a}) \mid \sigma(N) = 2m^2 + 1$ and therefore $d \equiv 1$ or $3 \pmod 8$. Finally, Cohen's result mentioned above yields that a cannot be a multiple of 4 and therefore $2a + 1 \equiv 3 \pmod 8$, which proves (b). □

No result similar to (C) above has been known for quasiperfect numbers and neither has the author been able to prove such a result. Instead, in [23], the author proved that if $N = p_1^{2a_1} p_2^{2a_2} \cdots p_t^{2a_t}$ is quasiperfect and there exists a finite set S of primes such that each $2a_i + 1$ is divisible by a prime in S, then N must have a prime factor below $C_0 = \exp(2173.5 \, |S|^2 \max\{8l, e^{13.3}\})$. More generally, the author proved that if $N = p_1^{2a_1} p_2^{2a_2} \cdots p_t^{2a_t}$ satisfies that $\sigma(N) \geq 2N$ has no prime factor congruent to 5 or 7 modulo 8, and there exists a finite set S of primes such that each $2a_i + 1$ is divisible by a prime in S, then N must have a prime factor below C_0. In this paper, we prove the following result.

Theorem 2. *If $N = (p_1 p_2 \cdots p_t)^{2a} = m^2$ is quasiperfect, then N must have a prime factor below $C = e^{721.85} < 3.129477 \cdot 10^{313}$.*

We shall give an outline of our proof here. From Theorem 1, we can see that each $p_i^2 + p_i + 1$ divides $\sigma(N) = 2m^2 + 1$ and therefore cannot have a prime factor $\equiv 7, 13 \pmod{24}$, which is implicit in the Note of Lemma 3 of [2]. Using some sieve argument, we shall prove that the number of prime $p \leq X$ such that $p^2 + p + 1$ has no prime factor $\equiv 7, 13 \pmod{24}$ is $< cX/\log^{3/2} X$ with an explicit constant c.

Our sieve argument is based on the argument used in [22] to prove that an odd perfect number $q^b p_1^{2a_1} p_2^{2a_2} \cdots p_t^{2a_t}$ with all p_i in a given finite set S must have a prime factor below an effectively computable constant C depending on S. This method was refined by Fletcher, Nielsen and Ochem [7] and the author [23] to prove (D).

Finally, we would like to show the following lower bounds concerning quasiperfect numbers of the form $N = m^2$ with m squarefree.

Theorem 3. *If $N = m^2$ with m squarefree is quasiperfect, then $N > e^{17840573219}$ and N must have at least 406550054 distinct prime factors, one of which is $\geq$ 9457308739.*

Our results lead us to conjecture that there exists no quasiperfect number of the form $N = (p_1 p_2 \cdots p_t)^{2a}$.

2. Proof of Theorem 2

From (a) of Theorem 1 and a remark in the introduction, we see that if p is a prime factor of N, then $p \equiv \pm 1 \pmod 8$ and $p^2 + p + 1$ cannot have a prime factor congruent to 7 or 13 (mod 24). Letting $Q^{\pm}$ be the set of prime numbers $p \equiv \pm 1 \pmod 8$ such that $p^2 + p + 1$ has no prime factor congruent to 7 or 13 (mod 24), we can say that any prime factor of N must be contained in either Q^+ or Q^-.

We would like to introduce some notations in order to apply sieve methods. For a given prime p, we let Ω_p denote a set of congruent classes modulo p and define $\rho(p)$ to be the number of such congruent classes. Under this setting, let $S(x, \Omega, y)$ denote the set of integers not exceeding x which does not belong to Ω_p for any prime $p \leq y$.

Hereafter, $\pm$ denotes a sign which is possibly different at each occurrence but the same in the same context. For our purpose, we put $\Omega_p^{\pm} = \{n \mid (8n \pm 1)((8n \pm 1)^2 + (8n \pm 1) + 1) \equiv 0 \pmod{p}\}$ for primes p congruent to 7 or 13 (mod 24) and $\Omega_p^{\pm} = \{n \mid 8n \pm 1 \equiv 0 \pmod{p}\}$ for other odd primes p. Then, we see that $\rho^{\pm}(2) = 0$, $\rho^{\pm}(p) = 3$ for primes $p \equiv 7, 13 \pmod{24}$ and $\rho^{\pm}(p) = 1$ for the other primes p. Moreover, it is clear that if $p = 8n \pm 1$ belongs to $Q^{\pm}$ and $y < n \leq x$, then n must be contained in $S(x, \Omega^{\pm}, y)$ and therefore

$$\pi^{\pm}(8x \pm 1) \leq y + \left|S(x, \Omega^{\pm}, y)\right|, \tag{1}$$

where $\pi^{\pm}(X)$ denotes the number of primes not exceeding X that belongs to $Q^{\pm}$.

In order to estimate $|S(x, \Omega^{\pm}, y)|$, we use the sieve argument mentioned in the introduction. Let us introduce the following notation:

$$B(z) = B^{\pm}(z) = \frac{1}{\log z} \sum_{p \leq z} \frac{\rho^{\pm}(p) \log p}{p}, \tag{2}$$

$$V(z) = V^{\pm}(z) = \prod_{p \leq z} \left(1 - \frac{\rho^{\pm}(p)}{p}\right) \tag{3}$$

and

$$\psi_0^{\pm}(v, u) = 1 - \exp(-\psi_1(B^{\pm}(x^{\frac{1}{v}}), v/u)), \tag{4}$$

where

$$\psi_1(K, t) = \max\left\{0, t \log \frac{t}{K} - t + K\right\}. \tag{5}$$

Now, we have the following sieve inequality.

Lemma 1. *For any real $v \geq u \geq 2$, we have*

$$\left|S(x, \Omega^{\pm}, x^{\frac{1}{u}})\right| \leq \frac{(x + x^{\frac{2}{u}}) V^{\pm}(x^{\frac{1}{u}})}{\psi_0^{\pm}(v, u)}. \tag{6}$$

Proof. This is essentially a special case of Lemma 2.2 of [23]. However, we would like to give the proof again for self-containedness.

Let $g(m) = g^{\pm}(m)$ be the multiplicative function over the squarefree integers m with $g^{\pm}(p) = \rho^{\pm}(p)/(p - \rho^{\pm}(p))$ for each prime p. We define $G(w) = \sum_{n \leq w} g(n)$ and $G_z(w) = \sum_{n \leq w, p|n \Rightarrow p \leq z} g(n)$. Observing that $0 \leq \rho(p) < p$ for any prime p, we use Theorem 7.14 in [11] to obtain

$$\left|S(x, \Omega, x^{\frac{1}{u}})\right| \leq \frac{(x + x^{\frac{2}{u}})}{G(x^{\frac{1}{u}})}. \tag{7}$$

Now Theorem 2.2.1 in [8] gives (6) with $\sqrt{D} = x^{1/u}$ and $B = \sup_t B(t)$ in place of $B(z)$. But we can see that this theorem still holds with B replaced by $B(z)$, whether the supremum B exists or not. Indeed, it follows from the argument in p.p. 53–54 in [8] that

$$1 - V(P(z))G_z(x^{1/u}) \leq \exp\left(-c\frac{\log x}{u \log z} + B(z)(e^c - 1)\right) \tag{8}$$

for any constant $c \geq 0$. Choosing $c = \log(v/u) - \log B(z)$, we obtain the lemma. □

Now we need to estimate $B(z)$ and $V(z)$. To this end, we need some explicit estimates for a sum and a product involving primes in arithmetic progressions. Our start points are the following error estimates in the prime number theorem for arithmetic progressions with difference 24, which can be obtained from some known explicit versions of the prime number theorem for arithmetic progressions and a recent numerical result for the Generalized Riemann Hypothesis of L-functions.

Lemma 2. *As in [4], we set $R = 9.645908801$. Let ℓ be any integer coprime to 6. Then,*

$$\frac{1}{z}\left|\psi(z; 24, \ell) - \frac{z}{8}\right| < \begin{cases} c_1 := 6.28782 \cdot 10^{-4} & \textit{for } z \geq 10^{10}, \\ c_2 := 2.04555 \cdot 10^{-5} & \textit{for } z \geq e^{30}, \\ c_3 := 1.24271 \cdot 10^{-7} & \textit{for } z \geq e^{60}. \end{cases} \tag{9}$$

Moreover, for any real $z \geq e^{250R}$, we have

$$\left|\psi(z; 24, \ell) - \frac{z}{8}\right| < \frac{0.001197z}{\log z} \tag{10}$$

and

$$\left|\psi(z; 24, \ell) - \frac{z}{8}\right| < \frac{2.88643z}{\log^2 z}. \tag{11}$$

Proof. Platt [17] confirmed that $L(s, \chi)$ has no nontrivial zero with $\Re s \neq 1/2$ and $|\Im s| \leq 10^8/24$ for any character χ modulo 24. From Theorem 3.6.3 of [18], we see that any nontrivial zero s of $L(s, \chi)$ satisfies $1 - \operatorname{Re} s \geq 1/R \log(24 |\operatorname{Im} s| / 68.9385)$ for any character χ modulo 24. Now (9) follows from Theorem 5.1.1 of [18] applied with $H_\chi = 10^8/24$ and $C_1(\chi) = 68.9385$ for each character χ modulo 24.

Moreover, we can apply Theorem 5 of [4] with $H = 1000, C_1(k) = 32$ (we note that H and $C_1(k)$ in this theorem can be taken smaller) and $X_4 \leq \sqrt{250}$ for any character modulo 24 to obtain (10) and (11). This proves the lemma. □

Remark 1. Kadiri [12] claimed to prove that any nontrivial zero s of $L(s, \chi)$ satisfies $1 - \operatorname{Re} s \geq 1/6.397 \log(24 |\operatorname{Im} s|)$ for all characters χ modulo 24. This would give better estimates.

Using these error estimates, we obtain the following bounds.

Lemma 3. *For any real $z \geq e^{60}$, we have*

$$\sum_{p\leq z, p\equiv 7,13 \pmod{24}} \frac{\log p}{p} < \frac{\log z}{4} \tag{12}$$

and

$$\prod_{p\leq z, p\equiv 7,13 \pmod{24}} \left(1-\frac{1}{p}\right)^{-1} > 0.952711 \log^{\frac{1}{4}} z. \tag{13}$$

Proof. In this proof of the lemma, we let ℓ be an integer congruent to 7 or 13 modulo 24. We begin by proving (12) for $z \geq e^{60}$. Lemma 2 gives

$$\begin{aligned}\int_{10^{10}}^{\infty} \frac{|E_\psi(t;24,l)|}{t^2}dt <&(30-10\log 10)c_1 + 30c_2 + (250R-60)c_3 + \frac{2.88643}{250R} \\ <&0.008 + 0.002 = 0.01,\end{aligned} \tag{14}$$

where $E_\psi(t;k,\ell) = \psi(t,k,\ell) - t/\varphi(k)$. By Theorem 5.2.1 of [18], we have

$$\int_{10^8}^{10^{10}} \frac{|E_\psi(t;24,l)|}{t^2}dt < 1.745\int_{10^8}^{10^{10}} \frac{dt}{t^{3/2}} = \frac{3.49}{10^4} - \frac{3.49}{10^5} < 0.00032. \tag{15}$$

Some calculations yield that

$$\sum_{p<10^8, p\equiv 7 \pmod{24}} \frac{\log p}{p} < \log 10 - 0.101846 \tag{16}$$

and

$$\sum_{p<10^8, p\equiv 13 \pmod{24}} \frac{\log p}{p} < \log 10 - 0.202137. \tag{17}$$

Moreover, other calculations give $\psi(10^8;24,7) > 12499496$ and $\psi(10^8;24,13) > 12499441$. We use partial summation and then apply these inequalities and (11) to obtain that, for $z \geq 10^8$ and $\ell = 7$ or 13,

$$\begin{aligned}\sum_{\substack{p\leq z,\\ p\equiv \ell \pmod{24}}} \frac{\log p}{p} \leq& \sum_{\substack{p\leq 10^8,\\ p\equiv \ell \pmod{24}}} \frac{\log p}{p} + \sum_{\substack{10^8<n\leq z,\\ n\equiv \ell \pmod{24}}} \frac{\Lambda(n)}{n} \\ =& \sum_{\substack{p\leq 10^8,\\ p\equiv \ell \pmod{24}}} \frac{\log p}{p} + \frac{\psi(z;24,\ell)}{z} - \frac{\psi(10^8;24,\ell)}{10^8} + \int_{10^8}^{z} \frac{\psi(t;24,\ell)}{t^2}dt \\ <& \frac{\log z}{8} - 0.101846 + \frac{0.01}{\log z} + 0.0004 + 0.01 \\ <& \frac{\log z}{8},\end{aligned} \tag{18}$$

which immediately gives (12) for $z \geq e^{60}$ (and even for $z \geq 10^8$).

Next we prove (13) for $z \geq e^{60}$. By (9) we have

$$\int_z^{e^{250R}} \frac{c_3(1+\log t)}{t\log^2 t}dt = c_3\left(\frac{1}{\log z} - \frac{1}{250R} + \log(250R) - \log\log z\right) \tag{19}$$

for $e^{60} \leq z < e^{250R}$ and, by (10), we have

$$\int_z^{\infty} \frac{0.001197(1+\log t)}{t\log^3 t}dt = 0.001197\left(\frac{1}{\log z} + \frac{1}{2\log^2 z}\right) < 4.9659 \times 10^{-7} \tag{20}$$

for $z \geq e^{250R}$.

Henceforth, we let $z \geq e^{60}$ as in the lemma. Using (19) and (20), we have

$$\begin{aligned}&\sum_{\substack{p\leq z,\\ p\equiv \ell \pmod{24}}} \frac{1}{p}\\ =&\frac{1}{8}\log\log z + M(24,\ell) + \frac{E_\theta(z;24,\ell)}{z\log z} - \int_z^{\infty}\frac{(1+\log t)E_\theta(t;24,\ell)}{t^2\log^2 t}dt \\ >&\frac{1}{8}\log\log z + M(24,\ell) - \frac{c_3}{60} - 4.61034\times 10^{-6} - 4.9659\times 10^{-7},\end{aligned} \tag{21}$$

where $E_\theta(z;k,\ell) = \theta(z;k,\ell) - z/\varphi(k)$ and $M(k,\ell)$ denotes the limit

$$\lim_{x\to\infty} \sum_{\substack{p\leq x,\\ p\equiv \ell \pmod{k}}} \frac{1}{p} - \frac{\log\log x}{\varphi(k)}. \tag{22}$$

Using $M(24,7) > 0.0038975\cdots$ and $M(24,13) > -0.068154106\cdots$, as is given by [14], we obtain

$$\sum_{\substack{p\leq z,\\ p\equiv 7,13 \pmod{24}}} \frac{1}{p} > \frac{1}{8}\log\log z - 0.064266, \tag{23}$$

and therefore

$$\begin{aligned}\prod_{\substack{p\leq z,\\ p\equiv 7,13 \pmod{24}}} \left(1-\frac{1}{p}\right)^{-1} &= \exp\left[\sum_{\substack{p\leq z,\\ p\equiv 7,13 \pmod{24}}}\left(\frac{1}{p}+\frac{1}{2p^2}+\cdots\right)\right]\\ &> \exp\left(0.015815 + \sum_{\substack{p\leq z,\\ p\equiv 7,13 \pmod{24}}}\frac{1}{p}\right)\\ &>0.952711\log^{\frac{1}{4}} z,\end{aligned} \tag{24}$$

which proves the lemma. □

Now $V(P(z))$ and $B(z)$ can be easily estimated from Lemma 3 and known error estimates for the ordinary prime number theorem. Indeed, combining (13) and the estimate

$$\prod_{p\leq z}\left(1-\frac{1}{p}\right)<\frac{e^{-\gamma}}{\log z}\left(1+\frac{1}{5\log^3 z}\right) \tag{25}$$

for $z \geq 2278382$ given in Theorem 5.9 of [5], we conclude that

$$\begin{aligned} V(P(z)) &< \prod_{2<p\leq z}\left(1-\frac{1}{p}\right)\prod_{\substack{p\leq z,\\ p\equiv 7,13 \pmod{24}}}\left(1-\frac{1}{p}\right)^2 \\ &< \frac{2e^{-\gamma}}{(0.952711)^2\log^{\frac{3}{2}} z}\left(1+\frac{1}{5\log^2 z}\right) \\ &< \frac{1.23273}{\log^{\frac{3}{2}} z}, \end{aligned} \tag{26}$$

for $z \geq e^{60}$. Moreover, by (3.22) of [19], we have

$$\sum_{p\leq z}\frac{\log p}{p}<\log z-1.33258+\frac{1}{2\log z}<\log z \tag{27}$$

for $z > 319$ and, together with (12), we immediately obtain

$$\sum_{p\leq z}\frac{\rho(p)\log(p)}{p}<\frac{3}{2}\log z \tag{28}$$

or, equivalently, $B(z) < 1.5$ for $z \geq e^{60}$.

Equations (26) and (28) allow us to take $B = 1.5, u = 2.0173$ and $v = 7.58$ in the sieve inequality (6), which gives

$$\left|S(x,\Omega^{\pm},y)\right| \leq \frac{37.13862x}{\log^{\frac{3}{2}} x} \tag{29}$$

for $x > e^{719.77}$ and therefore, by (1),

$$\pi^{\pm}(X) \leq \frac{4.65574X}{\log^{\frac{3}{2}} X} \tag{30}$$

for $X > C = e^{721.85}$.

Now we have

$$\begin{aligned}\prod_{p\geq C, p\in Q^+\cup Q^-}\frac{p}{p-1} &< \exp\left[\sum_{p\geq C, p\in Q^+\cup Q^-}\left(\frac{1}{p}+\frac{1}{2p^2}+\cdots\right)\right]\\ &< \exp\left(\frac{1}{C}+\sum_{p\geq C, p\in Q^+\cup Q^-}\frac{1}{p}\right)\\ &< \exp\left(\frac{2}{C}+\int_C^\infty \frac{9.31148dt}{t\log^{\frac{3}{2}} t}\right)\\ &= \exp\left(\frac{2}{C}+\frac{18.62296}{\log^{\frac{1}{2}} C}\right) < 2.\end{aligned} \tag{31}$$

However, as mentioned in the beginning of this section, any prime factor of N must belong to $Q^+\cup Q^-$. Hence, if N has no prime factor below C, then $\sigma(N) < N\prod_{p|N} p/(p-1) < 2N$, which contradicts the assumption that $\sigma(N) = 2N+1$. This proves Theorem 2.

3. Proof of Theorem 3

Assume that $N = m^2$ with m squarefree is quasiperfect and let P be the greatest prime factor of N. As noted in the previous section, any prime factor of N belongs to $Q^+\cup Q^-$.

Calculation gives

$$\prod_{p<2^{29}, p\in Q^+\cup Q^-}\frac{p^2+p+1}{p^2} < 1.75014319434, \tag{32}$$

$$\prod_{p<2^{29}, p\in Q^+\cup Q^-} p > e^{62460825.5} \tag{33}$$

and

$$\pi^+(2^{29})+\pi^-(2^{29}) = 3285696. \tag{34}$$

By Theorem 6.12 of [5], we have

$$\prod_{2^{29}<p\leq P}\frac{p}{p-1} < \frac{\log P}{29\log 2}\left(1+\frac{1}{121945\log^3 2}\right)\left(1+\frac{1}{5\log^3 P}\right). \tag{35}$$

It follows from (32) and (35) that

$$\prod_{p\leq 9457308738, p\in Q^+\cup Q^-}\frac{p^2+p+1}{p^2} < 2, \tag{36}$$

and therefore we must have $P \geq 9457308739$.

Now, let $P_0 = 9457308739$. Then we must have

$$m \geq \left(\prod_{p<2^{29}, p\in Q^+\cup Q^-} p \right) \left(\prod_{2^{29}<p\leq P_0} p \right). \tag{37}$$

Theorem 4.2 of [5] gives

$$\prod_{2^{29}<p\leq P_0} p = \exp(\theta(P) - \theta(2^{29})) > \exp\left(P_0 - \frac{P_0}{100\log^2 P_0} - 536842885.9 \right) \tag{38}$$

and, with the aid of (37), we see that $m > e^{8920286609.5}$ and $N > e^{17840573219}$.

Finally, we observe that, using (34) and Theorem 5.1 of [5],

$$\begin{aligned} &\pi^+(P) + \pi^-(P) \geq \pi(P_0) - \pi(2^{29}) + \pi^+(2^{29}) + \pi^-(2^{29}) \\ &\geq \frac{P}{\log P}\left(1 + \frac{1}{\log P} + \frac{2}{\log^2 P} + \frac{7.32}{\log^3 P}\right) - \pi(2^{29}) + 3285696 > 406550053.02 \end{aligned} \tag{39}$$

and therefore N must have at least 406550054 prime factors. This completes the proof of Theorem 3.

References

[1] Paolo Cattaneo, Sui numeri quasiperfetti, *Boll. Un. Mat. Ital.* (3) **6** (1951), 59–62.

[2] Graeme L. Cohen, The nonexistence of quasiperfect numbers of certain forms, *Fibonacci Quart.* **20** (1982), 81–84.

[3] G. L. Cohen and R. J. Williams, Extensions of some results concerning odd perfect numbers, *Fibonacci Quart.* **23** (1985), 70–76.

[4] P. Dusart, Estimates of $\theta(x; k, l)$ for large values of x, *Math. Comp.* **71** (2001), 1137–1166.

[5] P. Dusart, Explicit estimates of some functions over primes, *Ramanujan J.*, **45** (2018), 227–251.

[6] Ronald Evans and Jonathan Pearlman, Nonexistence of odd perfect numbers of a certain form, *Fibonacci Quart.* **45** (2007), 122–127.

[7] S. Adam Fletcher, Pace P. Nielsen and Pascal Ochem, Sieve methods for odd perfect numbers, *Math. Comp.* **81** (2012), 1753–1776.

[8] G. Greaves, *Sieves in Number Theory*, Springer-Verlag, Berlin, 2001.

[9] Peter Hagis, Jr. and Graeme L. Cohen, Some results concerning quasiperfect numbers, *J. Aust. Math. Soc. (Ser. A)* **33** (1982), 275–286.

[10] P. Hagis Jr. and Wayne L. McDaniel, A new result concerning the structure of odd perfect numbers, *Proc. Amer. Math. Soc.* **32** (1972), 13–15.

[11] H. Iwaniec and Kowalski, *Analytic Number Theory*, American Mathematical Society Colloquium Publications, vol. 53, American Mathematical Society, Providence, RI, 2004.

[12] Habiba Kadiri, An explicit zero-free region for the Dirichlet L-functions, preprint `http://arxiv.org/abs/0510570`.

[13] H.-J. Kanold, Untersuchungen über ungerade vollkommene Zahlen, *J. Reine Angew. Math.* **183** (1941), 98–109.

[14] A. Languasco and A. Zaccagnini, Computing the Mertens and Meissel-Mertens constants for sums over arithmetic progressions, *Experiment. Math.* **19** (2010), 279–284. The complete set of results is available online `http://www.math.unipd.it/ languasc/Mertens-comput.html`.

[15] W. L. McDaniel, The non-existence of odd perfect numbers of a certain form, *Arch. Math. (Basel)* **21** (1970), 52–53.

[16] W. L. McDaniel and P. Hagis Jr., Some results concerning the non-existence of odd perfect numbers of the form $p^a M^{2\beta}$, *Fibonacci Quart.* **13** (1975), 25–28.

[17] David J. Platt, Numerical computations concerning the GRH, *Math. Comp.* **85** (2016), 3009–3027.

[18] O. Ramare and R. Rumely, Primes in arithmetic progressions, *Math. Comp.* **65** (1996), 397–425.

[19] J. B. Rosser and L. Schoenfeld, Approximate formulas for some functions of prime numbers, *Illinois J. Math.* **6** (1962), 64–94.

[20] R. Steuerwald, Verschärfung einen notwendigen Bedingung für die Existenz einen ungeraden vollkommenen Zahl, *S.-B. Bayer. Akad. Wiss.* 1937, 69–72.

[21] T. Yamada, Odd perfect numbers of a special form, *Colloq. Math.* **103** (2005), 303–307.

[22] Tomohiro Yamada, On the divisibility of odd perfect numbers by a high power of a prime, preprint, `https://arxiv.org/abs/math/0511410`.

[23] Tomohiro Yamada, On the divisibility of odd perfect numbers by a high power of a prime, II, submitted, preprint, `https://arxiv.org/abs/1609.08795`.

10

ELLIPTIC CURVES ARISING FROM THE TRIANGULAR NUMBERS

Abhishek Juyal
Harish-Chandra Research Institute, HBNI, Allahabad, India
abhinfo1402@gmail.com

Shiv Datt Kumar
Department of Mathematics, Motilal Nehru National Institute of Technology, Allahabad, India
sdt@mnnit.ac.in

Dustin Moody
National Institute of Standards and Technology, Gaithersburg, Maryland
dustin.moody@nist.gov

Abstract

We study the Legendre family of elliptic curves $E_t : y^2 = x(x-1)(x-\Delta_t)$, parametrized by triangular numbers $\Delta_t = t(t+1)/2$. We prove that the rank of E_t over the function field $\overline{\mathbb{Q}}(t)$ is 1, while the rank is 0 over $\mathbb{Q}(t)$. We also produce some infinite subfamilies whose Mordell-Weil rank is positive, and find high rank curves from within these families.

1. Introduction and Main Result

The study of polygonal numbers is an ancient problem and has been widely studied. Among polygonal numbers, triangular numbers occupy a central place and they have drawn considerable attention from many researchers. For example, Legendre proved that a triangular number can never be a cube or a fourth power of an integer. As a second example, Gauss famously noted in his diary that every natural number can be expressed as a sum of at most three triangular numbers, and Euler proved that there are infinitely many squares among the triangular numbers and found all such numbers [1].

This brief paper is an attempt to establish some interesting connections between triangular numbers and elliptic curves. The specific curves belong to the Legendre family (see definition below), and are parameterized by triangular numbers. In this context, we first begin with a brief review. Let Δ_t be the tth triangular number, which is defined as the sum of the first t natural numbers:

$$\Delta_t = 1 + 2 + 3 + \cdots + t = \frac{t(t+1)}{2}.$$

For any field k with $\mathrm{char}(k) \neq 2$, an elliptic curve in Legendre form is one given by the equation

$$y^2 = x(x-1)(x-\lambda),$$

for some $\lambda \in k$, with $\lambda \neq 0, 1$. A Legendre curve always has three rational points of order two, namely the points $(0,0), (1,0)$, and $(\lambda, 0)$. Conversely, any elliptic curve E/k which has three rational points of order two can be given by an elliptic curve of the form $y^2 = x(x-\alpha)(x-\beta)$ with $\alpha, \beta \in k^*$. Investigating the possible transformations ([19, III, Sect. 1]) yields that E is isomorphic to a curve in Legendre form if and only if at least one of $\pm\alpha, \pm\beta, \pm(\alpha - \beta)$ is a square in k^*.

Over the years, several authors have given considerable effort to study the ranks of certain families of Legendre curves with connections to other interesting areas. In [8], the authors showed that there exist infinitely many Pythagorean triples (a, b, c) for which the rank of the family

$$E_{a,b} : y^2 = x(x-a^2)(x-b^2)$$

is positive. Naskręcki, Izadi, and Nabardi continued in this line and showed constructions of families with rank at least two [7, 12]. Other researchers have looked at congruent number curves [4, 14, 15], or families of Legendre curves associated to Heron and Brahmagupta quadrilaterals [5, 6]. It is conjectured that there exist elliptic curves with arbitrarily high rank, although the highest known rank is 28. On the other hand, it has also been conjectured that the rank is absolutely bounded, with only finitely many elliptic curves of rank greater than 21 [13]. Restricting to Legendre curves, the record is 15 for an individual curve and 8 for an infinite family (all these records are due to Elkies [2, 3]).

We will set $\lambda = \Delta_t$, and define the main object of study as the family of elliptic curves

$$E_t : y^2 = x(x-1)(x - t(t+1)/2),$$

parametrized by the triangular numbers. We extend the definition of Δ_t to allow for rational values of t. As far as we can tell, this is the first time triangular numbers have been used to define a family of elliptic curves. Other papers relating triangular numbers and elliptic curves have tended to prove properties about triangular numbers using elliptic curve techniques [1].

The main result of this paper is the following. Our approach is similar to [21].

Theorem 1. *Let E_t be an elliptic curve over $\mathbb{Q}(t)$ given by the equation*

$$E_t : y^2 = x(x-1)(x - t(t+1)/2).$$

Then

(i) The associated elliptic surface (denoted $\mathcal{E}$) is rational.

(ii) The rank of $E_t(\overline{\mathbb{Q}}(t))$ is 1, with $(x,y) = (-t/2, t(t+2)/(2\sqrt{-2}))$ being a generator of the free part of the group $E_t(\overline{\mathbb{Q}}(t))$.

(iii) The rank of $E_t(\mathbb{Q}(t)) = 0$.

(iv) The torsion subgroup of $E_t(\mathbb{Q}(t))$ is isomorphic to $\mathbb{Z}_2 \times \mathbb{Z}_2$.

In Section 2, we will introduce the notion of elliptic surfaces and certain results necessary for the proof of Theorem 1. We then give the detailed proof of Theorem 1 in Section 3, and construct infinite families of the curves E_t with positive rank in Section 4. We conclude in Section 5 by giving some examples of specific curves E_t with high rank, and end with directions for further study.

2. Elliptic Surfaces

Definition. Let C be a smooth, irreducible projective curve over an algebraically closed field k. An elliptic surface over C is a pair (S, f), where S is a smooth, irreducible, projective surface over k, and $f : S \longrightarrow C$ is a relatively minimal elliptic fibration having a singular fiber and a zero section. We often write $f : S \longrightarrow C$ to denote the elliptic surface (S, f) over C.

Let $k(C)$ denote the function field of the curve C. Given an elliptic curve E over $k(C)$, one can associate an elliptic surface $f : \mathcal{E} \longrightarrow C$ with generic fiber E, the existence and uniqueness of which is guaranteed by the work of Kodaira and Néron. This elliptic surface is known as the Kodaira-Néron model of the elliptic curve E over $k(C)$.

Given that all the relevant results needed to prove our main theorem are well known, we just give their statements and omit their proofs.

Theorem 2 (Corollary 2.2, [17]). *Let (S, f) be an elliptic surface over C. The Néron-Severi group, denoted $NS(S)$, is finitely generated and torsion-free.*

Recall the classical Shioda-Tate formula.

Theorem 3 (Corollary 5.3, [17]). *Let (S, f) be an elliptic surface over C. For each point v of C having singular fiber, let m_v denote the number of components of the singular fiber above v. Let E denote the generic fiber of S. The rank of the Néron-Severi group of S, denoted $\rho(S)$, can be obtained from the equality*

$$\rho(S) = \text{rank } \mathrm{E(k(C))} + 2 + \sum_{\mathrm{v}} (\mathrm{m_v} - 1),$$

where the summation ranges over the the points of C under singular fibers.

We also need the following lemma.

Lemma 1 (Theorem IV.8.2, [20] and Corollary 7.5, [18]). *Let E be an elliptic curve over $\overline{\mathbb{Q}}(t)$. Let $\Sigma \subset \mathbb{P}^1(\overline{\mathbb{Q}}(t))$ be the set of points of bad reduction of E. Let $G(F_v)$ denote the group generated by simple components of the fiber F_v at $v \in \Sigma$. There exists an injective homomorphism*

$$\phi : E(\overline{\mathbb{Q}}(t))_{tors} \longrightarrow \prod_{v \in \Sigma} G(F_v).$$

If F_v is of multiplicative type I_n in Kodaira notation, the corresponding group is $\mathbb{Z}/n\mathbb{Z}$. If F_v is of additive type I_{2n}^*, the group is $(\mathbb{Z}/2\mathbb{Z})^2$.

3. Proof of Main Theorem

In this section, we give the proof of Theorem 1.

Proof. The elliptic curve E_t over $\mathbb{Q}(t)$ can be written in short Weierstrass form as

$$y^2 = x^3 + A(t)x + B(t),$$

where

$$A(t) = -\frac{4}{3}(t^4 + 2t^3 - t^2 - 2t + 4),$$

$$B(t) = -\frac{16}{27}(t^6 + 3t^5 - 5t^3 - 9t^2 - 6t + 8).$$

The discriminant of E_t is given by

$$\Delta(t) = -256t^2(t-1)^2(t+1)^2(t+2)^2.$$

We now prove each of the parts of the theorem.

(i) Given an elliptic curve

$$y^2 + a_1(t)xy + a_3(t)y = x^3 + a_2(t)x^2 + a_4(t)x + a_6(t)$$

over $\mathbb{Q}(t)$ in long Weierstrass form, we know from [17, Equation 10.14] that if $\deg(a_i(t)) \leq i$ for each i, then the associated elliptic surface $\mathcal{E}$ is rational. In our case, since $\deg(A(t)) = 4$ and $\deg(B(t)) = 6$, therefore the underlying elliptic surface is rational.

(ii) From the expression of the discriminant of E_t, we see that E_t has singular fibers at the values $t = 0, -2, \pm 1$, and ∞. We determine the numbers m_v, of irreducible components of the fiber over v, from Kodaira types of singular fibers [10, section 4]:

v	coefficients				
	$ord_{t=v}(A)$	$ord_{t=v}(B)$	$ord_{t=v}(\Delta)$	Kodaira type	$m_v - 1$
0	0	0	2	I_2	1
-2	0	0	2	I_2	1
-1	0	0	2	I_2	1
1	0	0	2	I_2	1
∞	0	0	4	I_4	3

Since $\mathcal{E}$ is a rational surface, we have $\rho(\mathcal{E}) = 10$. Thus by Theorem (3) we get,

$$10 = \text{rank } \mathrm{E_t}(\overline{\mathbb{Q}}(\mathrm{t})) + 2 + 1 + 1 + 1 + 1 + 3,$$

and hence rank $\mathrm{E_t}(\overline{\mathbb{Q}}(\mathrm{t})) = 1$.

The group $E_t(\overline{\mathbb{Q}}(t))$ is generated with the points of the form $(a_2T^2 + a_1T + a_0,\ b_3T^3 + b_2T^2 + b_1T + b_0)$, $a_i, b_i \in \overline{\mathbb{Q}}$ (see [17, Equation 10.14]). A straightforward calculation shows that the point $P = (-t/2, t(t+2)/(2\sqrt{-2}))$ is a generator of $E_t(\overline{\mathbb{Q}}(t))$.

(iii) This part's proof follows the idea in Corollary 6.3 of [12]. Since rank $\mathrm{E_t}(\overline{\mathbb{Q}}(\mathrm{t}))=$ 1, then necessarily rank $\mathrm{E_t}(\mathbb{Q}(\mathrm{t})) \leq 1$. We claim that rank $\mathrm{E_t}(\mathbb{Q}(\mathrm{t})) = 0$. On the contrary, assume that rank $\mathrm{E_t}(\mathbb{Q}(\mathrm{t})) = 1$. Then $H = E_t(\mathbb{Q}(t))$ is a finite indexed subgroup of $G = E_t(\overline{\mathbb{Q}}(t))$, and $G_{\mathbb{Q}} = H_{\mathbb{Q}}$, where $G_{\mathbb{Q}} = G \otimes_{\mathbb{Z}} \mathbb{Q}$ and $H_{\mathbb{Q}} = H \otimes_{\mathbb{Z}} \mathbb{Q}$ are one dimensional vector spaces over $\mathbb{Q}$.

We have a canonical Galois representation

$$\rho : \mathrm{Gal}(\overline{\mathbb{Q}}/\mathbb{Q}) \longrightarrow \mathrm{G}_{\mathbb{Q}}^{\times} = \mathrm{Aut}(\mathrm{G}_{\mathbb{Q}})$$

which is defined as follows. For any $\sigma \in \mathrm{Gal}(\overline{\mathbb{Q}}/\mathbb{Q})$ and $g \in G$, we define $\sigma(g\otimes 1) = \sigma(g)\otimes 1$, where $\sigma(g)$ is obtained by the action of σ on the coefficients of rational functions in the coordinates of g. Let $\sigma \in \mathrm{Gal}(\overline{\mathbb{Q}}/\mathbb{Q})$ be the element such that $\sigma(\sqrt{-1}) = -\sqrt{-1}$ and $\sigma(\sqrt{2}) = \sqrt{2}$. Then we get $\sigma(P \otimes 1) = -(P \otimes 1)$. Therefore ρ is a non-trivial character of $\mathrm{Gal}(\overline{\mathbb{Q}}/\mathbb{Q})$. Clearly ρ acts trivially on $H_{\mathbb{Q}}$ which is a contradiction as $G_{\mathbb{Q}} = H_{\mathbb{Q}}$.

(iv) By Lemma 1 and the table in the proof of (ii) above, we see that the torsion subgroup of $E_t(\overline{\mathbb{Q}}(t))$ is embedded in $\mathbb{Z}_2 \times \mathbb{Z}_2 \times \mathbb{Z}_2 \times \mathbb{Z}_2 \times \mathbb{Z}_4$. From the well-known specialization theorem [20], the specialization homomorphism is injective on the torsion. Specializing at $t = 3$, the curve is $E_3 : y^2 = x^3 - 7x^2 + 6x$. The torsion group of E_3 is easily computed to be $\mathbb{Z}_2 \times \mathbb{Z}_2$. We have exactly three 2-torsion points, namely $(0,0), (1,0)$ and $(t(t+1)/2, 0)$, on the elliptic curve E_t. Since these 2-torsion points are also the points of $E_t(\mathbb{Q}(t))$, it implies that $E_t(\mathbb{Q}(t))_{tors} = \mathbb{Z}_2 \times \mathbb{Z}_2$.

□

4. Infinite Families With Positive Rank

4.1. Rank 1 Families

In this section we construct infinite families of the curve E_t which have positive rank. We first construct a couple of subfamilies of rank (at least) 1, followed by a family with rank (at least) 2.

In order to construct a subfamily of rank one, we consider the right hand side of the curve $y^2 = x(x-1)(x - t(t+1)/2)$ as a polynomial in t. The coefficient of t^2 is $-x(x-1)/2$. Putting

$$-x(x-1)/2 = m^2,$$

we can rewrite the equation E_t as $(y - mt)(y + mt) = m^2t - 2m^2x$. Now writing $y - mt = p$, we get $p(y+mt) = m^2t - 2m^2x$. These two equations may be considered as two linear equations in y and t, and solving we find

$$t = \frac{p^2 + 2m^2x}{m(m-2pm)}, \quad y = \frac{-p^2 + 2m^2x + pm}{m - 2p}.$$

We divide both the numerator and denominators by m^2, and setting $p/m = g$, $x = 2/(h^2+2)$, and $m = h/(h^2+2)$, we obtain

$$t(g,h) = -\frac{(h^2g^2 + 2g^2 + 4)}{(h^2+2)(2g-1)},$$

for which the point

$$(x, y) = \left(\frac{2}{h^2 + 2}, \frac{h(h^2g^2 - gh^2 + 2g^2 - 2g - 4)}{(h^2 + 2)^2(2g - 1)} \right)$$

always lies on the elliptic curve

$$y^2 = x(x - 1)(x - t(g, h))(x - (t(g, h) + 1)/2).$$

Specialization at $(g, h) = (1, 1)$ yields the curve $E_{t(1,1)} : y^2 = x^3 - \frac{23}{9}x^2 + \frac{14}{9}x$, which has rank 1 over $\mathbb{Q}$, with the point $(2/3, -4/9)$ of infinite order. Since the specialization map [20, Theorem 11.4] is a homomorphism, this implies that the rank of the curve is at least 1 over $\mathbb{Q}(g, h)$.

It is not hard to find other families with rank at least one. For example, if we set $t = 4/(m^2 + 2)$, then the point with x-coordinate $t + 1$ is rational and has infinite order. Similarly, if we set $t = (m^2 + 1152)/(8m^2 + 1024)$, or $t = (6 - 3m^2)/(m^2 + 2)$ yields curves with a point of infinite order given by the x-coordinate $x = t - 1/8$, $x = t + 3$ respectively. For a more general construction, setting

$$t = \frac{(2c - 1)m^2 + 2c}{m^2 + 2c^2}$$

then the point with x-coordinate $x = ct$ will be rational and have infinite order.

4.2. A Subfamily of Rank Two

To construct a family with rank (at least) two, we begin by forcing the point with x-coordinate $x = -2t - 2$ to be rational. This requires that the expression $-(t + 4)(2t + 3)$ be square. We can parameterize

$$t = -\frac{4m^2 + 3}{m^2 + 2},$$

to ensure the expression is square. It can be checked that this point has infinite order via specialization. We then try to increase the rank by forcing the point with x-coordinate $x = (t + 4)/5$ to be rational. Some simple algebra shows this condition is equivalent to requiring that $(12m^2 - 1)/2$ is square. We therefore set

$$m = -\frac{k^2 + 8k + 24}{2k^2 - 48}.$$

Substituting back in, we find that

$$t = -16\frac{k^4 + 4k^3 - 8k^2 + 96k + 576}{(9k^2 + 88k + 216)(k^2 - 8k + 24)},$$

with the two points having x-coordinates $x_1(t) = -2t - 2$ and $x_2(t) = (t + 4)/5$.

It is easy to show via specialization that these points are independent and have infinite order. Indeed, for $k = 4$ then $t = -336/89$, with points $P_1 = (494/89, 22230/7921)$ and $P_2 = (4/89, 3740/7921)$. As verified by SAGE [16], both P_1 and P_2 have infinite order (which is easy to see since the torsion group is $\mathbb{Z}_2 \times \mathbb{Z}_2$). The determinant of the height pairing matrix for P_1 and P_2 is $6.94053525377041 \neq 0$, which implies the points are independent. In fact, the curve E_t has rank 2, and P_1 and P_2 can be shown to be generators. We thus have constructed an infinite family with rank (at least) two.

It is possible to construct other families with rank (at least) 2 by using the above technique, with different x-coordinates. We note that we attempted to find a rank 3 subfamily, but were unsuccessful.

5. Examples of Elliptic Curves of High Rank

We searched for curves with high rank, and were able to find some elliptic curves of rank 6 in the family E_t. We use the sieving method based on Mestre-Nagao sums ([9], [11]). Let $E/\mathbb{Q}$ be an elliptic curve, and p be a prime. Set $a_p = a_p(E) = p + 1 - |E(\mathbf{F_p})|$. Given a fixed integer N, the Mestre-Nagao sum is defined by

$$S(N, E) = \sum_{\text{primes } p \leq N} \left(1 - \frac{p-1}{|E(\mathbf{F_p})|}\right) \log(p) = \sum_{\text{primes } p \leq N} \frac{-a_p + 2}{p + 1 - a_p} \log(p).$$

It has been conjectured that in general, larger values of $S(N, E)$ tend to correspond to curves with high rank. Provided N is not too large, $S(N, E)$ can be calculated using SAGE [16].

Searching through the curves E_t with $t = t_1/t_2$ with $-10000 \leq t_1 \leq 10000, 1 \leq t_2 \leq 10000$, we found that for the curves with $S(523, E) > 20$, the rank was frequently 3 or 4. In the above range, we found only one curve with rank 6, which occurred for $t = 2961/3116$. Concretely, the elliptic curve

$$y^2 = x^3 - 74825818x^2 + 1397695377085056x,$$

which has generators:

$$P_1 = (39037824, 4879728000),$$

$$P_2 = (86307147531/2209, 549837626486445/103823),$$

$$P_3 = (139324262094/3481, 2843260991496900/205379),$$

$$P_4 = (1023658496/25, 2586273190144/125),$$

$P_5 = (1364049981504/32041, 185551816648665600/5735339)$,

$P_6 = (45163475, 51198909265)$.

There were several values of $t = t_1/t_2$ which correspond to curves E_t which have rank 5. See Table 1 below.

rank	$t = t_1/t_2$
5	324/965, 1572/1390, 1602/935, 1696/801, 1984/619, 1142/2192, 1360/2043, 1558/1771, 1817/3029, 2135/2614, 2143/3034, 2408/3005, 2578/2761, 2631/2255, 2697/2910, 2817/2712, 3048/2345, 5057/5215, 5156/5522, 5232/5597, 5394/5820, 5634/5424, 5635/5289, 5178/6806, 5331/6499, 2512/1513, 5081/3497, 10177/6255.

Table 1: Rank 5 curves

We similarly searched the other families given in Section 4. For the rank 1 family, we did not find any additional rank 5 or 6 curves. For the family described in Section 4.2, we found curves of rank 5 for $k = -23/12, -15/7$ (corresponding to $t = -115438864/71696361$ and -1345744/838881), and rank 6 for $k = 20, 7/3$ (corresponding to $t = -47824/23001$ and $= -1084624/410601$ respectively).

6. Conclusion

One can easily prove that a similar result is true for any polygonal number. If s is the number of sides in a polygon, the formula for the nth s-gonal number $P(s,n)$ is

$$P(s,n) = (s-2)\frac{n(n-1)}{2} + n.$$

The case $s = 2$ corresponds to Legendre curves, while the case $s = 3$ are the triangular number curves treated in this paper. When $s = 4$, then $P(4,n) = n^2$, with the related curve family $y^2 = x(x-1)(x-n^2)$ closely related to the works mentioned earlier involving Pythagorean triples [7], [8], [12]. It would be interesting to further examine the elliptic curves that arise from other polygonal numbers.

Acknowledgement. The authors would like to thank the referee for carefully reading the paper and for his/her helpful comments. The first author sincerely thanks Prof. Andrew Bremner and Pradeep Das for some fruitful discussion and Prof. Ajai Choudhry for his help in getting the parametrization used to construct the first subfamily of rank 1. The first author also thanks the Harish-Chandra Research Institute, Allahabad, for providing research facilities to pursue his research work and is indebted to his co-supervisor Prof. Kalyan Chakraborty for his continuous support and encouragement.

References

[1] J. S. Chahal and J. Top, Triangular numbers and elliptic curves, *Rocky Mountain J. Math.*, **26** (3), (1996), 937–949.

[2] A. Dujella, High rank elliptic curves with prescribed torsion, `http://web.math.hr/~duje/tors/tors.html` (accessed January 2018).

[3] A. Dujella, Infinite families of elliptic curves with high rank and prescribed torsion, `https://web.math.pmf.unizg.hr/~duje/tors/generic.html` (accessed January 2018).

[4] A. Dujella, A. S. Janfada, S. Salami, A search for high rank congruent number elliptic curves, *J. Integer Seq.*, **12** (2009), Article 09.5.8.

[5] F. Izadi, F. Khoshnam, D. Moody, Heron quadrilaterals via elliptic curves, *Rocky Mountain J. Math.*, **47** (4), (2017) 1227-1258.

[6] F. Izadi, F. Khoshnam, D. Moody, A. S. Zargar, Elliptic curves arising from Brahmagupta quadrilaterals, *Bull. Aust. Math. Soc.* **90** (01) (2014) 47-56.

[7] Farzali Izadi and Kamran Nabardi, Elliptic curves and Pythagorean triples, *Eur. J. Pure Appl. Math.*, **7** (2), (2014), 131-139.

[8] F. A. Izadi, K. Nabardi, F. Khoshnam, A new family of elliptic curves with positive rank arising from Pythagorean triples, *ArXiv e-prints (2010), arXiv:1012.5837v4.*

[9] J.F. Mestre, Construction de courbes elliptiques sur $\mathbb{Q}$ de rang ≥ 12, *C. R. Acad. Sci. Paris Ser. I*, **295** (1982), 643–644.

[10] R. Miranda, "*An overview of algebraic surfaces, in: Algebraic Geometry* (Ankara,1995)", Lecture Notes in Pure and Appl. Math. 193, Dekker, New York, 1997, 197–217.

[11] K. Nagao, An example of elliptic curve over $\mathbb{Q}$ with rank ≥ 20, *Proc. Japan Acad. Ser. A Math. Sci.* **69** (1993), 291–293.

[12] B. Naskręcki, Mordell-weil ranks of families of elliptic curves associated to Pythagorean triples, *Acta Arith.* **160**: (2013), 159–183.

[13] J. Park, B. Poonen, J. Voight, M. M. Wood, A heuristic for boundedness of ranks of elliptic curves, *ArXiv e-prints (2017), arXiv:1602.01431v2.*

[14] N. F. Rogers, Rank computations for the congruent number elliptic curves, *Exp. Math.*, **9** (2000), no. 4, 591-594.

[15] K. Rubin, and A. Silverberg, Rank frequencies for quadratic twists of elliptic curves, *Exp. Math.* **10** (2001), no. 4, 559-569.

[16] SAGE software, Version 4.5.3 `http://www.SAGEmath.org`.

[17] T. Shioda, On the Mordell-Weil lattices, *Comment. Math. Univ. St. Pauli*, **39** (1990), 211–240.

[18] T. Shioda, M. Schütt, Elliptic surfaces. Algebraic geometry in East Asia-Seoul 2008, 51–160, *Adv. Stud. Pure Math.*, **60**, *Math. Soc. Japan, Tokyo*, 2010.

[19] J. H. Silverman, *The Arithmetic of Elliptic Curves*, Springer-Verlag, New York, (1986).

[20] J. H. Silverman, *Advanced Topics in the Arithmetic of Elliptic Curves, Graduate Texts in Mathematics*, **151** Springer-Verlag, New York (1994).

[21] P. Tadić, On the family of elliptic curves $Y^2 = X^3 - T^2X + 1$, *Glas. Mat. Ser. III* **47** (67) (2012), 81–93.

11

FINITE SEARCHES, CHOWLA'S COSINE PROBLEM, AND LARGE NEWMAN POLYNOMIALS

Idris Mercer
Department of Mathematics and Statistics, Florida International University, Miami, Florida
imercer@fiu.edu

Abstract

A length n cosine sum is an expression of the form $\cos a_1\theta + \cdots + \cos a_n\theta$ where $a_1 < \cdots < a_n$ are positive integers, and a length n Newman polynomial is an expression of the form $z^{a_1} + \cdots + z^{a_n}$ where $a_1 < \cdots < a_n$ are nonnegative integers. We define $-\lambda(n)$ to be the largest minimum of a length n cosine sum as $\{a_1, \ldots, a_n\}$ ranges over all sets of n positive integers, and we define $\mu(n)$ to be the largest minimum modulus on the unit circle of a length n Newman polynomial as $\{a_1, \ldots, a_n\}$ ranges over all sets of n nonnegative integers. Since there are infinitely many possibilities for the a_j, it is not obvious how to compute $\lambda(n)$ or $\mu(n)$ for a given n in finitely many steps. Campbell et al. found the value of $\mu(3)$ in 1983, and Goddard found the value of $\mu(4)$ in 1992. In this paper, we find the values of $\lambda(2)$ and $\lambda(3)$ and nontrivial bounds on $\mu(5)$. We also include further remarks on the seemingly difficult general task of reducing the computation of $\lambda(n)$ or $\mu(n)$ to a finite problem.

1. Introduction

We define a *length n cosine sum* to be any expression of the form

$$\cos a_1\theta + \cdots + \cos a_n\theta$$

where $a_1 < \cdots < a_n$ are positive integers. If $f(\theta)$ is a length n cosine sum, then $f(0) = n$ and $f(\theta) \geq -n$ for all θ. For $f(\theta) = \cos a_1\theta + \cdots + \cos a_n\theta$, we define

$$L(a_1, \ldots, a_n) = \min_{\theta} f(\theta) = \min_{0 \leq \theta \leq \pi} f(\theta)$$

which is negative, since $f(\theta)$ has average value 0 on $[0, \pi]$. More specifically, we have

$$-n \leq L(a_1, \ldots, a_n) < 0.$$

We then define

$$-\lambda(n) = \sup L(a_1, \ldots, a_n)$$

where $\{a_1, \ldots, a_n\}$ ranges over all of the infinitely many sets of n positive integers. Then $0 \leq \lambda(n) \leq n$. There is also a construction showing $\lambda(n) = O(\sqrt{n})$ (see Section 2).

We define a *length n Newman polynomial* to be any expression of the form

$$z^{a_1} + \cdots + z^{a_n}$$

where $a_1 < \cdots < a_n$ are nonnegative integers. If $F(z)$ is a length n Newman polynomial and $|z| = 1$, then $0 \leq |F(z)| \leq n$. We define

$$M(a_1, \ldots, a_n) = \min_{|z|=1} |z^{a_1} + \cdots + z^{a_n}|$$

so we have

$$0 \leq M(a_1, \ldots, a_n) \leq n,$$

and in fact one can show $M(a_1, \ldots, a_n) \leq \sqrt{n}$ by considering the L^2 norm of F. (We note that the case $n = 2$ is uninteresting because $z^{a_1} + z^{a_2} = z^{a_1}(1 + z^{a_2 - a_1})$ always has minimum modulus 0 at the $(a_2 - a_1)$th roots of -1.) We then define

$$\mu(n) = \sup M(a_1, \ldots, a_n)$$

where $\{a_1, \ldots, a_n\}$ ranges over all of the infinitely many sets of n nonnegative integers.

Then $\lambda(n)$ and $\mu(n)$ are well-defined functions of n because we are taking the supremum of a bounded set. However, since there are infinitely many possibilities for the a_j, it is not obvious how to compute $\lambda(n)$ or $\mu(n)$ for a given n in finitely many steps. Also, since $\lambda(n)$ and $\mu(n)$ are defined as suprema of infinite sets, it is not obvious if $\lambda(n)$ or $\mu(n)$ are always equal to $-L(a_1, \ldots, a_n)$ or $M(a_1, \ldots, a_n)$ for a specific choice of $a_1, \ldots, a_n$. In any case, some brute-force exploration of specific small examples leads to conjectures such as

$$\begin{aligned}
\lambda(2) &= -L(1, 2) = 9/8 = 1.125, \\
\lambda(3) &= -L(1, 2, 3) \approx 1.315565, \\
\lambda(4) &= -L(1, 2, 3, 4) \approx 1.519558, \\
\lambda(5) &= -L(1, 2, 4, 5, 6) \approx 1.627461, \\
\lambda(6) &= -L(1, 2, 4, 6, 7, 8) \approx 1.591832,
\end{aligned}$$

and

$$\begin{aligned}
\mu(3) &= M(0, 1, 3) \approx 0.607346, \\
\mu(4) &= M(0, 1, 2, 4) \approx 0.752394, \\
\mu(5) &= M(0, 1, 2, 6, 9) = 1.
\end{aligned}$$

In 1983, Campbell et al. [2] proved that $\mu(3) = M(0,1,3)$, and in 1992, Goddard [6] proved that $\mu(4) = M(0,1,2,4)$. To prove results of that type, one must reduce a potentially infinite search to a finite search. The main results of this paper are:

Theorem 1.1. *We have*

$$\lambda(2) = -L(1,2) = 9/8 = 1.125,$$

$$\lambda(3) = -L(1,2,3) = \frac{17+7\sqrt{7}}{27} \approx 1.315565.$$

Theorem 1.2. *We have* $1 \leq \mu(5) \leq 1 + \pi/6 \approx 1.5236$.

Theorem 1.1 is proved in Section 4 and Theorem 1.2 is proved in Section 6.

Various authors have investigated the growth rates of the functions $\lambda(n)$ and $\mu(n)$. Around the late 1940s, Ankeny and Chowla conjectured [3, 4] that $\lambda(n)$ approaches infinity with n. This was first proved by Uchiyama and Uchiyama [9] using results of Cohen [5]; their lower bound for $\lambda(n)$ was sublogarithmic. Over the years, better lower bounds for $\lambda(n)$ have been found. The best lower bound currently known is due to Ruzsa [8]; it is superlogarithmic but grows more slowly than any power of n. The best known upper bound for $\lambda(n)$ appears to be $O(\sqrt{n})$. Chowla conjectured [4] that this is the true rate of growth.

It seems that the growth of $\mu(n)$ is less studied than the growth of $\lambda(n)$. There is a construction showing $\mu(n)$ exceeds a power of n for infinitely many n; specifically, we have $\mu(n) \geq n^{0.14}$ when n is a power of 9 (See Section 2). Boyd [1] conjectured that $\mu(n) > 1$ for all $n \geq 6$ and that $\log \mu(n)/\log n$ approaches a positive constant as n approaches infinity. The current author [7] proved that $\mu(n) > 0$ for all $n > 2$. It appears that nobody has proved that $\mu(n)$ approaches infinity with n.

It would be interesting to show that for any n, the value of $\lambda(n)$ or $\mu(n)$ can be computed in a finite number of steps (even an impractically large finite number) and it would be interesting to know if $\lambda(n)$ and $\mu(n)$ are always equal to $-L(a_1,\ldots,a_n)$ or $M(a_1,\ldots,a_n)$ for a specific choice of the a_j. This appears to be difficult to prove.

2. Constructions Bounding $\lambda(n)$ and $\mu(n)$

There are some straightforward constructions that lead to bounds on $\lambda(n)$ and $\mu(n)$. These probably count as mathematical folklore. We include them here for completeness.

Lemma 2.1. *For each positive integer n, there exists a length n cosine sum $f(\theta)$ that satisfies $f(\theta) \geq -\sqrt{2n} - \frac{1}{2}$ for all θ. It follows that $\lambda(n) \leq \sqrt{2n} + \frac{1}{2}$.*

Proof. Let $k = \lceil \sqrt{2n} \rceil$, so $\sqrt{2n} \leq k < \sqrt{2n} + 1$. Let $\{b_1 < \ldots < b_k\}$ be a set of k nonnegative integers such that the $\binom{k}{2}$ positive differences $b_j - b_i$ are all distinct;

for instance, we can take $b_j = 2^{j-1}$. (A set with this property is sometimes called a 'Sidon set'; in this proof, we do not need our Sidon set to be optimal in any sense.) Then if $z = e^{i\theta}$, we have

$$\begin{aligned}0 \le \left|z^{b_1} + \cdots + z^{b_k}\right|^2 &= (z^{b_1} + \cdots + z^{b_k})(z^{-b_1} + \cdots + z^{-b_k}) \\ &= k + \sum_{i<j} 2\mathrm{Re}(z^{b_j - b_i}) = k + 2\sum_{i<j} \cos(b_j - b_i)\theta.\end{aligned}$$

Thus, if we define $g(\theta) = \sum_{i<j} \cos(b_j - b_i)\theta$, then $g(\theta)$ is a length $\binom{k}{2}$ cosine sum satisfying $g(\theta) \ge -k/2$ for all θ. Now observe

$$n - \sqrt{\frac{n}{2}} = \frac{\sqrt{2n}(\sqrt{2n}-1)}{2} \le \frac{k(k-1)}{2} < \frac{(\sqrt{2n}+1)\sqrt{2n}}{2} = n + \sqrt{\frac{n}{2}}.$$

If $\binom{k}{2} > n$, choose $h(\theta)$ to be a sum consisting of $\binom{k}{2} - n$ of the cosines in the sum $g(\theta)$, and choose $f(\theta) = g(\theta) - h(\theta)$. If $\binom{k}{2} < n$, choose $h(\theta)$ to be a sum of $n - \binom{k}{2}$ cosines not appearing in the sum $g(\theta)$, and choose $f(\theta) = g(\theta) + h(\theta)$. If $\binom{k}{2} = n$, define $h(\theta) = 0$ and choose $f(\theta) = g(\theta)$.

Then $h(\theta)$ is a sum of at most $\sqrt{n/2}$ cosines, so we have $\pm h(\theta) \ge -\sqrt{n/2}$. We then have

$$f(\theta) = g(\theta) \pm h(\theta) \ge -\frac{k}{2} - \sqrt{\frac{n}{2}} > -\frac{\sqrt{2n}+1}{2} - \sqrt{\frac{n}{2}} = -\sqrt{2n} - \frac{1}{2}$$

which completes the proof. □

Lemma 2.2. *Let F be a length n Newman polynomial satisfying $|F(z)| \ge K_1$ for all $|z| = 1$, and let G be a length m Newman polynomial satisfying $|G(z)| \ge K_2$ for all $|z| = 1$. Then there exists a length nm Newman polynomial H satisfying $|H(z)| \ge K_1 K_2$ for all $|z| = 1$.*

Proof. Suppose

$$\begin{aligned}F(z) &= z^{a_1} + \cdots + z^{a_n}, \\ G(z) &= z^{b_1} + \cdots + z^{b_m}.\end{aligned}$$

If k is a sufficiently large positive integer, then the product

$$H(z) = F(z^k)G(z) = (z^{ka_1} + \cdots + z^{ka_n})(z^{b_1} + \cdots + z^{b_m})$$

has the property that the nm exponents $ka_i + b_j$ are all distinct, and hence $H(z)$ is a Newman polynomial of length nm. If $|z| = 1$, then $|H(z)| = \left|F(z^k)\right| |G(x)| \ge K_1 K_2$. This completes the proof. □

As mentioned in [1], the length 9 Newman polynomial

$$F(z) = 1 + z + z^2 + z^3 + z^4 + z^7 + z^8 + z^{10} + z^{12}$$

has unusually high minimum modulus on the unit circle. Specifically, $|F(z)| \geq 1.362$ for all z on the unit circle. Then, by repeated application of Lemma 2.2, we can construct for each k a Newman polynomial $H(z)$ of length 9^k with the property that $|H(z)| \geq 1.362^k$ for all z on the unit circle. Since $1.362 > 9^{0.14}$, this means that for infinitely many values of n, we have a length n Newman polynomial $H(z)$ that satisfies $|H(z)| \geq n^{0.14}$ on the unit circle.

For example, there is a Newman polynomial of length $9^3 = 729$ that satisfies $|H(z)| \geq 1.362^3 \approx 2.53$ on the unit circle, so $\mu(729) \geq 2.53$. It would be interesting to know (for example) the least n such that $\mu(n) \geq 2$.

3. Some Notation and Lemmas

In this section, we establish some notation and some useful lemmas.

Let $f(\theta) = \cos a_1\theta + \cdots + \cos a_n\theta$, let $g = \gcd(a_1, \ldots, a_n)$, and let $a'_j = a_j/g$. Then $\cos a'_1\theta + \cdots + \cos a'_n\theta$ is a length n cosine sum taking on the same values (and hence having the same minimum) as $f(\theta)$. Therefore, in the definition of $\lambda(n)$, it suffices to consider only those $\{a_1, \ldots, a_n\}$ for which $\gcd(a_1, \ldots, a_n) = 1$.

Let $F(z) = z^{a_1} + \cdots + z^{a_n}$, where $a_1 < \cdots < a_n$ are nonnegative integers. Define $b_j = a_j - a_1$. If $|z| = 1$, then

$$F_1(z) = F(z)/z^{a_1} = 1 + z^{b_2} + \cdots + z^{b_n}$$

has the same modulus as $F(z)$. Next, define $g = \gcd(b_2, \ldots, b_n)$ and $b'_j = b_j/g$. If

$$F_2(z) = 1 + z^{b'_2} + \cdots + z^{b'_n}$$

then $F_2(z^g) = F_1(z)$, so $F_1(z)$ and $F_2(z)$ have the same set of outputs as z ranges over the unit circle. Therefore, in the definition of $\mu(n)$, it suffices to consider only those $\{a_1, \ldots, a_n\}$ of the form $\{0, a_2, \ldots, a_n\}$ where $\gcd(a_2, \ldots, a_n) = 1$. Furthermore, there is one more symmetry we exploit. If

$$F(z) = 1 + z^{a_2} + \cdots + z^{a_{n-1}} + z^{a_n}$$

then we define

$$H(z) = z^{a_n}F(z^{-1}) = z^{a_n} + z^{a_n - a_2} + \cdots + z^{a_n - a_{n-1}} + 1,$$

and note that $|F(z)|$ and $|H(z)|$ have the same set of outputs as z ranges over the unit circle. So we are free to choose between $F(z)$ and $H(z)$ and can hence assume $a_{n-1} \geq a_n - a_2$.

For the above reasons, we make the following definitions. Define

$$\mathbb{N}'_n = \{(a_1, \ldots, a_n) \mid 0 < a_1 < \cdots < a_n,\ \gcd(a_1, \ldots, a_n) = 1\},$$
$$\mathbb{N}''_n = \{(0, a_2, \ldots, a_n) \mid 0 < a_2 < \cdots < a_n,\ \gcd(a_2, \ldots, a_n) = 1,\ a_{n-1} \geq a_n - a_2\}$$

and note that if $L(a_1, \ldots, a_n)$ and $M(a_1, \ldots, a_n)$ are as defined in Section 1, then we have

$$-\lambda(n) = \sup L(a_1, \ldots, a_n)$$

where the supremum is taken over all $(a_1, \ldots, a_n) \in \mathbb{N}'_n$, and we have

$$\mu(n) = \sup M(0, a_2, \ldots, a_n)$$

where the supremum is taken over all $(0, a_2, \ldots, a_n) \in \mathbb{N}''_n$.

We also define $\mathbb{T}$ to be $\mathbb{R}$ mod 2π. Following are some definitions and lemmas regarding subsets of $\mathbb{T}$.

Definition. An *equispaced subset* of $\mathbb{T}$ of *order* m is any subset of $\mathbb{T}$ of the form

$$\Big\{\theta_0 + \frac{2k\pi}{m} \;\Big|\; k \in \mathbb{Z}\Big\}.$$

Note that if we fix $\xi \in \mathbb{T}$, then the set $\{\theta \in \mathbb{T} \mid m\theta = \xi\}$ is an equispaced set of order m.

Lemma 3.1. *Let S_1 and S_2 be equispaced subsets of $\mathbb{T}$ of order m_1 and m_2 respectively, and let $g = \gcd(m_1, m_2)$. Then there exists $\theta_1 \in S_1$ and $\theta_2 \in S_2$ such that $|\theta_1 - \theta_2| \leq \frac{\pi g}{m_1 m_2}$.*

Proof. Suppose

$$S_1 = \Big\{\xi_1 + \frac{2k\pi}{m_1} \;\Big|\; k \in \mathbb{Z}\Big\} \qquad \text{and} \qquad S_2 = \Big\{\xi_2 + \frac{2k\pi}{m_2} \;\Big|\; k \in \mathbb{Z}\Big\}.$$

The real number $\frac{m_1 m_2}{2\pi}(\xi_1 - \xi_2)$ must be within $\frac{g}{2}$ of an integer multiple of g, call it ag. Also, ag can be written in the form $km_1 - \ell m_2$ for some integers k and ℓ. Thus we have

$$\Big|\frac{m_1 m_2}{2\pi}(\xi_1 - \xi_2) - (km_1 - \ell m_2)\Big| \leq \frac{g}{2},$$

which, multiplying by $\frac{2\pi}{m_1 m_2}$ and rearranging, gives us

$$\Big|\Big(\xi_1 + \frac{2\pi\ell}{m_1}\Big) - \Big(\xi_2 + \frac{2\pi k}{m_2}\Big)\Big| \leq \frac{\pi g}{m_1 m_2},$$

completing the proof of the lemma. ◻

The following lemma is straightforward.

Lemma 3.2. *If θ satisfies $|\theta - \pi| = \varepsilon$, then $\cos\theta \le -1 + \frac{1}{2}\varepsilon^2$.*

We will also need other bounds on the cosine function, which need not be the best bounds possible. One can show the following.

Lemma 3.3. *If θ satisfies $\left|\theta - \frac{2\pi}{3}\right| = \varepsilon \le \frac{\pi}{6}$, then*

$$\cos\theta \le -\frac{1}{2} + \frac{3}{\pi}\varepsilon$$

and if θ satisfies $\left|\theta - \frac{4\pi}{3}\right| = \varepsilon \le \frac{\pi}{6}$, then

$$\cos\theta \le -\frac{1}{2} + \frac{3}{\pi}\varepsilon.$$

Definition. If $S \subset \mathbb{T}$ is an equispaced set of order m, and $f(\theta)$ and $g(\theta)$ are real-valued functions on $\mathbb{T}$, then we define

$$\big\langle f(\theta), g(\theta)\big\rangle_S = \frac{1}{m}\sum_{\theta\in S} f(\theta)g(\theta),$$

the average value of $f(\theta)g(\theta)$ over S, which we can think of as a kind of dot product of $f(\theta)$ and $g(\theta)$.

One can verify that this dot product has the following properties:

- $\big\langle f_1(\theta) + f_2(\theta), g(\theta)\big\rangle_S = \big\langle f_1(\theta), g(\theta)\big\rangle_S + \big\langle f_2(\theta), g(\theta)\big\rangle_S$
- $\big\langle f(\theta), g_1(\theta) + g_2(\theta)\big\rangle_S = \big\langle f(\theta), g_1(\theta)\big\rangle_S + \big\langle f(\theta), g_2(\theta)\big\rangle_S$
- $\big\langle 1, 1\big\rangle_S = 1$
- $\big\langle 1, \cos k\theta\big\rangle_S = \big\langle \cos k\theta, 1\big\rangle_S = 0$ if k is *not* a multiple of m
- $\big\langle \cos k\theta, \cos k\theta\big\rangle_S = \frac{1}{2}$ if $2k$ is *not* a multiple of m
- $\big\langle \cos k\theta, \cos \ell\theta\big\rangle_S = 0$ if $k+\ell$ and $k-\ell$ are *not* multiples of m

A function that is nonnegative on $\mathbb{T}$ can be used as a 'weight function'. Some examples of nonnegative weight functions are:

$$1 - \cos k\theta,$$
$$2(1-\cos k\theta)^2 = 3 - 4\cos k\theta + \cos 2k\theta,$$

as well as any sum of such functions.

Lemma 3.4. *Let $w(\theta)$ be a nonnegative weight function on $\mathbb{T}$, let $g(\theta)$ be any real-valued function on $\mathbb{T}$, and let $S \subset \mathbb{T}$ be an equispaced set. If we have $\big\langle w(\theta), g(\theta)\big\rangle_S \le 0$, then $g(\theta) \le 0$ for some $\theta \in \mathbb{T}$.*

4. The Values of $\lambda(2)$ and $\lambda(3)$

To prove $\lambda(2) = -L(1,2) = 9/8$, let $(a,b) \in \mathbb{N}_2'$, so $\gcd(a,b) = 1$. We must show that $\cos a\theta + \cos b\theta \le -9/8$ for some θ.

If $b \le 2$, then $\cos a\theta + \cos b\theta = \cos\theta + \cos 2\theta$, which one can verify has minimum value $-9/8$. So suppose $b \ge 3$. Define

$$S_1 = \{\theta \in \mathbb{T} \mid a\theta = \pi\} \qquad \text{and} \qquad S_2 = \{\theta \in \mathbb{T} \mid b\theta = \pi\},$$

which are equispaced sets of order a and b respectively. By Lemma 3.1, there exist $\theta_1 \in S_1$ and $\theta_2 \in S_2$ such that $|\theta_1 - \theta_2| \le \frac{\pi}{ab}$. Then $\cos b\theta_2 = \pi$ and $|a\theta_2 - \pi| = |a\theta_2 - a\theta_1| \le \frac{\pi}{b}$, so by Lemma 3.2, we have

$$\cos a\theta_2 \le -1 + \frac{\pi^2}{2b^2} \le -1 + \frac{\pi^2}{18} \approx -0.45,$$

implying $\cos a\theta_2 + \cos b\theta_2 \le -2 + \frac{\pi^2}{18} \approx -1.45 < -9/8$. This completes the evaluation of $\lambda(2)$.

Next, we will show that $\lambda(3) = -L(1,2,3)$. First, observe that trigonometric identities allow us to write

$$\begin{aligned}\cos\theta + \cos 2\theta + \cos 3\theta &= \cos\theta + (2\cos^2\theta - 1) + (4\cos^3\theta - 3\cos\theta) \\ &= 4\cos^3\theta + 2\cos^2\theta - 2\cos\theta - 1\end{aligned}$$

which is a polynomial in $\cos\theta$ of degree 3. One can verify that

$$\min_{-1 \le c \le 1} 4c^3 + 2c^2 - 2c - 1 = -\frac{17 + 7\sqrt{7}}{27} \approx -1.315565.$$

That is, $-L(1,2,3) = \frac{17+7\sqrt{7}}{27} \approx 1.315565$. For brevity, let $K = -L(1,2,3)$.

To prove that $\lambda(3) = K$, we will partition $\mathbb{N}_3'$ into four subsets:

$$\begin{aligned}M_1 &= \{(a,b,c) \in \mathbb{N}_3' \mid c = 2a\}, \\ M_2 &= \{(a,b,c) \in \mathbb{N}_3' \mid c = 2b\}, \\ M_3 &= \{(a,b,c) \in \mathbb{N}_3' \mid c = a + b\}, \\ M_0 &= \big\{(a,b,c) \in \mathbb{N}_3' \mid c \notin \{2a, 2b, a+b\}\big\}.\end{aligned}$$

Speaking very informally, we can think of M_1, M_2, M_3 as subsets of $\mathbb{N}_3'$ that have only two 'degrees of freedom'. For each $j \in \{0,1,2,3\}$, we will show that with finitely many exceptions, if $(a,b,c) \in M_j$ then $\cos a\theta + \cos b\theta + \cos c\theta \le -K$ for some θ.

Suppose $(a,b,c) \in M_0$. We then have $0 < a < b < c$, $\gcd(a,b,c) = 1$, and $c \notin \{2a, 2b, a+b\}$. Let $S = \{\theta \in \mathbb{T} \mid c\theta = \pi\}$, which is an equispaced set of order c. Note that $2 - \cos a\theta - \cos b\theta$ is a nonnegative weight function, and consider

$$\delta = \left\langle 2 - \cos a\theta - \cos b\theta,\ \frac{1}{2} + \cos a\theta + \cos b\theta \right\rangle_S.$$

Now observe that none of a, b, or $b-a$ is a multiple of c, and that any of $2a$, $2b$, or $a+b$ that is a multiple of c must equal c. It then follows from properties of the dot product that

$$\delta = \langle 2, \frac{1}{2} \rangle_S - \langle \cos a\theta, \cos a\theta \rangle_S - \langle \cos b\theta, \cos b\theta \rangle_S = 1 - \frac{1}{2} - \frac{1}{2} = 0$$

and then by Lemma 3.4, we conclude that $\frac{1}{2} + \cos a\theta + \cos b\theta \leq 0$ for some $\theta \in S$. But that θ satisfies $\cos c\theta = -1$ and therefore $\cos a\theta + \cos b\theta + \cos c\theta \leq -3/2 < -K$.

Next, suppose $(a, b, c) \in M_1$. If $a \leq 2$ then (a, b, c) must be $(2, 3, 4)$, and we observe that $\cos 2\theta + \cos 3\theta + \cos 4\theta = -2 < -K$ if $\theta = \pi/3$. We therefore assume $a \geq 3$. Now define

$$S_1 = \left\{\theta \in \mathbb{T} \mid a\theta = \frac{2\pi}{3}\right\}, \qquad S_2 = \{\theta \in \mathbb{T} \mid b\theta = \pi\},$$

which are equispaced sets of order a and b respectively. Note that $\gcd(a, b) = 1$ since any common divisor would divide $2a = c$. By Lemma 3.1, there exist $\theta_1 \in S_1$ and $\theta_2 \in S_2$ such that $|\theta_1 - \theta_2| \leq \frac{\pi}{ab}$. Then $\cos a\theta_1 = \cos 2\pi/3 = -1/2$ and $\cos c\theta_1 = \cos 2a\theta_1 = \cos 4\pi/3 = -1/2$. Also, we have $|b\theta_1 - \pi| = |b\theta_1 - b\theta_2| \leq \frac{\pi}{a}$, so by Lemma 3.2, we have

$$\cos b\theta_1 \leq -1 + \frac{\pi^2}{2a^2} \leq -1 + \frac{\pi^2}{18} \approx -0.45,$$

implying $\cos a\theta_1 + \cos b\theta_1 + \cos c\theta_1 \leq -2 + \frac{\pi^2}{18} \approx -1.45 < -K$.

Next, suppose $(a, b, c) \in M_2$. This case is very similar to $(a, b, c) \in M_1$. If $b \leq 2$ then (a, b, c) must be $(1, 2, 4)$, and we observe that $\cos\theta + \cos 2\theta + \cos 4\theta = -3/2 < -K$ if $\theta = 2\pi/3$. We therefore assume $b \geq 3$. We define

$$S_1 = \{\theta \in \mathbb{T} \mid a\theta = \pi\}, \qquad S_2 = \left\{\theta \in \mathbb{T} \mid b\theta = \frac{2\pi}{3}\right\},$$

which are equispaced sets of order a and b, and we note that $\gcd(a, b) = 1$, so there exist $\theta_1 \in S_1$ and $\theta_2 \in S_2$ such that $|\theta_1 - \theta_2| \leq \frac{\pi}{ab}$. Then $\cos b\theta_2 = \cos 2\pi/3 = -1/2$ and $\cos c\theta_2 = \cos 2b\theta_2 = \cos 4\pi/3 = -1/2$. Also, we have $|a\theta_2 - \pi| = |a\theta_2 - a\theta_1| \leq \frac{\pi}{b}$, so Lemma 3.2 gives us

$$\cos a\theta_2 \leq -1 + \frac{\pi^2}{2b^2} \leq -1 + \frac{\pi^2}{18} \approx -0.45,$$

implying $\cos a\theta_2 + \cos b\theta_2 + \cos c\theta_2 \leq -2 + \frac{\pi^2}{18} \approx -1.45 < -K$.

Finally, suppose $(a, b, c) \in M_3$. We will show the existence of a finite subset $M' \subset M_3$ such that if $(a, b, c) \in M_3 \setminus M'$, then $\cos a\theta + \cos b\theta + \cos c\theta \leq -K$ for some θ. That will reduce this case to checking the finitely many elements of M'. It will suffice to choose

$$M' = \{(a, b, c) \in M_3 \mid b \leq 32\} = \{(a, b, a+b) \in M_3 \mid 1 \leq a < b \leq 32\}$$

which has at most $\binom{32}{2} = 496$ elements. At the end of this section, we will give an alternative argument that avoids such a large finite set.

So suppose $(a, b, c) \in M_3 \setminus M'$, so $c = a + b$ and $b \geq 33$. We choose

$$S_1 = \Big\{\theta \in \mathbb{T} \mid a\theta = \frac{2\pi}{3}\Big\}, \qquad S_2 = \Big\{\theta \in \mathbb{T} \mid b\theta = \frac{2\pi}{3}\Big\},$$

which are equispaced sets of order a and b respectively. Note that $\gcd(a, b) = 1$ since any common divisor would divide $a+b = c$. By Lemma 3.1, there exist $\theta_1 \in S_1$ and $\theta_2 \in S_2$ such that $|\theta_1 - \theta_2| \leq \frac{\pi}{ab}$. Then $\cos b\theta_2 = \cos 2\pi/3 = -1/2$. Also, we have $|a\theta_2 - 2\pi/3| = |a\theta_2 - a\theta_1| \leq \frac{\pi}{b}$ and

$$\left|c\theta_2 - \frac{4\pi}{3}\right| = \left|\left(a\theta_2 - \frac{2\pi}{3}\right) + \left(b\theta_2 - \frac{2\pi}{3}\right)\right| = \left|\left(a\theta_2 - \frac{2\pi}{3}\right) + 0\right| \leq \frac{\pi}{b}.$$

By Lemma 3.3, we then have

$$\cos a\theta_2 \leq -\frac{1}{2} + \frac{3}{\pi}\frac{\pi}{b} \qquad \text{and} \qquad \cos c\theta_2 \leq -\frac{1}{2} + \frac{3}{\pi}\frac{\pi}{b}$$

which implies

$$\cos a\theta_2 + \cos b\theta_2 + \cos c\theta_2 \leq -\frac{3}{2} + \frac{6}{b} \leq -\frac{3}{2} + \frac{6}{33} \approx -1.318 < -K.$$

We remark that the case $(a, b, c) \in M_3$ can be dealt with more easily if we take as known the fact that

$$\mu(3) = M(0, 1, 3) = \sqrt{\frac{47 - 14\sqrt{7}}{27}} \approx 0.607346,$$

which was shown in Section 3 of [2]. Note that this result says that given any degree 3 Newman polynomial $F(z) = 1 + z^k + z^\ell$, we have $|F(z)| \leq M(0, 1, 3)$ for some $|z| = 1$. We apply this to the case $F(z) = 1 + z^a + z^{a+b}$. The statement

$$\left|1 + z^a + z^{a+b}\right| \leq \sqrt{\frac{47 - 14\sqrt{7}}{27}} \qquad \text{for some } |z| = 1$$

is equivalent to

$$\left|1 + z^a + z^{a+b}\right|^2 \leq \frac{47 - 14\sqrt{7}}{27} \qquad \text{for some } |z| = 1$$

but we also have, if $z = e^{i\theta}$,

$$\begin{aligned}
\left|1 + z^a + z^{a+b}\right|^2 &= (1 + z^a + z^{a+b})(1 + z^{-a} + z^{-a-b}) \\
&= 3 + (z^a + z^{-a}) + (z^b + z^{-b}) + (z^{a+b} + z^{-a-b}) \\
&= 3 + 2\big(\cos a\theta + \cos b\theta + \cos(a + b)\theta\big).
\end{aligned}$$

The condition

$$3 + 2\big(\cos a\theta + \cos b\theta + \cos(a+b)\theta\big) \leq \frac{47 - 14\sqrt{7}}{27}$$

is equivalent to

$$\cos a\theta + \cos b\theta + \cos(a+b)\theta \leq \frac{1}{2}\Big(\frac{47 - 14\sqrt{7}}{27} - 3\Big) = -\frac{17 + 7\sqrt{7}}{27} = L(1,2,3).$$

In [2], the fact that $\mu(3) = M(0,1,3)$ is proved by reducing to a finite number of cases, but their number of cases is much less than 496. Since we used the value of $\mu(3)$ just in the case where $(a,b,c) \in M_3$, it could be said that our evaluation of $\lambda(3)$ is a generalization of the evaluation of $\mu(3)$.

5. A Possible Strategy for $\lambda(4)$

The current author is unaware of how to evaluate $\lambda(4)$, but includes in this section a possible outline of a strategy where we reduce that problem to a finite set of problems that, speaking informally, have fewer 'degrees of freedom'.

We conjecture that $\lambda(4) = -L(1,2,3,4) \approx 1.519558$. To prove this, we must show that if $(a,b,c,d) \in \mathbb{N}_4'$, then $\cos a\theta + \cos b\theta + \cos c\theta + \cos d\theta \leq L(1,2,3,4)$ for some θ.

Let $(a,b,c,d) \in \mathbb{N}_4'$, and define $S = \{\theta \in \mathbb{T} \mid d\theta = \pi\}$, which is an equispaced set of order d. Define the nonnegative weight function

$$\begin{aligned} w(\theta) &= (1 - \cos a\theta) + (1 - \cos b\theta) + 2(1 - \cos c\theta)^2 \\ &= 1 - \cos a\theta + 1 - \cos b\theta + 3 - 4\cos c\theta + \cos 2c\theta \\ &= 5 - \cos a\theta - \cos b\theta - 4\cos c\theta + \cos 2c\theta. \end{aligned}$$

If we can show that

$$\Big\langle w(\theta),\ \frac{3}{5} + \cos a\theta + \cos b\theta + \cos c\theta \Big\rangle_S \leq 0,$$

then it will follow that $\cos a\theta + \cos b\theta + \cos c\theta \leq -3/5$ for some θ satisfying $\cos d\theta = -1$, so then $\cos a\theta + \cos b\theta + \cos c\theta + \cos d\theta \leq -8/5 = -1.6 < L(1,2,3,4)$.

Using previously stated properties of the dot product, we note that evaluating

$$\Big\langle 5 - \cos a\theta - \cos b\theta - 4\cos c\theta + \cos 2c\theta,\ \frac{3}{5} + \cos a\theta + \cos b\theta + \cos c\theta \Big\rangle_S$$

depends on properties of the set $\{0,a,b,c,2c\} \pm \{0,a,b,c\}$. This is illustrated in Table 1. If all the nonzero numbers in the body of Table 1 are nonmultiples of d,

	0	a	b	c	$2c$
0	0	a	b	c	$2c$
a	a	$2a$	$b \pm a$	$c \pm a$	$2c \pm a$
b	b	$b \pm a$	$2b$	$c \pm b$	$2c \pm b$
c	c	$c \pm a$	$c \pm b$	$2c$	$2c \pm c$

Table 1: Elements of $\{0, a, b, c, 2c\} \pm \{0, a, b, c\}$

then properties of the dot product give us

$$\left\langle 5 - \cos a\theta - \cos b\theta - 4\cos c\theta + \cos 2c\theta,\ \frac{3}{5} + \cos a\theta + \cos b\theta + \cos c\theta \right\rangle_S$$
$$= 5 \cdot \frac{3}{5} - \frac{1}{2} - \frac{1}{2} - 4 \cdot \frac{1}{2} = 0$$

and so in that case, the desired conclusion follows. We now observe:

- The positive numbers a, b, c, $b-a$, $c-a$, $c-b$ are all less than d, and are hence nonmultiples of d.
- The positive numbers $2a$, $2b$, $2c$, $b+a$, $c+a$, $c+b$, $2c-a$, $2c-b$ are all less than $2d$, and hence if any of them are multiples of d, they are equal to d.
- The positive numbers $2c+a$, $2c+b$, $2c+c$ are all less than $3d$, and hence if any of them are multiples of d, they are equal to d or $2d$.

It follows that if $(a, b, c, d) \in \mathbb{N}_4'$ has the property that d is *not* equal to any of these 14 linear combinations of a, b, c

$$2a, 2b, 2c, b+a, c+a, c+b, 2c-a, 2c-b, 2c+a, c+\frac{1}{2}a, 2c+b, c+\frac{1}{2}b, 3c, \frac{3}{2}c$$

then $\cos a\theta + \cos b\theta + \cos c\theta + \cos d\theta \le -8/5 < L(1, 2, 3, 4)$ for some θ. It then remains to deal with those (a, b, c, d) where d is equal to one of the 14 combinations listed above.

Enumerate those 14 linear combinations as $\varphi_1(a, b, c), \ldots, \varphi_{14}(a, b, c)$, and then for $1 \le j \le 14$, define

$$M_j = \{(a, b, c, d) \in \mathbb{N}_4' \mid d = \varphi_j(a, b, c)\}.$$

Then each M_j, speaking very informally, is a subset of $\mathbb{N}_4'$ having only 3 'degrees of freedom'. If we can prove for each j that if $(a, b, c, d) \in M_j$, then $\cos a\theta + \cos b\theta + \cos c\theta + \cos d\theta \le L(1, 2, 3, 4)$ for some θ, then that would complete the evaluation of $\lambda(4)$. Perhaps it is possible to reduce each M_j to a finite collection of subsets of $\mathbb{N}_4'$ that have 2 degrees of freedom.

6. Bounds on $\mu(5)$

It was observed in [6] that

$$M(0,1,2,6,9) = \min_{|z|=1} \left|1 + z + z^2 + z^6 + z^9\right| = 1$$

and it is suspected that $\mu(5) = 1$, although this appears to be difficult to prove. In this section, we give a short elementary argument showing that $\mu(5) \le \sqrt{3} \approx 1.732$, and a longer case analysis showing that $\mu(5) \le 1 + \pi/5 \approx 1.628$. More generally, we will show that for any positive integer m, the problem of showing $\mu(5) \le 1 + \pi/m$ can be reduced to checking a finite number of cases. In particular, we will sketch a proof that $\mu(5) \le 1 + \pi/6 \approx 1.524$.

To show $\mu(5) \le K$, we must show that for all of the infinitely many $(0, a, b, c, d)$ in $\mathbb{N}_5''$, we have $\left|1 + z^a + z^b + z^c + z^d\right| \le K$ for some $|z| = 1$.

Let $(0, a, b, c, d) \in \mathbb{N}_5''$. To show $\left|1 + z^a + z^b + z^c + z^d\right| \le \sqrt{3}$ for some $|z| = 1$, define $S = \{\theta \in \mathbb{T} \mid d\theta = \pi\}$, and let $z = e^{i\theta}$. Observe that we have

$$\begin{aligned}
\left|z^a + z^b + z^c\right|^2 &= (z^a + z^b + z^c)(z^{-a} + z^{-b} + z^{-c}) \\
&= 3 + (z^{b-a} + z^{a-b}) + (z^{c-a} + z^{a-c}) + (z^{c-b} + z^{b-c}) \\
&= 3 + \cos(b-a)\theta + \cos(c-a)\theta + \cos(c-b)\theta.
\end{aligned}$$

Now note that $b - a$, $c - a$, $c - b$ are positive integers less than d, so they are not multiples of d. It follows that the average of $\left|z^a + z^b + z^c\right|^2$ over S is 3. Therefore at least one $\theta \in S$ satisfies $\left|z^a + z^b + z^c\right|^2 \le 3$ and hence $\left|1 + z^a + z^b + z^c + z^d\right| = \left|z^a + z^b + z^c\right| \le \sqrt{3}$.

To get better bounds on $\mu(5)$, we will use some straightforward lemmas similar to those in Section 3.

Lemma 6.1. *If θ satisfies $|\theta - \pi| \le \delta$, then $\left|1 + e^{i\theta}\right| \le \delta$.*

Lemma 6.2. *Let k, ℓ, m be distinct positive integers with $\ell < m$, and let $g = \gcd(k, m - \ell)$. We then have*

$$\left|1 + z^k + z^\ell + z^m\right| \le \frac{\pi g}{k}$$

for some z on the unit circle.

Proof. Define the sets

$$\begin{aligned}
S_1 &= \{\theta \in \mathbb{T} \mid k\theta = \pi\}, \\
S_2 &= \{\theta \in \mathbb{T} \mid (m - \ell)\theta = \pi\},
\end{aligned}$$

which are equispaced sets of order k and $m-\ell$ respectively. By Lemma 3.1, there exist $\theta_1 \in S_1$ and $\theta_2 \in S_2$ such that $|\theta_1 - \theta_2| \le \frac{\pi g}{k(m-\ell)}$, and hence

$$|(m-\ell)\theta_1 - \pi| = |(m-\ell)\theta_1 - (m-\ell)\theta_2| \le \frac{\pi g}{k}.$$

Then, if $z = e^{i\theta_1}$, we have

$$\begin{aligned}\left|1+z^k+z^\ell+z^m\right| &= \left|1+z^k+z^\ell(1+z^{m-\ell})\right| = \left|1+e^{ik\theta_1}+z^\ell(1+z^{m-\ell})\right| \\ &= \left|1-1+z^\ell(1+z^{m-\ell})\right| = \left|z^\ell(1+z^{m-\ell})\right| = \left|1+z^{m-\ell}\right|\end{aligned}$$

which, by Lemma 6.1, is at most $\pi g/k$. □

Now to prove $\mu(5) \le 1+\pi/5$, let $(0,a,b,c,d) \in \mathbb{N}_5''$, and define $\alpha(z) = 1+z^a+z^b+z^c+z^d$. We must show that

$$|\alpha(z)| \le 1+\frac{\pi}{5}$$

for some z on the unit circle. From the definition of $\mathbb{N}_5''$, we have not only $0<a<b<c<d$, but also $c \ge d-a$, which implies $c > d-b$. We thus have the following inequalities, where both sides are positive integers:

$$\begin{aligned} d &> b-a \\ d &> c-b \\ d &> c-a \\ c &> b-a \\ c &> d-b \\ c &\ge d-a \end{aligned}$$

We now define

$$\begin{aligned} g_1 &= \gcd(d, b-a) && \text{so } g_1 \le b-a < d \\ g_2 &= \gcd(d, c-b) && \text{so } g_2 \le c-b < d \\ g_3 &= \gcd(d, c-a) && \text{so } g_3 \le c-a < d \\ g_4 &= \gcd(c, b-a) && \text{so } g_4 \le b-a < c \\ g_5 &= \gcd(c, d-b) && \text{so } g_5 \le d-b < c \\ g_6 &= \gcd(c, d-a) && \text{so } g_6 \le d-a \le c. \end{aligned}$$

Applying Lemma 6.2 to the case $(k,\ell,m) = (d,a,b)$, we have

$$\begin{aligned}|\alpha(z)| &\le |z^c| + \left|1+z^d+z^a+z^b\right| \\ &= 1+\left|1+z^d+z^a+z^b\right| \le 1+\frac{\pi g_1}{d} \qquad \text{for some } |z|=1.\end{aligned}$$

Similarly, applying Lemma 6.2 to each of the cases

$$\begin{aligned}(k,\ell,m)&=(d,b,c)\\(k,\ell,m)&=(d,a,c)\\(k,\ell,m)&=(c,a,b)\\(k,\ell,m)&=(c,b,d)\\(k,\ell,m)&=(c,a,d),\end{aligned}$$

we can conclude that we have

$$\begin{aligned}|\alpha(z)|&\le 1+\frac{\pi g_2}{d} &&\text{for some } |z|=1\\|\alpha(z)|&\le 1+\frac{\pi g_3}{d} &&\text{for some } |z|=1\\|\alpha(z)|&\le 1+\frac{\pi g_4}{c} &&\text{for some } |z|=1\\|\alpha(z)|&\le 1+\frac{\pi g_5}{c} &&\text{for some } |z|=1\\|\alpha(z)|&\le 1+\frac{\pi g_6}{c} &&\text{for some } |z|=1.\end{aligned}$$

If any of the six numbers

$$\frac{g_1}{d},\ \frac{g_2}{d},\ \frac{g_3}{d},\ \frac{g_4}{c},\ \frac{g_5}{c},\ \frac{g_6}{c} \tag{1}$$

is less than or equal to $1/5$, we are done. So for the remainder of the proof, we assume each of those six numbers is greater than $1/5$.

Note that d and $b-a$ are integer multiples of g_1, with $d > b-a$. The condition $g_1/d > 1/5$ is equivalent to $d < 5g_1$, which implies that $d \in \{2g_1, 3g_1, 4g_1\}$. Then since $b-a$ is an integer multiple of g_1 smaller than d, one of the following must be true:

$$\begin{aligned}d=2g_1 \text{ and } b-a=g_1 &\implies \frac{1}{2}d=b-a\\d=3g_1 \text{ and } b-a=g_1 &\implies \frac{1}{3}d=b-a\\d=3g_1 \text{ and } b-a=2g_1 &\implies \frac{2}{3}d=b-a\\d=4g_1 \text{ and } b-a=1g_1 &\implies \frac{1}{4}d=b-a\\d=4g_1 \text{ and } b-a=3g_1 &\implies \frac{3}{4}d=b-a\end{aligned}$$

So we have $rd = b - a$, where

$$r \in \left\{\frac{1}{4}, \frac{1}{3}, \frac{1}{2}, \frac{2}{3}, \frac{3}{4}\right\}.$$

We will now apply similar reasoning to the other five expressions in (1). Note that the following hold:

$$\begin{aligned}
&d \text{ and } c-b \text{ are integer multiples of } g_2 \text{ with } d > c-b, \text{ and } d < 5g_2\\
&d \text{ and } c-a \text{ are integer multiples of } g_3 \text{ with } d > c-a, \text{ and } d < 5g_3\\
&c \text{ and } b-a \text{ are integer multiples of } g_4 \text{ with } c > b-a, \text{ and } c < 5g_4\\
&c \text{ and } d-b \text{ are integer multiples of } g_5 \text{ with } c > d-b, \text{ and } c < 5g_5\\
&c \text{ and } d-a \text{ are integer multiples of } g_6 \text{ with } c \geq d-a, \text{ and } c < 5g_6.
\end{aligned}$$

Notice the slight difference in the last line. This means that our list of possibilities for c and $d-a$ will include $c = d - a = g_6$.

Using similar reasoning as before, we can conclude:

$$\begin{aligned}
rd &= b-a, \text{ where } r \in \{\tfrac{1}{4}, \tfrac{1}{3}, \tfrac{1}{2}, \tfrac{2}{3}, \tfrac{3}{4}\}\\
sd &= c-b, \text{ where } s \in \{\tfrac{1}{4}, \tfrac{1}{3}, \tfrac{1}{2}, \tfrac{2}{3}, \tfrac{3}{4}\}\\
td &= c-a, \text{ where } t \in \{\tfrac{1}{4}, \tfrac{1}{3}, \tfrac{1}{2}, \tfrac{2}{3}, \tfrac{3}{4}\}\\
uc &= b-a, \text{ where } u \in \{\tfrac{1}{4}, \tfrac{1}{3}, \tfrac{1}{2}, \tfrac{2}{3}, \tfrac{3}{4}\}\\
vc &= d-b, \text{ where } v \in \{\tfrac{1}{4}, \tfrac{1}{3}, \tfrac{1}{2}, \tfrac{2}{3}, \tfrac{3}{4}\}\\
wc &= d-a, \text{ where } w \in \{\tfrac{1}{4}, \tfrac{1}{3}, \tfrac{1}{2}, \tfrac{2}{3}, \tfrac{3}{4}, 1\}.
\end{aligned}$$

Thus there are finitely many possibilities for r, s, t, u, v, w. Notice that this remains true if we change our goal from showing $\mu(5) \leq 1 + \pi/5$ to showing $\mu(5) \leq 1 + \pi/m$. In that case, the possible values for r, s, t, u, v, w are fractions between 0 and 1 whose denominators are less than m.

We claim that for each of the finitely many possible triples r, s, u, there is at most one element of $\mathbb{N}_5''$ satisfying the three conditions

$$\begin{aligned}
rd &= b-a\\
sd &= c-b\\
uc &= b-a.
\end{aligned}$$

To verify this claim, we write these conditions as

$$\begin{aligned}
1a - 1b + 0c + rd &= 0\\
0a + 1b - 1c + sd &= 0\\
1a - 1b + uc + 0d &= 0,
\end{aligned}$$

which we temporarily regard as a system of equations in *rational* unknowns a, b, c, d. Writing as a matrix and row-reducing, we have

$$\begin{bmatrix} 1 & -1 & 0 & r \\ 0 & 1 & -1 & s \\ 1 & -1 & u & 0 \end{bmatrix} \to \begin{bmatrix} 1 & -1 & 0 & r \\ 0 & 1 & -1 & s \\ 0 & 0 & u & -r \end{bmatrix} \to \begin{bmatrix} 1 & -1 & 0 & r \\ 0 & 1 & -1 & s \\ 0 & 0 & 1 & -\frac{r}{u} \end{bmatrix}.$$

Note that u is always nonzero. Continuing, we have

$$\begin{bmatrix} 1 & -1 & 0 & r \\ 0 & 1 & -1 & s \\ 0 & 0 & 1 & -\frac{r}{u} \end{bmatrix} \to \begin{bmatrix} 1 & -1 & 0 & r \\ 0 & 1 & 0 & s-\frac{r}{u} \\ 0 & 0 & 1 & -\frac{r}{u} \end{bmatrix} \to \begin{bmatrix} 1 & 0 & 0 & r+s-\frac{r}{u} \\ 0 & 1 & 0 & s-\frac{r}{u} \\ 0 & 0 & 1 & -\frac{r}{u} \end{bmatrix}.$$

It follows that the system has an infinite one-parameter family of *rational* solutions, given by

$$\begin{aligned} a &= \left(\frac{r}{u} - r - s\right)q \\ b &= \left(\frac{r}{u} - s\right)q \\ c &= \left(\frac{r}{u}\right)q \\ d &= q, \end{aligned}$$

where q can be any rational number. In other words, the rational solutions (a, b, c, d) are precisely the rational multiples of the 4-tuple

$$\left(\frac{r}{u} - r - s,\ \frac{r}{u} - s,\ \frac{r}{u},\ 1\right).$$

However, at most one rational multiple of a given rational 4-tuple can consist of relatively prime nonnegative integers. That is, there is at most one possible $(0, a, b, c, d) \in \mathbb{N}_5''$ for each choice of r, s, u.

For the specific case of showing $\mu(5) \leq 1 + \pi/5$, it is possible to enumerate all the possibilities by hand. Recall that we have the restrictions

$$\begin{aligned} rd &= b - a, \text{ where } r \in \{\tfrac{1}{4}, \tfrac{1}{3}, \tfrac{1}{2}, \tfrac{2}{3}, \tfrac{3}{4}\} \\ sd &= c - b, \text{ where } s \in \{\tfrac{1}{4}, \tfrac{1}{3}, \tfrac{1}{2}, \tfrac{2}{3}, \tfrac{3}{4}\} \\ td &= c - a, \text{ where } t \in \{\tfrac{1}{4}, \tfrac{1}{3}, \tfrac{1}{2}, \tfrac{2}{3}, \tfrac{3}{4}\} \\ uc &= b - a, \text{ where } u \in \{\tfrac{1}{4}, \tfrac{1}{3}, \tfrac{1}{2}, \tfrac{2}{3}, \tfrac{3}{4}\} \\ vc &= d - b, \text{ where } v \in \{\tfrac{1}{4}, \tfrac{1}{3}, \tfrac{1}{2}, \tfrac{2}{3}, \tfrac{3}{4}\} \\ wc &= d - a, \text{ where } w \in \{\tfrac{1}{4}, \tfrac{1}{3}, \tfrac{1}{2}, \tfrac{2}{3}, \tfrac{3}{4}, 1\}. \end{aligned}$$

Notice that

$$(r + s)d = rd + sd = (b - a) + (c - b) = c - a = td$$

r	s	t
1/4	1/4	1/2
1/4	1/2	3/4
1/3	1/3	2/3
1/2	1/4	3/4

Table 2: Permissible values of r, s, t

u	v	w
1/4	1/4	1/2
1/4	1/2	3/4
1/4	3/4	1
1/3	1/3	2/3
1/3	2/3	1
1/2	1/4	3/4
1/2	1/2	1
2/3	1/3	1
3/4	1/4	1

Table 3: Permissible values of u, v, w

so $r + s = t$, and also notice that

$$(u + v)c = uc + vc = (b - a) + (d - b) = d - a = wc$$

so $u + v = w$. The only permissible values of r, s, t that satisfy this are shown in Table 2, and the only permissible values of u, v, w that satisfy this are shown in Table 3. Next, notice that we have $rd = b - a = uc$, which together with $c < d$ implies that $u > r$. Also notice that we have

$$sd = c - b \text{ implies } b = c - sd$$
$$vc = d - b \text{ implies } b = d - vc$$
$$c - sd = d - vc \text{ implies } c + vc = d + sd \text{ implies } (1 + v)c = (1 + s)d,$$

which together with $c < d$ implies $1 + v > 1 + s$, so $v > s$. The only permissible values of r, s, u, v satisfying both $u > r$ and $v > s$ are shown in Table 4. We now use the conditions $rd = uc$ and $(1 + v)c = (1 + s)d$ to conclude

$$\frac{r}{u} = \frac{c}{d} = \frac{1+s}{1+v}, \text{ implying } r(1+v) = u(1+s)$$

which only some of our possible values of r, s, u, v will satisfy. We see in Table 5 that only one of our possibilities satisfies $r(1 + v) = u(1 + s)$. We conclude that $(r, s, u, v) = (\frac{1}{4}, \frac{1}{4}, \frac{1}{3}, \frac{2}{3})$. We then have

$$\left(\frac{r}{u} - r - s,\ \frac{r}{u} - s,\ \frac{r}{u},\ 1\right) = \left(\frac{1}{4},\ \frac{1}{2},\ \frac{3}{4},\ 1\right),$$

r	s	u	v
1/4	1/4	1/3	1/3
1/4	1/4	1/3	2/3
1/4	1/4	1/2	1/2
1/4	1/4	2/3	1/3
1/4	1/2	1/3	2/3
1/3	1/3	1/2	1/2
1/2	1/4	2/3	1/3

Table 4: Permissible values of r, s, u, v satisfying $u > r$ and $v > s$

r	s	u	v	$1+s$	$1+v$	$r(1+v)$	$u(1+s)$
1/4	1/4	1/3	1/3	5/4	4/3	1/3	5/12
1/4	1/4	1/3	2/3	5/4	5/3	5/12	5/12
1/4	1/4	1/2	1/2	5/4	3/2	3/8	5/8
1/4	1/4	2/3	1/3	5/4	4/3	1/3	5/6
1/4	1/2	1/3	2/3	3/2	5/3	5/12	1/2
1/3	1/3	1/2	1/2	4/3	3/2	1/2	2/3
1/2	1/4	2/3	1/3	5/4	4/3	2/3	5/6

Table 5: Checking whether $r(1+v) = u(1+s)$

implying $(a, b, c, d) = (1, 2, 3, 4)$. But the polynomial $f(z) = 1 + z + z^2 + z^3 + z^4$ certainly satisfies $|f(z)| \le 1 + \pi/5$ for some $|z| = 1$, because it has zeros at the nontrivial fifth roots of unity. This completes the proof that $\mu(5) \le 1 + \pi/5$.

The above argument can be modified to show that $\mu(5) \le 1+\pi/m$ for some other positive integers m, but doing so by hand is cumbersome and computer assistance is helpful. We will give a rough outline of an argument that $\mu(5) \le 1 + \pi/6$.

This time, the argument involves finding r, s, t, u, v, w that satisfy

$$\begin{aligned} rd &= b - a \\ sd &= c - b \\ td &= c - a \\ uc &= b - a \\ vc &= d - b \\ wc &= d - a, \end{aligned}$$

where $0 < r, s, t, u, v < 1$, $0 < w \le 1$, and r, s, t, u, v, w are fractions with denominators strictly less than 6. They must further satisfy

$$r + s = t, \qquad u + v = w, \qquad r < u, \qquad s < v, \qquad r(1+v) = u(1+s).$$

A finite search (aided by computer) reveals that the only eligible values of r, s, u, v

r	s	u	v	$1+s$	$1+v$
1/3	1/3	2/5	3/5	4/3	8/5
1/4	1/4	1/3	2/3	5/4	5/3
1/5	1/5	1/4	1/2	6/5	3/2
1/5	2/5	1/4	3/4	7/5	7/4
2/5	1/5	1/2	1/2	6/5	3/2
3/5	1/5	2/3	1/3	6/5	4/3

Table 6: Eligible r, s, u, v with denominators less than 6

$\frac{r}{u} - r - s$	$\frac{r}{u} - s$	$\frac{r}{u}$	(a, b, c, d)
1/6	3/6	5/6	(1, 3, 5, 6)
1/4	2/4	3/4	(1, 2, 3, 4)
2/5	3/5	4/5	(2, 3, 4, 5)
1/5	2/5	4/5	(1, 2, 4, 5)
1/5	3/5	4/5	(1, 3, 4, 5)
1/10	7/10	9/10	(1, 7, 9, 10)

Table 7: Eligible 4-tuples for verifying $\mu(5) \leq 1 + \pi/6$

are those shown in Table 6. (The values of $1 + s$ and $1 + v$ are included for convenience.) As before, each eligible (r, s, u, v) gives us a 4-tuple

$$\left(\frac{r}{u} - r - s,\ \frac{r}{u} - s,\ \frac{r}{u},\ 1\right)$$

of which precisely one integer multiple is an eligible (a, b, c, d). This gives us six possibilities, shown in Table 7. One can verify that for each of these six possibilities for (a, b, c, d), we have

$$\left|1 + z^a + z^b + z^c + z^d\right| \leq 1 + \pi/6$$

for some $|z| = 1$. This completes the sketch of the proof that $\mu(5) \leq 1 + \pi/6$.

In conclusion, note that the contributions of Campbell et al. and Goddard appeared gradually. The results in this paper can be regarded as extensions of that work. It appears that evaluating $\lambda(n)$ or $\mu(n)$ in finitely many steps is a genuinely subtle problem.

References

[1] D.W. Boyd, Large Newman polynomials, in *Diophantine Analysis (Kensington, 1985)*, 159–170, London Math. Soc. Lecture Note Ser., 109, Cambridge Univ. Press, Cambridge, 1986.

[2] D.M. Campbell, H.R.P. Ferguson and R.W. Forcade, Newman polynomials on $|z| = 1$, *Indiana Univ. Math. J.* **32** (1983), 517–525.

[3] S. Chowla, The Riemann zeta and allied functions, *Bull. Amer. Math. Soc.* **58** (1952), 287–305.

[4] S. Chowla, Some applications of a method of A. Selberg, *J. Reine Angew. Math.* **217** (1965), 128–132.

[5] P.J. Cohen, On a conjecture of Littlewood and idempotent measures, *Amer. J. Math.* **82** (1960), 191–212.

[6] B. Goddard, Finite exponential series and Newman polynomials, *Proc. Amer. Math. Soc.* **116** (1992), 313–320.

[7] I. Mercer, Newman polynomials not vanishing on the unit circle, *Integers* **12** (2012), A67, 7 pp.

[8] I.Z. Ruzsa, Negative values of cosine sums, *Acta Arith.* **111** (2004), 179–186.

[9] M. Uchiyama (née Katayama) and S. Uchiyama, On the cosine problem, *Proc. Japan Acad.* **36** (1960), 475–479.

12

ON A DIOPHANTINE EQUATION WITH THREE PRIME VARIABLES

Min Zhang
School of Applied Science, Beijing Information Science and Technology University, Beijing, P. R. China
min.zhang.math@gmail.com

Jinjiang Li[1]
Department of Mathematics, China University of Mining and Technology, Beijing, P. R. China
jinjiang.li.math@gmail.com

Abstract

Let $[x]$ denote the integral part of the real number x, and N be a sufficiently large integer. In this paper, it is proved that, for $1 < c < \frac{3113}{2703}$, the Diophantine equation $N = [p_1^c] + [p_2^c] + [p_3^c]$ is solvable in prime variables p_1, p_2, p_3, which constitutes an improvement upon that of Cai.

1. Introduction and Main Result

Let $[x]$ be the integral part of the real number x. In 1933–1934, Segal [13, 14] first considered the Waring's problem with non–integer degrees, who showed that for any sufficiently large integer N and $c > 1$ being not an integer, there exists a integer $k_0 = k_0(c) > 0$ such that the equation

$$N = [x_1^c] + [x_2^c] + \cdots + [x_k^c]$$

is solvable for $k \geqslant k_0(c)$. Later, Segal's bound for $k_0(c)$ was improved by Deshouillers [4] and by Arkhilov and Zhitkov [1], respectively. Let $G(c)$ be the least of the integers $k_0(c)$ such that every sufficiently large integer N can be written as a sum of not more than $k_0(c)$ numbers with the form $[n^c]$. In particular, Deshouillers [5] and Gritsenko [8] considered the case $k = 2$ and gave $G(c) = 2$ for $1 < c < 4/3$ and $1 < c < 55/41$, respectively.

In 1937, Vinogradov [15] solved asymptotic form of the ternary Goldbach problem. He proved that, for sufficiently large integer N satisfying $N \equiv 1 \pmod 2$,

[1]Corresponding author.

the equation $N = p_1 + p_2 + p_3$ is solvable in primes p_1, p_2, p_3. As an analogue of the ternary Goldbach problem, in 1995, Laporta and Tolev [12] investigated the solvability of the equation $N = [p_1^c] + [p_2^c] + [p_3^c]$ in prime variables p_1, p_2, p_3. Define

$$\mathcal{R}_3(N) = \sum_{N=[p_1^c]+[p_2^c]+[p_3^c]} (\log p_1)(\log p_2)(\log p_3).$$

Laporta and Tolev [12] showed that the sum $\mathcal{R}_3(N)$ has an asymptotic formula for $1 < c < 17/16$, and gave

$$\mathcal{R}_3(N) = \frac{\Gamma^3(1+1/c)}{\Gamma(3/c)} N^{3/c-1} + O\Big(N^{3/c-1} \exp\big(-(\log N)^{1/3-\delta}\big)\Big)$$

for any $0 < \delta < 1/3$. Later, Kumchev and Nedeva [11] improved the result of Laporta and Tolev [12], and enlarged the range of c to $12/11$. Afterwards, Zhai and Cao [17] refined the result of Kumchev and Nedeva [11], who extended the range of c to $258/235$. In 2018, Cai [3] enhanced the result of Zhai and Cao [17] and gave the upper bound of c as $137/119$.

In this paper, we shall continue to improve the result of Cai [3] and establish the following result.

Theorem 1. *Let $1 < c < \frac{3113}{2703}$, and N be a sufficiently large integer. Then we have*

$$\mathcal{R}_3(N) = \frac{\Gamma^3(1+1/c)}{\Gamma(3/c)} N^{3/c-1} + O\Big(N^{3/c-1} \exp\big(-(\log N)^{1/3-\delta}\big)\Big)$$

for any $0 < \delta < 1/3$, where the implied constant in the O-term depends only on c.

Notation. Throughout this paper, we suppose that $1 < c < \frac{3113}{2703}$. Let p, with or without subscripts, always denote a prime number; and let ε always denote arbitrary small positive constant, which may not be the same at different occurrences. As usual, we use $[x]$, $\{x\}$ and $\|x\|$ to denote the integral part of x, the fractional part of x and the distance from x to the nearest integer, respectively. Also, we write $e(x) = e^{2\pi i x}$; $f(x) \ll g(x)$ means that $f(x) = O(g(x))$; and $f(x) \asymp g(x)$ means that $f(x) \ll g(x) \ll f(x)$.

We also define

$$P = N^{1/c}, \qquad \tau = P^{1-c-\varepsilon}, \qquad S(\alpha) = \sum_{p \leqslant P} (\log p) e\big([p^c]\alpha\big),$$

$$S(\alpha, X) = \sum_{X < p \leqslant 2X} (\log p) e\big([p^c]\alpha\big), \qquad \mathcal{T}(\alpha, X) = \sum_{X < n \leqslant 2X} e\big([n^c]\alpha\big).$$

2. Preliminary Lemmas

In this section, we shall state some preliminary lemmas, which are required in the proof of Theorem 1.

Lemma 1. *Let $L, Q \geqslant 1$ and z_ℓ be complex numbers. Then we have*

$$\left|\sum_{L<\ell\leqslant 2L} z_\ell\right|^2 \leqslant \left(2+\frac{L}{Q}\right)\sum_{|q|<Q}\left(1-\frac{|q|}{Q}\right)\sum_{L<\ell+q,\ell-q\leqslant 2L} z_{\ell+q}\overline{z_{\ell-q}}.$$

Proof. See Lemma 2 of Fouvry and Iwaniec [6]. □

Lemma 2. *Suppose that $f(x):[a,b]\to\mathbb{R}$ has continuous derivatives of arbitrary order on $[a,b]$, where $1\leqslant a<b\leqslant 2a$. Suppose further that*

$$\left|f^{(j)}(x)\right| \asymp \lambda_1 a^{1-j}, \qquad j\geqslant 1, \qquad x\in[a,b].$$

Then for any exponential pair (κ,λ), we have

$$\sum_{a<n\leqslant b} e(f(n)) \ll \lambda_1^\kappa a^\lambda + \lambda_1^{-1}.$$

Proof. See (3.3.4) of Graham and Kolesnik [7]. □

Lemma 3. *Let x be a non-integer, $\alpha\in(0,1)$, and $H\geqslant 3$. Then we have*

$$e(-\alpha\{x\}) = \sum_{|h|\leqslant H} c_h(\alpha)e(hx) + O\left(\min\left(1,\frac{1}{H\|x\|}\right)\right),$$

where

$$c_h(\alpha) = \frac{1-e(-\alpha)}{2\pi i(h+\alpha)}.$$

Proof. See Lemma 12 of Buriev [2] or Lemma 3 of Kumchev and Nedeva [11]. □

Lemma 4. *Let $3<U<V<Z<X$ and suppose that $Z-\frac{1}{2}\in\mathbb{N}$, $X\geqslant 64Z^2U$, $Z\geqslant 4U^2$, $V^3\geqslant 32X$. Assume further that $F(n)$ is a complex-valued function such that $|F(n)|\leqslant 1$. Then the sum*

$$\sum_{X<n\leqslant 2X}\Lambda(n)F(n)$$

may be decomposed into $O(\log^{10}X)$ sums, each of which either of Type I:

$$\sum_{M<m\leqslant 2M} a(m)\sum_{K<k\leqslant 2K} F(mk)$$

with $K\gg Z$, where $a(m)\ll m^\varepsilon$, $MK\asymp X$, or of Type II:

$$\sum_{M<m\leqslant 2M} a(m)\sum_{K<k\leqslant 2K} b(k)F(mk)$$

with $U\ll M\ll V$, where $a(m)\ll m^\varepsilon$, $b(k)\ll k^\varepsilon$, $MK\asymp X$.

Proof. See Lemma 3 of Heath–Brown [9]. □

Lemma 5. *For any $\varepsilon > 0$, the pair $\left(\frac{32}{205}+\varepsilon, \frac{269}{410}+\varepsilon\right)$ is an exponential pair.*

Proof. See the Corollary of Theorem 1 of Huxley [10]. □

Lemma 6. *Let $\alpha, \beta \in \mathbb{R}$ with $\alpha\beta(\alpha-1)(\beta-1) \neq 0$. Define the bilinear sums of Type I as follows*

$$S_I(M,K) := \sum_{M<m\leqslant 2M} \sum_{k\in\mathcal{I}(m)} a(m) e\left(F\frac{m^\alpha k^\beta}{M^\alpha K^\beta}\right),$$

where $F > 0, M \geqslant 1, K \geqslant 1, |a(m)| \ll 1$, and $\mathcal{I}(m)$ is a subinterval of $(K, 2K]$. Then for any exponential pair (κ, λ), we have

$$S_I(M,K) \ll \left((F^{1+2\kappa}M^{4+4\kappa}K^{3+2\lambda})^{\frac{1}{6+4\kappa}} + M^{\frac{1}{2}}K + MK^{\frac{1}{2}} + F^{-1}MK\right)\log(2+FMK).$$

Proof. See Theorem 2 of Wu [16]. □

Lemma 7. *For any real number θ, there holds*

$$\min\left(1, \frac{1}{H\|\theta\|}\right) = \sum_{h=-\infty}^{+\infty} a_h e(h\theta),$$

where

$$a_h \ll \min\left(\frac{\log 2H}{H}, \frac{1}{|h|}, \frac{H}{h^2}\right).$$

Proof. See p.245 of Heath–Brown [9]. □

Lemma 8. *Let $1 < c < \frac{3113}{2703}, P^{\frac{5}{6}} \ll X \ll P, H = X^{\frac{2047}{27030}}$ and $c_h(\alpha)$ denote complex numbers such that $|c_h(\alpha)| \ll (1+|h|)^{-1}$. Then, for any $\alpha \in (\tau, 1-\tau)$, if $M \ll X^{\frac{7781}{13515}}$, we have*

$$\mathcal{S}_I(\alpha) := \sum_{|h|\leqslant H} c_h(\alpha) \sum_{M<m\leqslant 2M} a(m) \sum_{K<k\leqslant 2K} e\big((h+\alpha)(mk)^c\big) \ll X^{\frac{24983}{27030}+\varepsilon},$$

where $a(m) \ll m^\varepsilon$ and $MK \asymp X$.

Proof. Obviously, we have

$$\big|\mathcal{S}_I(\alpha)\big| \ll X^\varepsilon \max_{|\xi|\in(\tau,H+1)} \sum_{M<m\leqslant 2M} \left|\sum_{K<k\leqslant 2K} e\big(\xi(mk)^c\big)\right|. \tag{1}$$

If $M \ll X^{\frac{3959}{9010}}$, then we use Lemma 2 to estimate the inner sum over k in (1) with the exponential pair $(\kappa,\lambda) = AB(0,1) = (\frac{1}{6},\frac{2}{3})$ and derive that

$$\begin{aligned}\mathcal{S}_I(\alpha) &\ll X^{\varepsilon} \max_{|\xi|\in(\tau,H+1)} \sum_{M<m\leqslant 2M} \left((|\xi|X^c K^{-1})^{\frac{1}{6}} K^{\frac{2}{3}} + \frac{K}{|\xi|X^c} \right) \\ &\ll X^{\varepsilon} \max_{|\xi|\in(\tau,H+1)} \left(|\xi|^{\frac{1}{6}} X^{\frac{c}{6}} K^{\frac{1}{2}} M + \frac{MK}{|\xi|X^c} \right) \\ &\ll X^{\varepsilon} \big(H^{\frac{1}{6}} X^{\frac{c}{6}+\frac{1}{2}} M^{\frac{1}{2}} + X^{1-c}\tau^{-1} \big) \ll X^{\frac{24983}{27030}+\varepsilon}.\end{aligned}$$

If $X^{\frac{3959}{9010}} \ll M \ll X^{\frac{7781}{13515}}$, we use Lemma 6 to estimate the inner sum over k in (1) with the exponential pair $AB(0,1) = (\frac{1}{6},\frac{2}{3})$ and obtain that

$$\begin{aligned}\mathcal{S}_I(\alpha) &\ll X^{\varepsilon} \max_{|\xi|\in(\tau,H+1)} \left((|\xi|X^c)^{\frac{1}{5}} M^{\frac{7}{10}} K^{\frac{13}{20}} + M^{\frac{1}{2}} K + MK^{\frac{1}{2}} + |\xi|^{-1} X^{-c} MK \right) \\ &\ll X^{\varepsilon} \left(H^{\frac{1}{5}} M^{\frac{1}{20}} X^{\frac{c}{5}+\frac{13}{20}} + XM^{-\frac{1}{2}} + X^{\frac{1}{2}} M^{\frac{1}{2}} + X^{1-c}\tau^{-1} \right) \ll X^{\frac{24983}{27030}+\varepsilon},\end{aligned}$$

which completes the proof of Lemma 8. □

Lemma 9. *Let $1 < c < \frac{3113}{2703}$, $P^{\frac{5}{6}} \ll X \ll P$, $H = X^{\frac{2047}{27030}}$ and $c_h(\alpha)$ denote complex numbers such that $|c_h(\alpha)| \ll (1+|h|)^{-1}$. Then, for any $\alpha \in (\tau, 1-\tau)$, if there holds $X^{\frac{2047}{13515}} \ll M \ll X^{\frac{355}{901}}$, then we have*

$$\mathcal{S}_{II}(\alpha) := \sum_{|h|\leqslant H} c_h(\alpha) \sum_{M<m\leqslant 2M} a(m) \sum_{K<k\leqslant 2K} b(k) e\big((h+\alpha)(mk)^c\big) \ll X^{\frac{24983}{27030}+\varepsilon},$$

where $a(m) \ll m^{\varepsilon}$, $b(k) \ll k^{\varepsilon}$ and $MK \asymp X$.

Proof. Let $Q = X^{\frac{2047}{13515}}(\log X)^{-1}$. From Lemma 1 and Cauchy's inequality, we derive that

$$
\begin{aligned}
|S_{II}(\alpha)| &\ll \sum_{|h|\leqslant H} |c_h(\alpha)| \left| \sum_{K<k\leqslant 2K} b(k) \sum_{M<m\leqslant 2M} a(m)e\big((h+\alpha)(mk)^c\big)\right| \\
&\ll \sum_{|h|\leqslant H} |c_h(\alpha)| \left(\sum_{K<k\leqslant 2K} |b(k)|^2\right)^{\frac{1}{2}} \left(\sum_{K<k\leqslant 2K} \left| \sum_{M<m\leqslant 2M} a(m)e\big((h+\alpha)(mk)^c\big)\right|^2\right)^{\frac{1}{2}} \\
&\ll K^{\frac{1}{2}+\varepsilon} \sum_{|h|\leqslant H} |c_h(\alpha)| \left(\sum_{K<k\leqslant 2K} \frac{M}{Q} \sum_{0\leqslant q<Q} \left(1-\frac{q}{Q}\right)\right. \\
&\qquad \left. \times \sum_{M+q<m\leqslant 2M-q} a(m+q)\overline{a(m-q)}e\big((h+\alpha)k^c\Delta_c(m,q)\big)\right)^{\frac{1}{2}} \\
&\ll K^{\frac{1}{2}+\varepsilon} \sum_{|h|\leqslant H} |c_h(\alpha)| \left(\frac{M}{Q} \sum_{K<k\leqslant 2K} \left(M^{1+\varepsilon} + \sum_{1\leqslant q<Q} \left(1-\frac{q}{Q}\right)\right.\right. \\
&\qquad \left.\left. \times \sum_{M+q<m\leqslant 2M-q} a(m+q)\overline{a(m-q)}e\big((h+\alpha)k^c\Delta_c(m,q)\big)\right)\right)^{\frac{1}{2}} \\
&\ll X^{\varepsilon} \sum_{|h|\leqslant H} |c_h(\alpha)| \left(\frac{X^2}{Q} + \frac{X}{Q} \sum_{1\leqslant q<Q} \sum_{M<m\leqslant 2M} \left| \sum_{K<k\leqslant 2K} e\big((h+\alpha)k^c\Delta_c(m,q)\big)\right|\right)^{\frac{1}{2}},
\end{aligned} \tag{2}
$$

where $\Delta_c(m,q) = (m+q)^c - (m-q)^c$. Thus, it is sufficient to estimate the sum

$$S_0 := \sum_{K<k\leqslant 2K} e\big((h+\alpha)k^c\Delta_c(m,q)\big).$$

By Lemma 2 with the exponential pair $(\kappa,\lambda) = BA^2B(0,1) = (\frac{2}{7},\frac{4}{7})$, we have

$$S_0 \ll \big(|h+\alpha|X^{c-1}q\big)^{\frac{2}{7}} K^{\frac{4}{7}} + \frac{1}{|h+\alpha|X^{c-1}q}.$$

Putting the above estimate into (2), we obtain that

$$\begin{aligned}
&\mathcal{S}_{II}(\alpha)\\
&\ll X^{\varepsilon}\sum_{|h|\leqslant H}|c_h(\alpha)|\Bigg(\frac{X^2}{Q}+\frac{X}{Q}\sum_{1\leqslant q<Q}\sum_{M<m\leqslant 2M}\\
&\qquad\qquad\times\bigg(\big(|h+\alpha|X^{c-1}q\big)^{\frac{2}{7}}K^{\frac{4}{7}}+\frac{1}{|h+\alpha|X^{c-1}q}\bigg)\Bigg)^{\frac{1}{2}}\\
&\ll X^{\varepsilon}\sum_{|h|\leqslant H}|c_h(\alpha)|\bigg(\frac{X^2}{Q}+\frac{X}{Q}\Big(H^{\frac{2}{7}}X^{\frac{2}{7}(c-1)}MK^{\frac{4}{7}}Q^{\frac{9}{7}}+X^{1-c}M\tau^{-1}\log Q\Big)\bigg)^{\frac{1}{2}}\\
&\ll X^{1+\varepsilon}Q^{-\frac{1}{2}}\sum_{|h|\leqslant H}|c_h(\alpha)|\ll X^{1+\varepsilon}Q^{-\frac{1}{2}}\sum_{|h|\leqslant H}\frac{1}{1+|h|}\ll X^{\frac{24983}{27030}+\varepsilon},
\end{aligned}$$

which completes the proof of Lemma 9. □

Lemma 10. *For $\alpha\in(\tau,1-\tau)$, there holds*

$$S(\alpha)\ll P^{\frac{24983}{27030}+\varepsilon}.$$

Proof. First, we have

$$S(\alpha)=\mathcal{U}(\alpha)+O(P^{1/2}),$$

where

$$\mathcal{U}(\alpha)=\sum_{n\leqslant P}\Lambda(n)e([n^c]\alpha).$$

By a splitting argument, it is sufficient to prove that, for $P^{5/6}\ll X\ll P$ and $\alpha\in(\tau,1-\tau)$, there holds

$$\mathcal{U}^*(\alpha):=\sum_{X<n\leqslant 2X}\Lambda(n)e([n^c]\alpha)\ll X^{\frac{24983}{27030}+\varepsilon}.$$

By Lemma 3 with $H=X^{\frac{2047}{27030}}$, we have

$$\begin{aligned}
\mathcal{U}^*(\alpha)&=\sum_{X<n\leqslant 2X}\Lambda(n)e\big(n^c\alpha-\{n^c\}\alpha\big)=\sum_{X<n\leqslant 2X}\Lambda(n)e\big(n^c\alpha\big)e\big(-\{n^c\}\alpha\big)\\
&=\sum_{X<n\leqslant 2X}\Lambda(n)e\big(n^c\alpha\big)\Bigg(\sum_{|h|\leqslant H}c_h(\alpha)e(hn^c)+O\bigg(\min\Big(1,\frac{1}{H\|n^c\|}\Big)\bigg)\Bigg)\\
&=\sum_{|h|\leqslant H}c_h(\alpha)\sum_{X<n\leqslant 2X}\Lambda(n)e\big((h+\alpha)n^c\big)+O\Bigg(\log X\cdot\sum_{X<n\leqslant 2X}\min\Big(1,\frac{1}{H\|n^c\|}\Big)\Bigg).
\end{aligned}\tag{3}$$

By Lemma 7 and Lemma 2 with the exponential pair $(\kappa, \lambda) = B(0,1) = (\frac{1}{2}, \frac{1}{2})$, we derive that

$$\begin{aligned}
&\sum_{X<n\leqslant 2X} \min\left(1, \frac{1}{H\|n^c\|}\right) \\
&= \sum_{X<n\leqslant 2X} \sum_{\ell=-\infty}^{+\infty} a_\ell e(\ell n^c) \ll \sum_{\ell=-\infty}^{+\infty} |a_\ell| \left| \sum_{X<n\leqslant 2X} e(\ell n^c) \right| \\
&\ll \frac{X \log 2H}{H} + \sum_{1\leqslant \ell \leqslant H} \frac{1}{\ell} \left| \sum_{X<n\leqslant 2X} e(\ell n^c) \right| + \sum_{\ell > H} \frac{H}{\ell^2} \left| \sum_{X<n\leqslant 2X} e(\ell n^c) \right| \\
&\ll \frac{X \log 2H}{H} + \sum_{1\leqslant \ell \leqslant H} \frac{1}{\ell} \left((X^{c-1}\ell)^{\frac{1}{2}} X^{\frac{1}{2}} + \frac{1}{\ell X^{c-1}} \right) \\
&\qquad + \sum_{\ell > H} \frac{H}{\ell^2} \left((X^{c-1}\ell)^{\frac{1}{2}} X^{\frac{1}{2}} + \frac{1}{\ell X^{c-1}} \right) \\
&\ll X^{\frac{24983}{27030}} \log X + H^{\frac{1}{2}} X^{\frac{c}{2}} + X^{1-c} \ll X^{\frac{24983}{27030}} \log X. \qquad (4)
\end{aligned}$$

Taking $U = X^{\frac{2047}{13515}}, V = X^{\frac{355}{901}}$, and $Z = [X^{\frac{5734}{13515}}] + \frac{1}{2}$ in Lemma 4, it is easy to see that the sum

$$\sum_{|h|\leqslant H} c_h(\alpha) \sum_{X<n\leqslant 2X} \Lambda(n) e\big((h+\alpha)n^c\big)$$

can be represented as $O(\log^{10} X)$ sums, each of which either of Type I

$$\mathcal{S}_I(\alpha) = \sum_{|h|\leqslant H} c_h(\alpha) \sum_{M<m\leqslant 2M} a(m) \sum_{K<k\leqslant 2K} e\big((h+\alpha)(mk)^c\big)$$

with $K \gg Z, a(m) \ll m^\varepsilon, MK \asymp X$, or of Type II

$$\mathcal{S}_{II}(\alpha) = \sum_{|h|\leqslant H} c_h(\alpha) \sum_{M<m\leqslant 2M} a(m) \sum_{K<k\leqslant 2K} b(k) e\big((h+\alpha)(mk)^c\big)$$

with $U \ll M \ll V, a(m) \ll m^\varepsilon, b(k) \ll k^\varepsilon, MK \asymp X$. For the Type I sums, by noting the fact that $K \gg Z$ and $MK \asymp X$, we deduce that $M \ll X^{\frac{7781}{13515}}$. From Lemma 8, we have $\mathcal{S}_I(\alpha) \ll X^{\frac{24983}{27030}+\varepsilon}$. For the Type II sums, by Lemma 9, we have $\mathcal{S}_{II}(\alpha) \ll X^{\frac{24983}{27030}+\varepsilon}$. Therefore, we conclude that

$$\sum_{|h|\leqslant H} c_h(\alpha) \sum_{X<n\leqslant 2X} \Lambda(n) e\big((h+\alpha)n^c\big) \ll X^{\frac{24983}{27030}+\varepsilon}. \qquad (5)$$

From (3)–(5), we derive the desired result. This completes the proof of Lemma 10. □

Lemma 11. *For $\alpha \in (0,1)$, we have*

$$\mathcal{T}(\alpha, X) \ll X^{\frac{269c+538}{1217}+\varepsilon} + \frac{1}{\alpha X^{c-1}}.$$

Proof. Taking $H_1 = X^{\frac{679-269c}{1217}}$, and by Lemma 3, we deduce that

$$\begin{aligned}
\mathcal{T}(\alpha, X) &= \sum_{X<n\leqslant 2X} e\big((n^c - \{n^c\})\alpha\big) \\
&= \sum_{X<n\leqslant 2X} e(n^c\alpha)\left(\sum_{|h|\leqslant H_1} c_h(\alpha) e(hn^c) + O\left(\min\left(1, \frac{1}{H_1\|n^c\|}\right)\right)\right) \\
&= \sum_{|h|\leqslant H_1} c_h(\alpha) \sum_{X<n\leqslant 2X} e((h+\alpha)n^c) + O\left(\sum_{X<n\leqslant 2X} \min\left(1, \frac{1}{H_1\|n^c\|}\right)\right). \qquad (6)
\end{aligned}$$

From Lemma 7, we get

$$\begin{aligned}
\sum_{X<n\leqslant 2X} \min\left(1, \frac{1}{H_1\|n^c\|}\right) &= \sum_{X<n\leqslant 2X} \sum_{k=-\infty}^{+\infty} a_k e(kn^c) \\
&\ll \sum_{k=-\infty}^{+\infty} |a_k| \left|\sum_{X<n\leqslant 2X} e(kn^c)\right|. \qquad (7)
\end{aligned}$$

According to Lemma 5, the pair $BA(\frac{32}{205}+\varepsilon, \frac{269}{410}+\varepsilon) = (\frac{269}{948}+\varepsilon, \frac{269}{474}+\varepsilon)$ is an exponential pair. Then we shall use Lemma 2 with the exponential pair $(\frac{269}{948}+\varepsilon, \frac{269}{474}+\varepsilon)$ to estimate the sum over n on the right–hand side in (7), and derive that

$$\begin{aligned}
&\sum_{X<n\leqslant 2X} \min\left(1, \frac{1}{H_1\|n^c\|}\right) \\
&\ll \frac{X\log 2H_1}{H_1} + \sum_{1\leqslant k\leqslant H_1} \frac{1}{k}\left|\sum_{X<n\leqslant 2X} e(kn^c)\right| + \sum_{k>H_1} \frac{H_1}{k^2}\left|\sum_{X<n\leqslant 2X} e(kn^c)\right| \\
&\ll \frac{X\log 2H_1}{H_1} + \sum_{1\leqslant k\leqslant H_1} \frac{1}{k}\left((X^{c-1}k)^{\frac{269}{948}+\varepsilon} X^{\frac{269}{474}+\varepsilon} + \frac{1}{kX^{c-1}}\right) \\
&\qquad + \sum_{k>H_1} \frac{H_1}{k^2}\left((X^{c-1}k)^{\frac{269}{948}+\varepsilon} X^{\frac{269}{474}+\varepsilon} + \frac{1}{kX^{c-1}}\right) \\
&\ll X^{\frac{269c+538}{1217}} \log X + H_1^{\frac{269}{948}+\varepsilon} X^{\frac{269(c+1)}{948}+\varepsilon} + X^{1-c} \ll X^{\frac{269c+538}{1217}+\varepsilon}. \qquad (8)
\end{aligned}$$

Similarly, for the first term in (6), we have

$$\begin{aligned}
&\sum_{|h|\leqslant H_1} c_h(\alpha) \sum_{X<n\leqslant 2X} e\big((h+\alpha)n^c\big)\\
&= c_0(\alpha)\sum_{X<n\leqslant 2X} e(\alpha n^c) + \sum_{1\leqslant |h|\leqslant H_1} c_h(\alpha) \sum_{X<n\leqslant 2X} e\big((h+\alpha)n^c\big)\\
&\ll \frac{1}{\alpha X^{c-1}} + \sum_{1\leqslant |h|\leqslant H_1} \frac{1}{h}\bigg(\big((h+\alpha)X^{c-1}\big)^{\frac{269}{948}+\varepsilon} X^{\frac{269}{474}+\varepsilon} + \frac{1}{(h+\alpha)X^{c-1}}\bigg)\\
&\ll \frac{1}{\alpha X^{c-1}} + H_1^{\frac{269}{948}+\varepsilon} X^{\frac{269(c+1)}{948}+\varepsilon} + X^{1-c}\\
&\ll \frac{1}{\alpha X^{c-1}} + X^{\frac{269c+538}{1217}+\varepsilon}. \qquad (9)
\end{aligned}$$

Combining (6)–(9), we obtain the desired result. This completes the proof of Lemma 11. □

3. Proof of Theorem 1

By the definition of $\mathcal{R}_3(N)$, we have

$$\begin{aligned}
\mathcal{R}_3(N) &= \int_0^1 S^3(\alpha)e(-N\alpha)\mathrm{d}\alpha = \int_{-\tau}^{1-\tau} S^3(\alpha)e(-N\alpha)\mathrm{d}\alpha\\
&= \int_{-\tau}^{\tau} S^3(\alpha)e(-N\alpha)\mathrm{d}\alpha + \int_{\tau}^{1-\tau} S^3(\alpha)e(-N\alpha)\mathrm{d}\alpha\\
&= \mathcal{R}_3^{(1)}(N) + \mathcal{R}_3^{(2)}(N), \qquad (10)
\end{aligned}$$

say. By the argument of Laporta and Tolev [12], we can know that, for $1 < c < 3/2$, there holds

$$\mathcal{R}_3^{(1)}(N) = \frac{\Gamma^3(1+1/c)}{\Gamma(3/c)} N^{3/c-1} + O\Big(N^{3/c-1}\exp\big(-(\log N)^{1/3-\delta}\big)\Big) \qquad (11)$$

for any $0 < \delta < 1/3$, where the implied constant in the O–term depends only on c. Thus, it is sufficient to estimate $\mathcal{R}_3^{(2)}(N)$. First, we have

$$S(\alpha) = \sum_{p\leqslant P^{5/6}} (\log p)e\big([p^c]\alpha\big) + \sum_{P^{5/6}<p\leqslant P} (\log p)e\big([p^c]\alpha\big). \qquad (12)$$

By a splitting argument and (12), we deduce that

$$\begin{aligned}
\mathcal{R}_3^{(2)}(N) &\ll (\log P)\max_{P^{5/6}\ll X\ll P}\left|\int_{\tau}^{1-\tau} S^2(\alpha)S(\alpha,X)e(-N\alpha)\mathrm{d}\alpha\right| + P^{\frac{5}{6}}\int_0^1 \big|S(\alpha)\big|^2\mathrm{d}\alpha\\
&\ll (\log P)\max_{P^{5/6}\ll X\ll P}\left|\int_{\tau}^{1-\tau} S^2(\alpha)S(\alpha,X)e(-N\alpha)\mathrm{d}\alpha\right| + P^{\frac{11}{6}}\log P. \qquad (13)
\end{aligned}$$

For $P^{5/6} \ll X \ll P$, we have

$$\begin{aligned}
&\left|\int_{\tau}^{1-\tau} S^2(\alpha)S(\alpha,X)e(-N\alpha)\mathrm{d}\alpha\right| \\
&= \left|\sum_{X<p\leqslant 2X} (\log p)\int_{\tau}^{1-\tau} S^2(\alpha)e\big(([p^c]-N)\alpha\big)\mathrm{d}\alpha\right| \\
&\leqslant \sum_{X<p\leqslant 2X} (\log p)\left|\int_{\tau}^{1-\tau} S^2(\alpha)e\big(([p^c]-N)\alpha\big)\mathrm{d}\alpha\right| \\
&\ll (\log X) \sum_{X<n\leqslant 2X} \left|\int_{\tau}^{1-\tau} S^2(\alpha)e\big(([n^c]-N)\alpha\big)\mathrm{d}\alpha\right|.
\end{aligned}$$

By Cauchy's inequality, we deduce that

$$\begin{aligned}
&\left|\int_{\tau}^{1-\tau} S^2(\alpha)S(\alpha,X)e(-N\alpha)\mathrm{d}\alpha\right| \\
&\ll X^{\frac{1}{2}+\varepsilon} \sum_{X<n\leqslant 2X} \left|\int_{\tau}^{1-\tau} S^2(\alpha)e\big(([n^c]-N)\alpha\big)\mathrm{d}\alpha\right|^2\Bigg)^{\frac{1}{2}} \\
&= X^{\frac{1}{2}+\varepsilon} \sum_{X<n\leqslant 2X} \int_{\tau}^{1-\tau} S^2(\alpha)e\big(([n^c]-N)\alpha\big)\mathrm{d}\alpha \cdot \int_{\tau}^{1-\tau} \overline{S^2(\beta)e\big(([n^c]-N)\beta\big)}\mathrm{d}\beta\Bigg)^{\frac{1}{2}} \\
&= X^{\frac{1}{2}+\varepsilon}\left(\int_{\tau}^{1-\tau} \overline{S^2(\beta)e(-N\beta)}\mathrm{d}\beta \int_{\tau}^{1-\tau} S^2(\alpha)\mathcal{T}(\alpha-\beta,X)e(-N\alpha)\mathrm{d}\alpha\right)^{\frac{1}{2}} \\
&\ll X^{\frac{1}{2}+\varepsilon}\left(\int_{\tau}^{1-\tau} \big|S(\beta)\big|^2\mathrm{d}\beta \int_{\tau}^{1-\tau} \big|S^2(\alpha)\mathcal{T}(\alpha-\beta,X)\big|\mathrm{d}\alpha\right)^{\frac{1}{2}}. \qquad (14)
\end{aligned}$$

For the inner integral in (14), we have

$$\begin{aligned}
&\int_{\tau}^{1-\tau} \big|S^2(\alpha)\mathcal{T}(\alpha-\beta,X)\big|\mathrm{d}\alpha \\
&\ll \left(\int_{(\tau,1-\tau)\cap\{\alpha:\,|\alpha-\beta|\leqslant X^{-c}\}} + \int_{(\tau,1-\tau)\cap\{\alpha:\,|\alpha-\beta|>X^{-c}\}}\right)\big|S^2(\alpha)\mathcal{T}(\alpha-\beta,X)\big|\mathrm{d}\alpha. \qquad (15)
\end{aligned}$$

For the first term on the right–hand side of (15), we use Lemma 10 and the trivial estimate $\mathcal{T}(\alpha-\beta,X) \ll X$ to deduce that

$$\begin{aligned}
&\int_{(\tau,1-\tau)\cap\{\alpha:\,|\alpha-\beta|\leqslant X^{-c}\}} \big|S^2(\alpha)\mathcal{T}(\alpha-\beta,X)\big|\mathrm{d}\alpha \\
&\ll X\cdot \sup_{\alpha\in(\tau,1-\tau)} \big|S(\alpha)\big|^2 \times \int_{|\alpha-\beta|\leqslant X^{-c}} \mathrm{d}\alpha \ll P^{\frac{49966}{27030}+\varepsilon}X^{1-c}. \qquad (16)
\end{aligned}$$

For the second term on the right–hand side of (15), by Lemma 10 and Lemma 11, we obtain

$$
\begin{aligned}
&\int_{(\tau,1-\tau)\cap\{\alpha:\,|\alpha-\beta|>X^{-c}\}} \left|S^2(\alpha)\mathcal{T}(\alpha-\beta,X)\right|\mathrm{d}\alpha \\
&\ll \int_{(\tau,1-\tau)\cap\{\alpha:\,|\alpha-\beta|>X^{-c}\}} \left|S(\alpha)\right|^2\left(X^{\frac{269c+538}{1217}+\varepsilon}+\frac{1}{|\alpha-\beta|X^{c-1}}\right)\mathrm{d}\alpha \\
&\ll X^{\frac{269c+538}{1217}+\varepsilon}\times\int_0^1|S(\alpha)|^2\mathrm{d}\alpha+\sup_{\alpha\in(\tau,1-\tau)}\left|S(\alpha)\right|^2\times\int_{|\alpha-\beta|>X^{-c}}\frac{\mathrm{d}\alpha}{|\alpha-\beta|X^{c-1}} \\
&\ll X^{\frac{269c+538}{1217}+\varepsilon}P\log P+P^{\frac{49966}{27030}+\varepsilon}X^{1-c}. \qquad (17)
\end{aligned}
$$

Combining (15) and (17), we conclude that

$$
\int_\tau^{1-\tau}\left|S^2(\alpha)\mathcal{T}(\alpha-\beta,X)\right|\mathrm{d}\alpha \ll X^{\frac{269c+538}{1217}+\varepsilon}P\log P+P^{\frac{49966}{27030}+\varepsilon}X^{1-c}. \qquad (18)
$$

Inserting (18) into (14), we obtain

$$
\left|\int_\tau^{1-\tau}S^2(\alpha)S(\alpha,X)e(-N\alpha)\mathrm{d}\alpha\right| \ll X^{\frac{1}{2}+\varepsilon}\left(X^{\frac{269c+538}{2434}}P+P^{\frac{24983}{27030}}X^{\frac{1-c}{2}}\right)\ll P^{3-c-\varepsilon}, \qquad (19)
$$

provided that $1<c<\frac{3113}{2703}$. From (13) and (19), we obatin

$$
\mathcal{R}_3^{(2)}(N)\ll P^{3-c-\varepsilon}. \qquad (20)
$$

By (10), (11) and (20), we get the desired result. This completes the proof of Theorem 1.

Acknowledgement The authors would like to express the most sincere gratitude to Professor Wenguang Zhai for his valuable advice and constant encouragement. This work is supported by the Fundamental Research Funds for the Central Universities (Grant No. 2019QS02).

References

[1] G. I. Arkhipov and A. N. Zhitkov, On Waring's problem with non–integer degrees, *Izv. Akad. Nauk SSSR* **48** (1984), 1138–1150.

[2] K. Buriev, *Additive problems with prime numbers*, Thesis, Moscow University, 1989.

[3] Y. C. Cai, On a Diophantine equation involving primes, *Ramamujan J.* DOI: 10.1007/s11139-018-0027-6.

[4] J. M. Deshouillers, Problème de Waring avec exposants non entiers, *Bull. Soc. Math. France* **101** (1973), 285–295.

[5] J. M. Deshouillers, Un problème binaire en théorie additive, *Acta Arith.* **25** (1974), no. 4, 393–403.

[6] E. Fouvry and H. Iwaniec, Exponential sums with monomials, *J. Number Theory* **33** (1989), no. 3, 311–333.

[7] S. W. Graham and G. Kolesnik, *Van der Corput's Method of Exponential Sums*, Cambridge University Press, New York, 1991.

[8] S. A. Gritsenko, Three additive problems, *Russian Acad. Sci. Izv. Math.* **41** (1993), no. 3, 447–464.

[9] D. R. Heath–Brown, The Pjateckiĭ-Šapiro prime number theorem, *J. Number Theory* **16** (1983), no. 2, 242–266.

[10] M. N. Huxley, Exponential sums and the Riemann zeta function V, *Proc. London Math. Soc. (3)* **90** (2005), no. 1, 1–41.

[11] A. Kumchev and T. Nedeva, On an equation with prime numbers, *Acta Arith.* **83** (1998), no. 2, 117–126.

[12] M. B. S. Laporta and D. I. Tolev, On an equation with prime numbers, *Math. Notes* **57** (1995), no. 5–6, 654–657.

[13] B. I. Segal, On a theorem similar to the Waring theorem, *Dokl. Akad. Nauk. SSSR* **1** (1933), 47–49.

[14] B. I. Segal, The Waring theorem with fractional and irrational degrees, *Trudy Mat. Inst. Steklov.* **5** (1934), 73–86.

[15] I. M. Vinogradov, I. M. Vinogradov, Representation of an odd number as the sum of three primes, *Dokl. Akad. Nauk. SSSR* **15** (1937), 291–294.

[16] J. Wu, Double exponential sums and some applications, *Monatsh. Math.* **128** (1999), no. 3, 255–262.

[17] W. G. Zhai and X. D. Cao, A Diophantine equation with prime numbers, *Acta Math. Sinica (Chin. Ser.)* **45** (2002), no. 3, 443–454.

13

ON BINOMIAL COEFFICIENTS MODULO SQUARES OF PRIMES

Darij Grinberg
School of Mathematics, University of Minnesota, Minneapolis, Minnesota
darijgrinberg@gmail.com

Abstract

We give elementary proofs for the Apagodu-Zeilberger-Stanton-Amdeberhan-Tauraso congruences

$$\sum_{n=0}^{rp-1} \binom{2n}{n} \equiv \eta_p \sum_{n=0}^{r-1} \binom{2n}{n} \bmod p^2,$$
$$\sum_{n=0}^{rp-1} \sum_{m=0}^{sp-1} \binom{n+m}{m}^2 \equiv \eta_p \sum_{m=0}^{r-1} \sum_{n=0}^{s-1} \binom{n+m}{m}^2 \bmod p^2,$$

where p is an odd prime, r and s are nonnegative integers, and

$$\eta_p = \begin{cases} 0, & \text{if } p \equiv 0 \bmod 3; \\ 1, & \text{if } p \equiv 1 \bmod 3; \\ -1, & \text{if } p \equiv 2 \bmod 3. \end{cases}$$

1. Introduction

In this note, we prove that any odd prime p and any $r, s \in \mathbb{N}$ satisfy

$$\sum_{n=0}^{rp-1} \binom{2n}{n} \equiv \eta_p \sum_{n=0}^{r-1} \binom{2n}{n} \bmod p^2 \qquad \text{(Theorem 3)};$$
$$\sum_{n=0}^{rp-1} \sum_{m=0}^{sp-1} \binom{n+m}{m}^2 \equiv \eta_p \sum_{m=0}^{r-1} \sum_{n=0}^{s-1} \binom{n+m}{m}^2 \bmod p^2 \qquad \text{(Theorem 4)},$$

where

$$\eta_p = \begin{cases} 0, & \text{if } p \equiv 0 \bmod 3; \\ 1, & \text{if } p \equiv 1 \bmod 3; \\ -1, & \text{if } p \equiv 2 \bmod 3. \end{cases}$$

These two congruences are (slightly extended) versions of the "Super-Conjectures" $1''$ and $4'$ stated by Apagodu and Zeilberger in [3][1]. Our proofs are more elementary than previous proofs by Stanton [16], and Amdeberhan and Tauraso [1].

A more detailed version of the present paper is available on the arXiv [10].

1.1. Binomial Coefficients

Let us first recall the definition of binomial coefficients.[2]

Definition 1. Let $n \in \mathbb{Z}$ and $m \in \mathbb{Z}$. Then, the *binomial coefficient* $\binom{m}{n}$ is a rational number defined by

$$\binom{m}{n} = \begin{cases} \dfrac{m(m-1)\cdots(m-n+1)}{n!}, & \text{if } n \in \mathbb{N}; \\ 0, & \text{if } n \notin \mathbb{N}. \end{cases}$$

This is the definition used in [7] and [9]. Some authors follow other conventions instead.

The following proposition is well-known (see, e.g., [9, Proposition 1.9]), and will be tacitly used below (as we study congruences involving binomial coefficients).

Proposition 1. *We have* $\binom{m}{n} \in \mathbb{Z}$ *for any* $m \in \mathbb{Z}$ *and* $n \in \mathbb{Z}$.

1.2. Classical Congruences

The behavior of binomial coefficients modulo primes and prime powers is a classical subject of research; see [14, §2.1] for a survey of much of it. Let us state two of the most basic results in this subject.

Theorem 1. *Let p be a prime. Let a and b be two integers. Let c and d be two elements of* $\{0, 1, \ldots, p-1\}$. *Then,*

$$\binom{ap+c}{bp+d} \equiv \binom{a}{b}\binom{c}{d} \bmod p.$$

Theorem 1 is known under the name of *Lucas's theorem*, and is proven in many places (e.g., [14, §2.1], [11, Proof of §4], [2, proof of Lucas's theorem] and [7, Exercise 5.61]) at least in the case when a and b are nonnegative integers. The standard proof of Theorem 1 in this case uses generating functions; this proof applies (mutatis mutandis) in the general case as well. See [9, Theorem 1.11] for an elementary proof of Theorem 1.

Another fundamental result is the following.

[1]In the arXiv preprint version of [3] (`https://arxiv.org/abs/1606.03351v2`arXiv:1606.03351v2), these congruences appear as "Super-Conjectures" 1" and $5'$, respectively.

[2]We use the notation $\mathbb{N}$ for the set $\{0, 1, 2, \ldots\}$.

Theorem 2. *Let p be a prime. Let a and b be two integers. Then,*

$$\binom{ap}{bp} \equiv \binom{a}{b} \bmod p^2.$$

Theorem 2 is a known result, often attributed to Charles Babbage. It appears with proof in [9, Theorem 1.12]; again, many sources prove it for nonnegative a and b (such as [15, Exercise 1.14 **c**] or [7, Exercise 5.62]). Notice that if $p \geq 5$, then the modulus p^2 can be replaced by p^3 or (depending on a, b and p) by even higher powers of p; see [14, (22) and (23)] for the details. See also [17, Lemma 2.1] for another strengthening of Theorem 2.

1.3. Modulo-p^2 Congruences

Definition 2. For any $p \in \mathbb{Z}$, we define an integer $\eta_p \in \{-1, 0, 1\}$ by

$$\eta_p = \begin{cases} 0, & \text{if } p \equiv 0 \bmod 3; \\ 1, & \text{if } p \equiv 1 \bmod 3; \\ -1, & \text{if } p \equiv 2 \bmod 3. \end{cases}$$

Notice that η_p is the so-called *Legendre symbol* $\left(\frac{p}{3}\right)$ known from number theory.

We are now ready to state two conjectures by Apagodu and Zeilberger, which we shall prove in the sequel. The first one is [3, Super-Conjecture 1″]:[3]

Theorem 3. *Let p be an odd prime. Let $r \in \mathbb{N}$. Set*

$$\alpha_r = \sum_{n=0}^{r-1} \binom{2n}{n}.$$

Then,

$$\sum_{n=0}^{rp-1} \binom{2n}{n} \equiv \eta_p \alpha_r \bmod p^2.$$

Theorem 3 has been proven by Dennis Stanton [16] using Laurent series (in the case when $p \geq 5$), and by Liu [12, (1.3)] using harmonic numbers. We shall reprove it elementarily. Note that we can apply Theorem 3 to $r = 1$, and obtain the congruence $\sum_{n=0}^{p-1} \binom{2n}{n} \equiv \eta_p \bmod p^2$ for each odd prime p; this is [3, Super-Conjecture 1].

The next conjecture that we shall prove is [3, Super-Conjecture 5′].

[3] To be precise (and boastful), our Theorem 3 is somewhat stronger than [3, Super-Conjecture 1″], since we only require p to be odd (rather than $p \geq 5$). The same remark applies to Theorem 4. That said, the $p = 3$ case may well fall prey to simpler methods.

Theorem 4. *Let p be an odd prime. Let $r \in \mathbb{N}$ and $s \in \mathbb{N}$. Set*

$$\epsilon_{r,s} = \sum_{m=0}^{r-1} \sum_{n=0}^{s-1} \binom{n+m}{m}^2 .$$

Then,

$$\sum_{n=0}^{rp-1} \sum_{m=0}^{sp-1} \binom{n+m}{m}^2 \equiv \eta_p \epsilon_{r,s} \bmod p^2 .$$

A proof of Theorem 4 has been found by Amdeberhan and Tauraso, and was outlined in [1, §6]; we give a different, elementary proof.

1.4. Bailey's Congruence and Analogues

On our way to proving the above two theorems, we shall show a modulo-p^2 congruence for certain binomial coefficients that can be regarded as a counterpart to Theorem 2.

Theorem 5. *Let p be a prime. Let $N \in \mathbb{Z}$ and $K \in \mathbb{Z}$ and $i \in \{1, 2, \ldots, p-1\}$. Then we have*

(a)

$$\binom{Np}{Kp+i} \equiv N \binom{N-1}{K} \binom{p}{i} \bmod p^2;$$

(b)

$$\binom{Np}{Kp-i} \equiv N \binom{N-1}{K-1} \binom{p}{i} \bmod p^2;$$

(c)

$$\binom{Np}{Kp+i} + \binom{Np}{Kp-i} \equiv N \binom{N}{K} \binom{p}{i} \bmod p^2 .$$

Theorem 5 (a) is essentially the result [4, Theorem 4] by Bailey (see also [14, (26)]); in fact, it transforms into [4, Theorem 4] if we rewrite $N\binom{N-1}{K}$ as $(K+1)\binom{N}{K+1}$ (using Proposition 8 below). We shall nevertheless give our own proof for it.

1.5. Polynomial Summation

Let us state two further lemmas that will be crucial to our proofs of Theorems 3 and 4, but are likely to have other uses as well.

Let $\mathbb{Z}[X]$ be the ring of all polynomials in one indeterminate X with integer coefficients. It is well-known that all integers p, c and l and every polynomial $P \in$

$\mathbb{Z}[X]$ satisfy $P(cp+l) \equiv P(l) \bmod p$. Thus, polynomials in $\mathbb{Z}[X]$ can be applied to modulo-p congruences, yielding new modulo-p congruences. More interesting is the fact that sometimes polynomials can be used to turn modulo-p congruences into modulo-p^2 congruences. Two vehicles for such "lifting" of congruences are the following two lemmas (proven further below).

Lemma 1. *Let p be a prime. Let $c \in \mathbb{Z}$. Let $P \in \mathbb{Z}[X]$ be a polynomial of degree $< 2p-1$. Then, $\sum_{l=0}^{p-1} (P(cp+l) - P(l)) \equiv 0 \bmod p^2$.*

Lemma 2. *Let p be an odd prime. Let $c \in \mathbb{Z}$. Let $P \in \mathbb{Z}[X]$ be a polynomial of degree $\leq p-1$. Then,*

$$\sum_{l=0}^{p-1} (P(cp+l) - P(l)) P(l) \equiv 0 \bmod p^2.$$

2. The Proofs

2.1. Identities and Congruences From the Literature

Before we come to the proofs of the above-listed results, let us collect various well-known facts that will prove useful.

We assume that the reader is familiar with standard properties of binomial coefficients (see, e.g., [8, §3.1], [7, Chapter 5] or [9, §1]):

Proposition 2. *We have $\binom{m}{n} = 0$ for every $m \in \mathbb{N}$ and $n \in \mathbb{N}$ satisfying $m < n$.*

Proposition 3. *Let $m \in \mathbb{N}$ and $n \in \mathbb{N}$ satisfy $m \geq n$. Then, $\binom{m}{n} = \binom{m}{m-n}$.*

Proposition 4. *Let $m \in \mathbb{Z}$ and $n \in \mathbb{N}$. Then, $\binom{m}{n} = (-1)^n \binom{n-m-1}{n}$.*

Proposition 5. *Let $m \in \mathbb{Z}$ and $n \in \mathbb{Z}$. Then, $\binom{m}{n} = \binom{m-1}{n-1} + \binom{m-1}{n}$.*

Proposition 6. *For every $x \in \mathbb{Z}$ and $y \in \mathbb{Z}$ and $n \in \mathbb{N}$, we have*

$$\binom{x+y}{n} = \sum_{k=0}^{n} \binom{x}{k} \binom{y}{n-k}.$$

Proposition 6 is the so-called *Vandermonde convolution identity*, and is a particular case of [8, Theorem 3.29].

Proposition 7. *For each $n \in \mathbb{N}$, we have*

$$\sum_{i=0}^{n-1} (-1)^i \binom{n-1-i}{i} = (-1)^n \cdot \begin{cases} 0, & \text{if } n \equiv 0 \bmod 3; \\ -1, & \text{if } n \equiv 1 \bmod 3; \\ 1, & \text{if } n \equiv 2 \bmod 3. \end{cases}$$

Proposition 7 is [8, Corollary 8.68]. Apart from that, Proposition 7 can be easily derived from [7, §5.2, Problem 3], [5, Identity 172] or [6].

Proposition 8. *Let $n \in \mathbb{Z}$ and $k \in \mathbb{Z}$. Then, $k\binom{n}{k} = n\binom{n-1}{k-1}$.*

Proposition 8 (sometimes known as the "absorption identity") appears in [7, (5.6)], and is easily proven just from the definition of binomial coefficients.

Let us also recall a result from elementary number theory:

Theorem 6. *Let p be a prime. Let $k \in \mathbb{N}$. Assume that k is not a positive multiple of $p-1$. Then,*

$$\sum_{l=0}^{p-1} l^k \equiv 0 \bmod p.$$

Theorem 6 is proven, e.g., in [9, Theorem 3.1] and (in a slightly rewritten form) in [13, Theorem 1].

2.2. Variants and Consequences of Vandermonde Convolution

We are now going to state a number of identities that are restatements or particular cases of the Vandermonde convolution identity (Proposition 6). We begin with the following one.

Corollary 1. *Let $u \in \mathbb{Z}$ and $l \in \mathbb{N}$ and $w \in \mathbb{N}$. Then,*

$$\sum_{m=0}^{l} \binom{u}{w+m}\binom{l}{m} = \binom{u+l}{w+l}.$$

Proof of Corollary 1. Proposition 6 (applied to $x = u$, $y = l$ and $n = w + l$) yields

$$\binom{u+l}{w+l} = \sum_{k=0}^{w+l} \binom{u}{k}\binom{l}{w+l-k} = \sum_{k=0}^{w-1} \binom{u}{k} \underbrace{\binom{l}{w+l-k}}_{\substack{=0 \\ \text{(by Proposition 2} \\ \text{(since } l<w+l-k \\ \text{(because } k<w\text{)))}}} + \sum_{k=w}^{w+l} \binom{u}{k}\binom{l}{w+l-k}$$

$$= \sum_{k=w}^{w+l} \binom{u}{k}\binom{l}{w+l-k} = \sum_{m=0}^{l} \binom{u}{w+m} \underbrace{\binom{l}{w+l-(w+m)}}_{\substack{=\binom{l}{l-m}=\binom{l}{m} \\ \text{(by Proposition 3)}}}$$

(here, we have substituted $w+m$ for k in the sum)

$$= \sum_{m=0}^{l} \binom{u}{w+m} \binom{l}{m}.$$

This proves Corollary 1. □

Let us also state another corollary of Proposition 6.

Corollary 2. *Let $x \in \mathbb{Z}$ and $y \in \mathbb{N}$ and $n \in \mathbb{Z}$. Then,*

$$\binom{x+y}{n} = \sum_{i=0}^{y} \binom{x}{n-i} \binom{y}{i}.$$

See [9, Corollary 2.2] for a proof of Corollary 2.

Lemma 3. *Let $u \in \mathbb{Z}$ and $w \in \mathbb{N}$ and $l \in \mathbb{N}$. Then,*

$$\binom{u+2l}{w+l} = \binom{u}{w}\binom{2l}{l} + \sum_{i=1}^{l} \left(\binom{u}{w+i} + \binom{u}{w-i} \right) \binom{2l}{l-i}.$$

Proof of Lemma 3. Corollary 2 (applied to $x = u$, $y = 2l$ and $n = w+l$) yields

$$\binom{u+2l}{w+l} = \sum_{i=0}^{2l} \binom{u}{w+l-i} \binom{2l}{i} = \sum_{i=-l}^{l} \binom{u}{w+i} \binom{2l}{l-i}$$

(here, we have substituted $l-i$ for i in the sum)

$$= \sum_{\substack{i \in \{-l, -l+1, \ldots, l\}; \\ i \neq 0}} \binom{u}{w+i} \binom{2l}{l-i} + \binom{u}{w} \binom{2l}{l}$$

(here, we have split off the addend for $i = 0$ from the sum). Hence,

$$\begin{aligned}
\binom{u+2l}{w+l} - \binom{u}{w}\binom{2l}{l} &= \sum_{\substack{i \in \{-l, -l+1, \ldots, l\}; \\ i \neq 0}} \binom{u}{w+i} \binom{2l}{l-i} \\
&= \sum_{i=1}^{l} \binom{u}{w+i} \binom{2l}{l-i} + \sum_{i=-l}^{-1} \binom{u}{w+i} \binom{2l}{l-i} \\
&= \sum_{i=1}^{l} \binom{u}{w+i} \binom{2l}{l-i} + \sum_{i=1}^{l} \binom{u}{w-i} \underbrace{\binom{2l}{l+i}}_{\substack{= \binom{2l}{l-i} \\ \text{(by Proposition 3)}}}
\end{aligned}$$

$$\left(\begin{array}{c}\text{here, we have substituted } -i \text{ for } i\\ \text{in the second sum}\end{array}\right)$$

$$= \sum_{i=1}^{l} \left(\binom{u}{w+i} + \binom{u}{w-i} \right) \binom{2l}{l-i}.$$

Adding $\binom{u}{w}\binom{2l}{l}$ to both sides yields the claim of Lemma 3. □

Lemma 4. *Let $p \in \mathbb{N}$. Let $c \in \mathbb{Z}$. Let $l \in \{0, 1, \ldots, p-1\}$. Then,*

$$\binom{cp+2l}{l} = \sum_{k=0}^{p-1} \binom{cp+l}{k}\binom{l}{k}.$$

Proof of Lemma 4. Corollary 2 (applied to $x = cp + l$, $y = l$ and $n = l$) yields

$$\binom{cp+2l}{l} = \sum_{i=0}^{l} \binom{cp+l}{l-i}\binom{l}{i} = \sum_{k=0}^{l} \binom{cp+l}{k} \underbrace{\binom{l}{l-k}}_{\substack{=\binom{l}{k} \\ \text{(by Proposition 3)}}}$$

(here, we have substituted k for $l-i$ in the sum)

$$= \sum_{k=0}^{l} \binom{cp+l}{k}\binom{l}{k}.$$

Comparing this with

$$\sum_{k=0}^{p-1} \binom{cp+l}{k}\binom{l}{k} = \sum_{k=0}^{l} \binom{cp+l}{k}\binom{l}{k} + \sum_{k=l+1}^{p-1} \binom{cp+l}{k} \underbrace{\binom{l}{k}}_{\substack{=0 \\ \text{(by Proposition 2} \\ \text{(applied to } m=l \text{ and } n=k) \\ \text{(since } l<k))}}$$

$$= \sum_{k=0}^{l} \binom{cp+l}{k}\binom{l}{k},$$

we obtain $\sum_{k=0}^{p-1} \binom{cp+l}{k}\binom{l}{k} = \binom{cp+2l}{l}$. This proves Lemma 4. □

Lemma 5. *Let $p \in \mathbb{N}$. Let $l \in \mathbb{N}$. Then,*

$$\sum_{i=1}^{l} \binom{p}{i}\binom{2l}{l-i} = \binom{p+2l}{l} - \binom{2l}{l}.$$

Proof of Lemma 5. Proposition 6 (applied to $x = p$, $y = 2l$ and $n = l$) yields

$$\binom{p+2l}{l} = \sum_{k=0}^{l} \binom{p}{k}\binom{2l}{l-k} = \sum_{i=0}^{l} \binom{p}{i}\binom{2l}{l-i}$$

(here, we have renamed the summation index k as i)

$$= \underbrace{\binom{p}{0}}_{=1}\underbrace{\binom{2l}{l-0}}_{=\binom{2l}{l}} + \sum_{i=1}^{l}\binom{p}{i}\binom{2l}{l-i} = \binom{2l}{l} + \sum_{i=1}^{l}\binom{p}{i}\binom{2l}{l-i}.$$

Solving this equality for $\sum_{i=1}^{l}\binom{p}{i}\binom{2l}{l-i}$, we obtain Lemma 5. □

2.3. Proof of Bailey's Congruence

Proof of Theorem 5. From $i \in \{1, 2, \ldots, p-1\}$, we conclude that both $i - 1$ and $p - i$ are elements of $\{0, 1, \ldots, p - 1\}$. Notice also that i is not divisible by p (since $i \in \{1, 2, \ldots, p - 1\}$); hence, i is coprime to p (since p is a prime). Therefore, i is also coprime to p^2.

(a) Proposition 8 (applied to $n = Np$ and $k = Kp + i$) yields

$$(Kp+i)\binom{Np}{Kp+i} = Np\binom{Np-1}{Kp+i-1} = Np \underbrace{\binom{(N-1)p+(p-1)}{Kp+(i-1)}}_{\substack{\equiv \binom{N-1}{K}\binom{p-1}{i-1} \bmod p \\ \text{(by Theorem 1, applied to} \\ a=N-1,\ b=K,\ c=p-1 \text{ and } d=i-1)}}$$

$$\equiv Np\binom{N-1}{K}\binom{p-1}{i-1} \bmod p^2 \tag{1}$$

(notice that the presence of the p factor has turned a congruence modulo p into a congruence modulo p^2). Thus,

$$(Kp+i)\binom{Np}{Kp+i} \equiv Np\binom{N-1}{K}\binom{p-1}{i-1} \equiv 0 \bmod p,$$

so that $0 \equiv \underbrace{(Kp+i)}_{\equiv i \bmod p}\binom{Np}{Kp+i} \equiv i\binom{Np}{Kp+i} \bmod p$. We can cancel i from this congruence (since i is coprime to p), and thus obtain $0 \equiv \binom{Np}{Kp+i} \bmod p$. Hence,

$\binom{Np}{Kp+i}$ is divisible by p. Thus, $p\binom{Np}{Kp+i}$ is divisible by p^2. In other words,

$$p\binom{Np}{Kp+i} \equiv 0 \bmod p^2. \tag{2}$$

Now,

$$(Kp+i)\binom{Np}{Kp+i} = K\underbrace{p\binom{Np}{Kp+i}}_{\substack{\equiv 0 \bmod p^2 \\ \text{(by (2))}}} + i\binom{Np}{Kp+i} \equiv i\binom{Np}{Kp+i} \bmod p^2.$$

Hence,

$$\begin{aligned} i\binom{Np}{Kp+i} &\equiv (Kp+i)\binom{Np}{Kp+i} \equiv Np\binom{N-1}{K}\binom{p-1}{i-1} \qquad \text{(by (1))} \\ &= N\binom{N-1}{K}\underbrace{p\binom{p-1}{i-1}}_{\substack{=i\binom{p}{i} \\ \text{(by Proposition 8)}}} = N\binom{N-1}{K}i\binom{p}{i} \bmod p^2. \end{aligned}$$

We can cancel i from this congruence (since i is coprime to p^2), and thus obtain

$$\binom{Np}{Kp+i} \equiv N\binom{N-1}{K}\binom{p}{i} \bmod p^2.$$

This proves Theorem 5 **(a)**.

(b) We have $i \in \{1, 2, \ldots, p-1\}$ and thus $p - i \in \{1, 2, \ldots, p-1\}$. Hence, Theorem 5 **(a)** (applied to $K-1$ and $p-i$ instead of K and i) yields

$$\binom{Np}{(K-1)p+(p-i)} \equiv N\binom{N-1}{K-1}\underbrace{\binom{p}{p-i}}_{\substack{=\binom{p}{i} \\ \text{(by Proposition 3)}}} = N\binom{N-1}{K-1}\binom{p}{i} \bmod p^2.$$

In view of $(K-1)p + (p-i) = Kp - i$, this can be rewritten as

$$\binom{Np}{Kp-i} \equiv N\binom{N-1}{K-1}\binom{p}{i} \bmod p^2.$$

This proves Theorem 5 **(b)**.

(c) We have

$$\underbrace{\binom{Np}{Kp+i}}_{\substack{\equiv N\binom{N-1}{K}\binom{p}{i} \bmod p^2 \\ \text{(by Theorem 5 (a))}}} + \underbrace{\binom{Np}{Kp-i}}_{\substack{\equiv N\binom{N-1}{K-1}\binom{p}{i} \bmod p^2 \\ \text{(by Theorem 5 (b))}}}$$

$$\equiv N\binom{N-1}{K}\binom{p}{i} + N\binom{N-1}{K-1}\binom{p}{i}$$

$$= N\underbrace{\left(\binom{N-1}{K-1} + \binom{N-1}{K}\right)}_{\substack{=\binom{N}{K} \\ \text{(by Proposition 5)}}}\binom{p}{i} = N\binom{N}{K}\binom{p}{i} \bmod p^2.$$

This proves Theorem 5 **(c)**. □

2.4. Proofs of Lemmas 1 and 2

Proof of Lemma 1. WLOG assume that $P = X^k$ for some $k \in \{0, 1, \ldots, 2p-2\}$ (since the congruence we are proving depends $\mathbb{Z}$-linearly on P). If $k = 0$, then Lemma 1 is easily checked. Thus, WLOG assume that $k \neq 0$. Hence, $k - 1 \in \mathbb{N}$.

We have $P = X^k$. Thus, each $l \in \{0, 1, \ldots, p-1\}$ satisfies

$$P(cp+l) = (cp+l)^k = \sum_{i=0}^{k} \binom{k}{i} (cp)^i l^{k-i} \qquad \text{(by the binomial formula)}$$

$$= \underbrace{(cp)^0 l^{k-0}}_{=l^k} + k\underbrace{(cp)^1}_{=cp} l^{k-1} + \sum_{i=2}^{k} \binom{k}{i} \underbrace{(cp)^i}_{\substack{\equiv 0 \bmod p^2 \\ (\text{since } i \geq 2)}} l^{k-i}$$

$$\equiv l^k + kcpl^{k-1} \bmod p^2$$

and $P(l) = l^k$ (since $P = X^k$). Thus,

$$\sum_{l=0}^{p-1} \left(\underbrace{P(cp+l)}_{\equiv l^k + kcpl^{k-1} \bmod p^2} - \underbrace{P(l)}_{=l^k} \right) \equiv \sum_{l=0}^{p-1} \underbrace{\left(l^k + kcpl^{k-1} - l^k\right)}_{=kcpl^{k-1}} = kcp \sum_{l=0}^{p-1} l^{k-1} \bmod p^2.$$

The claim of Lemma 1 now becomes obvious if $k = p$ (because if $k = p$, then kcp is already divisible by p^2); thus, we WLOG assume that $k \neq p$. Hence, $k - 1 \neq p - 1$.

If $k - 1$ was a positive multiple of $p - 1$, then we would have $k - 1 = p - 1$ (since $k \in \{0, 1, \ldots, 2p-2\}$), which would contradict $k - 1 \neq p - 1$. Hence, $k - 1$ is not a positive multiple of $p-1$. Thus, Theorem 6 (applied to $k-1$ instead of k) yields $\sum_{l=0}^{p-1} l^{k-1} \equiv 0 \bmod p$. Thus, $p \sum_{l=0}^{p-1} l^{k-1} \equiv 0 \bmod p^2$, so that

$$\sum_{l=0}^{p-1} \left(P\left(cp+l\right) - P\left(l\right)\right) \equiv kc \underbrace{p \sum_{l=0}^{p-1} l^{k-1}}_{\equiv 0 \bmod p^2} \equiv 0 \bmod p^2.$$

This proves Lemma 1. □

Lemma 6. *Let p, a and b be integers such that $a-b$ is divisible by p. Then, $a^2 - b^2 \equiv 2\left(a-b\right) b \bmod p^2$.*

Proof of Lemma 6. The difference $\left(a^2-b^2\right) - 2\left(a-b\right)b = \left(a-b\right)^2$ is divisible by p^2 (since $a-b$ is divisible by p). In other words, $a^2-b^2 \equiv 2\left(a-b\right) b \bmod p^2$. Lemma 6 is proven. □

Proof of Lemma 2. Fix $l \in \mathbb{Z}$. We have $P \in \mathbb{Z}\left[X\right]$. Thus, $P\left(u\right) - P\left(v\right)$ is divisible by $u-v$ whenever u and v are two integers[4]. Applying this to $u = cp+l$ and $v = l$, we conclude that $P\left(cp+l\right) - P\left(l\right)$ is divisible by $\left(cp+l\right) - l = cp$, and thus also divisible by p. Hence, Lemma 6 (applied to $a = P\left(cp+l\right)$ and $b = P\left(l\right)$) yields

$$\left(P\left(cp+l\right)\right)^2 - \left(P\left(l\right)\right)^2 \equiv 2\left(P\left(cp+l\right) - P\left(l\right)\right) P\left(l\right) \bmod p^2. \tag{3}$$

Now, forget that we fixed l. We thus have proven (3) for each $l \in \mathbb{Z}$.

The polynomial P has degree $\leq p-1$. Hence, the polynomial P^2 has degree $\leq 2\left(p-1\right) < 2p-1$. Thus, Lemma 1 (applied to P^2 instead of P) shows that

$$\sum_{l=0}^{p-1} \left(P^2\left(cp+l\right) - P^2\left(l\right)\right) \equiv 0 \bmod p^2.$$

Thus,

$$0 \equiv \sum_{l=0}^{p-1} \underbrace{\left(P^2\left(cp+l\right) - P^2\left(l\right)\right)}_{\substack{=\left(P\left(cp+l\right)\right)^2 - \left(P\left(l\right)\right)^2 \\ \equiv 2\left(P\left(cp+l\right) - P\left(l\right)\right) P\left(l\right) \bmod p^2 \\ \text{(by (3))}}} \equiv 2 \sum_{l=0}^{p-1} \left(P\left(cp+l\right) - P\left(l\right)\right) P\left(l\right) \bmod p^2.$$

We can cancel 2 from this congruence (since p is odd), and conclude that

$$0 \equiv \sum_{l=0}^{p-1} \left(P\left(cp+l\right) - P\left(l\right)\right) P\left(l\right) \bmod p^2.$$

This proves Lemma 2. □

[4]This is a well-known fact, and easily proven.

2.5. Applying Lemma 2

Now, let us prepare for the proofs of our results by showing several lemmas.

Lemma 7. *Let p be an odd prime. Let $c \in \mathbb{Z}$. Let $k \in \{0, 1, \ldots, p-1\}$. Then,*

$$\sum_{l=0}^{p-1} \left(\binom{cp+l}{k} - \binom{l}{k} \right) \binom{l}{k} \equiv 0 \mod p^2.$$

Proof of Lemma 7. Notice that $k!$ is coprime to p (since $k \leq p-1$), and thus $k!^2$ is coprime to p^2.

Define a polynomial $P \in \mathbb{Z}[X]$ by $P = X(X-1)\cdots(X-k+1)$. Then, P has degree $k \leq p-1$. Thus, Lemma 2 yields

$$\sum_{l=0}^{p-1} (P(cp+l) - P(l)) P(l) \equiv 0 \mod p^2.$$

Since each $n \in \mathbb{Z}$ satisfies $P(n) = n(n-1)\cdots(n-k+1) = k!\binom{n}{k}$, this can be rewritten as

$$\sum_{l=0}^{p-1} \left(k!\binom{cp+l}{k} - k!\binom{l}{k} \right) k!\binom{l}{k} \equiv 0 \mod p^2.$$

We can cancel $k!^2$ from this congruence (since $k!^2$ is coprime to p^2), and thus obtain

$$\sum_{l=0}^{p-1} \left(\binom{cp+l}{k} - \binom{l}{k} \right) \binom{l}{k} \equiv 0 \mod p^2.$$

This proves Lemma 7. □

Lemma 8. *Let p be an odd prime. Let $c \in \mathbb{Z}$. Then,*

$$\sum_{l=0}^{p-1} \left(\binom{cp+2l}{l} - \binom{2l}{l} \right) \equiv 0 \mod p^2.$$

Proof of Lemma 8. For each $l \in \{0, 1, \ldots, p-1\}$, we have

$$\underbrace{\binom{cp+2l}{l}}_{\substack{=\sum_{k=0}^{p-1}\binom{cp+l}{k}\binom{l}{k} \\ \text{(by Lemma 4)}}} - \underbrace{\binom{2l}{l}}_{\substack{=\sum_{k=0}^{p-1}\binom{l}{k}\binom{l}{k} \\ \text{(by Lemma 4,} \\ \text{applied to 0 instead of } c)}} = \sum_{k=0}^{p-1} \left(\binom{cp+l}{k} - \binom{l}{k} \right) \binom{l}{k}.$$

Summing these equalities over all $l \in \{0, 1, \ldots, p-1\}$, we find

$$\sum_{l=0}^{p-1} \left(\binom{cp+2l}{l} - \binom{2l}{l} \right) = \sum_{l=0}^{p-1} \sum_{k=0}^{p-1} \left(\binom{cp+l}{k} - \binom{l}{k} \right) \binom{l}{k}$$
$$= \sum_{k=0}^{p-1} \underbrace{\sum_{l=0}^{p-1} \left(\binom{cp+l}{k} - \binom{l}{k} \right) \binom{l}{k}}_{\substack{\equiv 0 \bmod p^2 \\ \text{(by Lemma 7)}}} \equiv \sum_{k=0}^{p-1} 0 = 0 \bmod p^2.$$

This proves Lemma 8. □

2.6. η_p Appears

Let us now prove the particular case of Theorem 3 for $r = 1$:

Theorem 7. *Let p be an odd prime. Then,*

$$\sum_{n=0}^{p-1} \binom{2n}{n} \equiv \eta_p \bmod p^2.$$

Proof of Theorem 7. Lemma 8 (applied to $c = -1$) yields

$$\sum_{l=0}^{p-1} \left(\binom{-p+2l}{l} - \binom{2l}{l} \right) \equiv 0 \bmod p^2.$$

Moving all the $\binom{2l}{l}$ addends to the right-hand side of this congruence, we obtain

$$\sum_{l=0}^{p-1} \binom{-p+2l}{l} \equiv \sum_{l=0}^{p-1} \binom{2l}{l} \bmod p^2. \tag{4}$$

Now,

$$\sum_{n=0}^{p-1} \binom{2n}{n} = \sum_{l=0}^{p-1} \binom{2l}{l} \equiv \sum_{l=0}^{p-1} \underbrace{\binom{-p+2l}{l}}_{\substack{=(-1)^l \binom{l-(-p+2l)-1}{l} \\ \text{(by Proposition 4)}}} \qquad \text{(by (4))}$$
$$= \sum_{l=0}^{p-1} (-1)^l \underbrace{\binom{l-(-p+2l)-1}{l}}_{=\binom{p-1-l}{l}} = \sum_{l=0}^{p-1} (-1)^l \binom{p-1-l}{l}$$

$$= \sum_{i=0}^{p-1} (-1)^i \binom{p-1-i}{i} = \underbrace{(-1)^p}_{\substack{=-1 \\ \text{(since } p \text{ is odd)}}} \cdot \begin{cases} 0, & \text{if } p \equiv 0 \bmod 3; \\ -1, & \text{if } p \equiv 1 \bmod 3; \\ 1, & \text{if } p \equiv 2 \bmod 3 \end{cases}$$

(by Proposition 7, applied to $n = p$)

$$= - \begin{cases} 0, & \text{if } p \equiv 0 \bmod 3; \\ -1, & \text{if } p \equiv 1 \bmod 3; \\ 1, & \text{if } p \equiv 2 \bmod 3 \end{cases} = \begin{cases} 0, & \text{if } p \equiv 0 \bmod 3; \\ 1, & \text{if } p \equiv 1 \bmod 3; \\ -1, & \text{if } p \equiv 2 \bmod 3 \end{cases} = \eta_p \bmod p^2.$$

□

2.7. Proving Theorem 3

Lemma 9. *Let $N \in \mathbb{Z}$ and $K \in \mathbb{N}$. Let p be a prime. Let $l \in \{0, 1, \ldots, p-1\}$. Then,*

$$\binom{Np+2l}{Kp+l} - \binom{N}{K}\binom{2l}{l} \equiv N\binom{N}{K}\left(\binom{p+2l}{l} - \binom{2l}{l}\right) \bmod p^2.$$

Proof of Lemma 9. Theorem 2 yields $\binom{Np}{Kp} \equiv \binom{N}{K} \bmod p^2$.

Lemma 3 (applied to $u = Np$ and $w = Kp$) yields

$$\binom{Np+2l}{Kp+l} = \underbrace{\binom{Np}{Kp}}_{\equiv \binom{N}{K} \bmod p^2} \binom{2l}{l} + \sum_{i=1}^{l} \underbrace{\left(\binom{Np}{Kp+i} + \binom{Np}{Kp-i}\right)}_{\substack{\equiv N\binom{N}{K}\binom{p}{i} \bmod p^2 \\ \text{(by Theorem 5 (c))}}} \binom{2l}{l-i}$$

$$\equiv \binom{N}{K}\binom{2l}{l} + \sum_{i=1}^{l} N\binom{N}{K}\binom{p}{i}\binom{2l}{l-i}$$

$$= \binom{N}{K}\binom{2l}{l} + N\binom{N}{K} \underbrace{\sum_{i=1}^{l} \binom{p}{i}\binom{2l}{l-i}}_{\substack{=\binom{p+2l}{l} - \binom{2l}{l} \\ \text{(by Lemma 5)}}}$$

$$= \binom{N}{K}\binom{2l}{l} + N\binom{N}{K}\left(\binom{p+2l}{l} - \binom{2l}{l}\right) \bmod p^2.$$

Subtracting $\binom{N}{K}\binom{2l}{l}$ from both sides of this congruence, we obtain the exact claim of Lemma 9. □

Lemma 10. *Let p be an odd prime. Let $N \in \mathbb{Z}$ and $K \in \mathbb{N}$. Then,*

$$\sum_{l=0}^{p-1} \binom{Np+2l}{Kp+l} \equiv \binom{N}{K} \eta_p \mod p^2.$$

Proof of Lemma 10. For any $l \in \{0, 1, \ldots, p-1\}$, we have (by Lemma 9)

$$\binom{Np+2l}{Kp+l} \equiv \binom{N}{K}\binom{2l}{l} + N\binom{N}{K}\left(\binom{p+2l}{l} - \binom{2l}{l}\right) \mod p^2.$$

Summing these congruences over all $l \in \{0, 1, \ldots, p-1\}$, we prove Lemma 10:

$$\begin{aligned}
\sum_{l=0}^{p-1} \binom{Np+2l}{Kp+l} &\equiv \sum_{l=0}^{p-1} \left(\binom{N}{K}\binom{2l}{l} + N\binom{N}{K}\left(\binom{p+2l}{l} - \binom{2l}{l}\right)\right) \\
&= \binom{N}{K} \sum_{l=0}^{p-1} \binom{2l}{l} + N\binom{N}{K} \underbrace{\sum_{l=0}^{p-1} \left(\binom{p+2l}{l} - \binom{2l}{l}\right)}_{\substack{\equiv 0 \mod p^2 \\ \text{(by Lemma 8, applied to } c=1)}} \\
&\equiv \binom{N}{K} \underbrace{\sum_{l=0}^{p-1} \binom{2l}{l}}_{\substack{= \sum_{n=0}^{p-1} \binom{2n}{n} \equiv \eta_p \mod p^2 \\ \text{(by Theorem 7)}}} \equiv \binom{N}{K} \eta_p \mod p^2.
\end{aligned}$$

□

Proof of Theorem 3. The map $\{0, 1, \ldots, p-1\} \times \{0, 1, \ldots, r-1\} \to \{0, 1, \ldots, rp-1\}$, $(l, K) \mapsto Kp + l$

is a bijection (since each element of $\{0, 1, \ldots, rp-1\}$ can be uniquely divided by p with remainder, and said remainder will belong to $\{0, 1, \ldots, r-1\}$). Thus, we can substitute $Kp + l$ for n in the sum $\sum_{n=0}^{rp-1} \binom{2n}{n}$. This sum thus can be rewritten as

$$\begin{aligned}
\sum_{n=0}^{rp-1} \binom{2n}{n} &= \underbrace{\sum_{(l,K) \in \{0,1,\ldots,p-1\} \times \{0,1,\ldots,r-1\}}}_{= \sum_{K=0}^{r-1} \sum_{l=0}^{p-1}} \underbrace{\binom{2(Kp+l)}{Kp+l}}_{= \binom{2Kp+2l}{Kp+l}} = \sum_{K=0}^{r-1} \underbrace{\sum_{l=0}^{p-1} \binom{2Kp+2l}{Kp+l}}_{\substack{\equiv \binom{2K}{K} \eta_p \mod p^2 \\ \text{(by Lemma 10,} \\ \text{applied to } N=2K)}} \\
&\equiv \underbrace{\sum_{K=0}^{r-1} \binom{2K}{K}}_{= \sum_{n=0}^{r-1} \binom{2n}{n} = \alpha_r} \eta_p = \alpha_r \eta_p = \eta_p \alpha_r \mod p^2.
\end{aligned}$$

This proves Theorem 3. □

2.8. Proving Theorem 4

Lemma 11. *Let p be an odd prime. Let $N \in \mathbb{Z}$ and $K \in \mathbb{N}$. Then,*

$$\sum_{l=0}^{p-1} \sum_{m=0}^{l} \left(\binom{Np+l}{Kp+m} - \binom{N}{K} \binom{l}{m} \right) \binom{l}{m} \equiv 0 \bmod p^2.$$

Proof of Lemma 11. We have

$$\begin{aligned}
&\sum_{l=0}^{p-1} \sum_{m=0}^{l} \left(\binom{Np+l}{Kp+m} - \binom{N}{K} \binom{l}{m} \right) \binom{l}{m} \\
&= \sum_{l=0}^{p-1} \underbrace{\sum_{m=0}^{l} \binom{Np+l}{Kp+m} \binom{l}{m}}_{\substack{= \binom{Np+2l}{Kp+l} \\ \text{(by Corollary 1,} \\ \text{applied to } u=Np+l \text{ and } w=Kp)}} - \binom{N}{K} \sum_{l=0}^{p-1} \underbrace{\sum_{m=0}^{l} \binom{l}{m} \binom{l}{m}}_{\substack{= \binom{2l}{l} \\ \text{(by Corollary 1,} \\ \text{applied to } u=l \text{ and } w=0)}} \\
&= \sum_{l=0}^{p-1} \binom{Np+2l}{Kp+l} - \binom{N}{K} \sum_{l=0}^{p-1} \binom{2l}{l} = \sum_{l=0}^{p-1} \underbrace{\left(\binom{Np+2l}{Kp+l} - \binom{N}{K} \binom{2l}{l} \right)}_{\substack{\equiv N \binom{N}{K} \left(\binom{p+2l}{l} - \binom{2l}{l} \right) \bmod p^2 \\ \text{(by Lemma 9)}}} \\
&\equiv N \binom{N}{K} \underbrace{\sum_{l=0}^{p-1} \left(\binom{p+2l}{l} - \binom{2l}{l} \right)}_{\substack{\equiv 0 \bmod p^2 \\ \text{(by Lemma 8, applied to } c=1)}} \equiv 0 \bmod p^2.
\end{aligned}$$

This proves Lemma 11. □

Lemma 12. *Let p be an odd prime. Let $N \in \mathbb{Z}$ and $K \in \mathbb{N}$. Then,*

$$\sum_{l=0}^{p-1} \sum_{m=0}^{l} \binom{Np+l}{Kp+m}^2 \equiv \binom{N}{K}^2 \eta_p \bmod p^2.$$

Proof of Lemma 12. Fix $l \in \{0, 1, \ldots, p-1\}$ and $m \in \{0, 1, \ldots, p-1\}$. Then, Theorem 1 (applied to $a = N$, $b = K$, $c = l$ and $d = m$) yields that $\binom{Np+l}{Kp+m} \equiv \binom{N}{K} \binom{l}{m} \bmod p$. In other words, $\binom{Np+l}{Kp+m} - \binom{N}{K} \binom{l}{m}$ is divisible by p. Hence,

Lemma 6 (applied to $a = \binom{Np+l}{Kp+m}$ and $b = \binom{N}{K}\binom{l}{m}$) shows that

$$\binom{Np+l}{Kp+m}^2 - \left(\binom{N}{K}\binom{l}{m}\right)^2 \equiv 2\left(\binom{Np+l}{Kp+m} - \binom{N}{K}\binom{l}{m}\right)\binom{N}{K}\binom{l}{m} \mod p^2. \tag{5}$$

Now, forget that we fixed l and m. We thus have proven (5) for all $l \in \{0, 1, \ldots, p-1\}$ and $m \in \{0, 1, \ldots, p-1\}$. Now,

$$\begin{aligned}
&\sum_{l=0}^{p-1}\sum_{m=0}^{l}\binom{Np+l}{Kp+m}^2 - \sum_{l=0}^{p-1}\sum_{m=0}^{l}\left(\binom{N}{K}\binom{l}{m}\right)^2 \\
&= \sum_{l=0}^{p-1}\sum_{m=0}^{l} \underbrace{\left(\binom{Np+l}{Kp+m}^2 - \left(\binom{N}{K}\binom{l}{m}\right)^2\right)}_{\substack{\equiv 2\left(\binom{Np+l}{Kp+m} - \binom{N}{K}\binom{l}{m}\right)\binom{N}{K}\binom{l}{m} \mod p^2 \\ \text{(by (5))}}} \\
&\equiv 2\binom{N}{K}\underbrace{\sum_{l=0}^{p-1}\sum_{m=0}^{l}\left(\binom{Np+l}{Kp+m} - \binom{N}{K}\binom{l}{m}\right)\binom{l}{m}}_{\substack{\equiv 0 \mod p^2 \\ \text{(by Lemma 11)}}} \equiv 0 \mod p^2.
\end{aligned}$$

Thus,

$$\begin{aligned}
\sum_{l=0}^{p-1}\sum_{m=0}^{l}\binom{Np+l}{Kp+m}^2 &\equiv \sum_{l=0}^{p-1}\sum_{m=0}^{l}\left(\binom{N}{K}\binom{l}{m}\right)^2 = \binom{N}{K}^2 \sum_{l=0}^{p-1} \underbrace{\sum_{m=0}^{l}\binom{l}{m}^2}_{\substack{=\sum_{m=0}^{l}\binom{l}{m}\binom{l}{m}=\binom{2l}{l} \\ \text{(by Corollary 1,} \\ \text{applied to } u=l \text{ and } w=0)}} \\
&= \binom{N}{K}^2 \underbrace{\sum_{l=0}^{p-1}\binom{2l}{l}}_{\substack{=\sum_{n=0}^{p-1}\binom{2n}{n} \equiv \eta_p \mod p^2 \\ \text{(by Theorem 7)}}} \\
&\equiv \binom{N}{K}^2 \eta_p \mod p^2.
\end{aligned}$$

This proves Lemma 12. □

Lemma 13. *Let p be a prime. Let $N \in \mathbb{Z}$ and $K \in \mathbb{Z}$. Let u and v be two elements of $\{0, 1, \dots, p-1\}$ satisfying $u + v \geq p$. Then, $p \mid \binom{Np+u+v}{Kp+u}$.*

Proof of Lemma 13. We have $u + v \geq p$. Thus, $u + v = p + c$ for some $c \in \mathbb{N}$. Consider this c. From $v \in \{0, 1, \dots, p-1\}$, we obtain $v < p$. Thus, $c + p = p + c = u + \underbrace{v}_{<p} < u + p$, so that $c < u \leq p - 1$ (since $u \in \{0, 1, \dots, p-1\}$). Thus, $c \in \{0, 1, \dots, p-1\}$ (since $c \in \mathbb{N}$). Also, $c < u$. Hence, Proposition 2 (applied to $m = c$ and $n = u$) yields $\binom{c}{u} = 0$.

Theorem 1 (applied to $a = N + 1$, $b = K$ and $d = u$) yields

$$\binom{(N+1)p+c}{Kp+u} \equiv \binom{N+1}{K} \underbrace{\binom{c}{u}}_{=0} = 0 \bmod p.$$

In other words, $p \mid \binom{(N+1)p+c}{Kp+u}$. In view of $(N+1)p + c = Np + \underbrace{p+c}_{=u+v} = Np + u + v$, this can be rewritten as $p \mid \binom{Np+u+v}{Kp+u}$. This proves Lemma 13. $\square$

Lemma 14. *Let p be an odd prime. Let $N \in \mathbb{Z}$ and $K \in \mathbb{N}$. Then,*

$$\sum_{u=0}^{p-1} \sum_{v=0}^{p-1} \binom{Np+u+v}{Kp+u}^2 \equiv \binom{N}{K}^2 \eta_p \bmod p^2.$$

Proof of Lemma 14. If u and v are two elements of $\{0, 1, \dots, p-1\}$ satisfying $v \geq p - u$, then

$$\binom{Np+u+v}{Kp+u}^2 \equiv 0 \bmod p^2 \tag{6}$$

[5].

Hence, any $u \in \{0, 1, \dots, p-1\}$ satisfies

$$\sum_{v=0}^{p-1} \binom{Np+u+v}{Kp+u}^2 = \sum_{v=0}^{p-u-1} \binom{Np+u+v}{Kp+u}^2 + \underbrace{\sum_{v=p-u}^{p-1} \binom{Np+u+v}{Kp+u}^2}_{\substack{\equiv 0 \bmod p^2 \\ \text{(by (6))}}}$$

[5] *Proof of (6):* Let u and v be two elements of $\{0, 1, \dots, p-1\}$ satisfying $v \geq p - u$. From $v \geq p - u$, we obtain $u + v \geq p$. Thus, Lemma 13 yields $p \mid \binom{Np+u+v}{Kp+u}$. Hence, $p^2 \mid \binom{Np+u+v}{Kp+u}^2$. This proves (6).

$$\equiv \sum_{v=0}^{p-u-1} \binom{Np+u+v}{Kp+u}^2 = \sum_{l=u}^{p-1} \binom{Np+l}{Kp+u}^2 \mod p^2$$

(here, we have substituted l for $u+v$ in the sum). Summing up these congruences for all $u \in \{0, 1, \ldots, p-1\}$, we obtain

$$\begin{aligned}
&\sum_{u=0}^{p-1}\sum_{v=0}^{p-1} \binom{Np+u+v}{Kp+u}^2 \\
&\equiv \underbrace{\sum_{u=0}^{p-1}\sum_{l=u}^{p-1}}_{=\sum_{l=0}^{p-1}\sum_{u=0}^{l}} \binom{Np+l}{Kp+u}^2 = \sum_{l=0}^{p-1}\sum_{u=0}^{l} \binom{Np+l}{Kp+u}^2 = \sum_{l=0}^{p-1}\sum_{m=0}^{l} \binom{Np+l}{Kp+m}^2 \\
&\qquad \text{(here, we have renamed the index } u \text{ as } m \text{ in the second sum)} \\
&\equiv \binom{N}{K}^2 \eta_p \mod p^2
\end{aligned}$$

(by Lemma 12). This proves Lemma 14. □

Proof of Theorem 4. First, let us observe that

$$\begin{aligned}
\epsilon_{r,s} &= \sum_{m=0}^{r-1}\sum_{n=0}^{s-1} \binom{n+m}{m}^2 = \sum_{n=0}^{s-1}\sum_{m=0}^{r-1} \binom{n+m}{m}^2 = \sum_{K=0}^{s-1}\sum_{L=0}^{r-1} \binom{K+L}{L}^2 \\
&= \sum_{K=0}^{s-1}\sum_{L=0}^{r-1} \binom{K+L}{K}^2
\end{aligned} \tag{7}$$

(since Proposition 3 yields $\binom{K+L}{L} = \binom{K+L}{K}$ for all $K \in \mathbb{N}$ and $L \in \mathbb{N}$).

Each $n \in \mathbb{N}$ satisfies

$$\sum_{m=0}^{sp-1} \binom{n+m}{m}^2 = \sum_{u=0}^{p-1}\sum_{K=0}^{s-1} \binom{n+Kp+u}{Kp+u}^2$$

(here, we have substituted $Kp+u$ for m in the sum, since the map

$$\{0, 1, \ldots, p-1\} \times \{0, 1, \ldots, s-1\} \to \{0, 1, \ldots, sp-1\}, \quad (u, K) \mapsto Kp+u$$

is a bijection). Summing up this equality over all $n \in \{0, 1, \ldots, rp-1\}$, we obtain

$$\sum_{n=0}^{rp-1}\sum_{m=0}^{sp-1} \binom{n+m}{m}^2 = \sum_{n=0}^{rp-1}\sum_{u=0}^{p-1}\sum_{K=0}^{s-1} \binom{n+Kp+u}{Kp+u}^2$$

$$= \sum_{v=0}^{p-1} \sum_{L=0}^{r-1} \sum_{u=0}^{p-1} \sum_{K=0}^{s-1} \binom{Lp + v + Kp + u}{Kp + u}^2$$

(here, we have substituted $Lp + v$ for n in the sum, since the map

$$\{0, 1, \ldots, p-1\} \times \{0, 1, \ldots, r-1\} \to \{0, 1, \ldots, rp-1\},$$
$$(v, L) \mapsto Lp + v$$

is a bijection).

Thus,

$$\sum_{n=0}^{rp-1} \sum_{m=0}^{sp-1} \binom{n+m}{m}^2 = \underbrace{\sum_{v=0}^{p-1} \sum_{L=0}^{r-1} \sum_{u=0}^{p-1} \sum_{K=0}^{s-1}}_{= \sum_{K=0}^{s-1} \sum_{L=0}^{r-1} \sum_{u=0}^{p-1} \sum_{v=0}^{p-1}} \underbrace{\binom{Lp + v + Kp + u}{Kp + u}^2}_{= \binom{(K+L)p + u + v}{Kp + u}^2}$$

$$= \sum_{K=0}^{s-1} \sum_{L=0}^{r-1} \underbrace{\sum_{u=0}^{p-1} \sum_{v=0}^{p-1} \binom{(K+L)p + u + v}{Kp + u}^2}_{\substack{\equiv \binom{K+L}{K}^2 \eta_p \bmod p^2 \\ \text{(by Lemma 14, applied to } N=K+L)}}$$

$$\equiv \underbrace{\sum_{K=0}^{s-1} \sum_{L=0}^{r-1} \binom{K+L}{K}^2}_{\substack{= \epsilon_{r,s} \\ \text{(by (7))}}} \eta_p = \epsilon_{r,s} \eta_p = \eta_p \epsilon_{r,s} \bmod p^2.$$

This proves Theorem 4. □

Acknowledgments. Thanks to Doron Zeilberger and Roberto Tauraso for alerting me to [1] and [17].

References

[1] Tewodros Amdeberhan, Roberto Tauraso, Two triple binomial sum supercongruences, *J. Number Theory* **175** (2017), 140-157. A preprint is available at `https://arxiv.org/abs/1607.02483v1`arXiv:1607.02483v1.

[2] Peter G. Anderson, Arthur T. Benjamin and Jeremy A. Rouse, Combinatorial proofs of Fermat's, Lucas's, and Wilson's theorems, *Amer. Math. Monthly* **112** (2005), 266-268.

[3] Moa Apagodu, Doron Zeilberger, Using the "Freshman's Dream" to prove combinatorial congruences, *Amer. Math. Monthly* **124** (2017), 597-608. (A preprint can be found at `https://arxiv.org/abs/1606.03351v2`, but is less up-to-date and uses a different numbering of the conjectures.)

[4] D. F. Bailey, Some binomial coefficient congruences, *Appl. Math. Lett.* **4** (1991), 1-5. `https://doi.org/10.1016/0893-9659(91)90043-U`

[5] Arthur T. Benjamin and Jennifer J. Quinn, *Proofs that Really Count: The Art of Combinatorial Proof*, The Mathematical Association of America, 2003.

[6] Arthur T. Benjamin and Jennifer J. Quinn, An alternate approach to alternating sums: A method to DIE for, *College Math. J.* **39** (2008) 191-202.

[7] Ronald L. Graham, Donald E. Knuth, Oren Patashnik, *Concrete Mathematics, Second Edition*, Addison-Wesley 1994.

[8] Darij Grinberg, *Notes on the combinatorial fundamentals of algebra*, available at `https://github.com/darijgr/detnotes/releases/tag/2018-11-07` (accessed 7 November 2018). See also `http://www.cip.ifi.lmu.de/~grinberg/ primes2015/sols.pdf` for a version that is getting updates.

[9] Darij Grinberg, *The Lucas and Babbage congruences*, available at `http://www.cip.ifi.lmu.de/~grinberg/lucascong.pdf` (accessed 3 October 2018).

[10] Darij Grinberg, *On binomial coefficients modulo squares of primes*, `arXiv:1712.02095v2` (more detailed version of this paper).

[11] Melvin Hausner, Applications of a simple counting technique, *Amer. Math. Monthly* **90** (1983), 127-129.

[12] Ji-Cai Liu, *On two conjectural supercongruences of Apagodu and Zeilberger*, `arXiv:1606.08432v3`.

[13] Kieren MacMillan and Jonathan Sondow, Proofs of power sum and binomial coefficient congruences via Pascal's identity, *Amer. Math. Monthly* **118** (2011), 549-551, `arXiv:1011.0076v1`.

[14] Romeo Meštrović, *Lucas' theorem: its generalizations, extensions and applications (1878–2014)*, `arXiv:1409.3820v1`.

[15] Richard Stanley, *Enumerative Combinatorics, volume 1*, 2nd edition, Cambridge University Press 2012. A preprint is available at `http://math.mit.edu/~rstan/ec/` .

[16] Dennis Stanton, *Addendum to "Using the "Freshman's Dream" to prove combinatorial congruences"*, `http://www.math.rutgers.edu/~zeilberg/mamarim/mamarimhtml/freshmanDennisStanton.pdf`.

[17] Zhi-Wei Sun, Roberto Tauraso, On some new congruences for binomial coefficients, *Int. J. Number Theory* **7**, (2011), 645–662. A preprint is available at `https://arxiv.org/abs/0709.1665v10`.

14

DIVISIBILITY PROPERTIES OF COEFFICIENTS OF MODULAR FUNCTIONS IN GENUS ZERO LEVELS

Victoria Iba

Paul Jenkins[1]
Department of Mathematics, Brigham Young University, Provo Utah
jenkins@math.byu.edu

Merrill Warnick
Department of Mathematics, Brigham Young University, Provo Utah
merrill.warnick@gmail.com

Abstract

We prove divisibility results for the Fourier coefficients of canonical basis elements for the spaces of weakly holomorphic modular forms of weight 0 and levels 6, 10, 12, and 18 with poles only at the cusp at infinity. In addition, we show that these Fourier coefficients satisfy Zagier duality in all weights, and give a general formula for the generating functions of such canonical bases for all genus zero levels.

1. Introduction

In 1949, Lehner [12, 13] showed that many of the Fourier coefficients $c(n)$ of the classical j-function $j(z) = q^{-1} + 744 + \sum c(n)q^n$, where $q = e^{2\pi iz}$ as usual, are highly divisible by small primes dividing n. For instance, he showed that $c(2^a 3^b 5^c 7^d n) \equiv 0 \pmod{2^{3a+8} 3^{2b+3} 5^{c+1} 7^d}$ for any positive integers a, b, c, d. These divisibility results were generalized to stronger congruences modulo larger powers of these primes by Aas [1] and Kolberg [10, 11], and such congruences were shown to hold for all elements of the canonical basis $\{q^{-m} + O(q)\}$ for the space of weakly holomorphic modular functions for $\mathrm{SL}_2(\mathbb{Z})$ by Griffin [6].

Let N be a positive integer such that the group $\Gamma_0(N)$ has genus zero; thus, $N \in \{1, 2, 3, 4, 5, 6, 7, 8, 9, 10, 12, 13, 16, 18, 25\}$. The space of modular functions of genus zero level N with poles only at the cusp at ∞ has a similar canonical basis, and analogous divisibility results have been shown for the Fourier coefficients of the basis elements in the genus zero prime power levels $N \in \{2, 3, 4, 5, 7, 8, 9, 13, 16, 25\}$

[1]This work was partially supported by a grant from the Simons Foundation (#281876 to Paul Jenkins). The authors thank the referee for many helpful suggestions.

in [2, 7, 8, 9]. In this paper, we prove divisibility results for the Fourier coefficients of canonical basis elements in the remaining genus zero levels $N \in \{6, 10, 12, 18\}$. Additionally, we show that the coefficients of these basis elements satisfy Zagier duality, and we explicitly construct their generating functions.

Let $M_k^!(N)$ be the space of weakly holomorphic modular forms for $\Gamma_0(N)$ of even integer weight k, and let $M_k^\sharp(N)$ be the subspace of such forms with poles allowed only at the cusp at ∞. If n_0 is the maximal order of vanishing at the cusp at ∞ for a modular form in $M_k^\sharp(N)$, then the space $M_k^\sharp(N)$ has a countably infinite canonical basis $\{f_{k,m}^{(N)}(z)\}_{m\geq -n_0}$ indexed by the order of vanishing $-m$ of $f_{k,m}^{(N)}(z)$ at the cusp at ∞. Specifically, define $f_{k,m}^{(N)}(z)$ to be the unique form in $M_k^\sharp(N)$ with Fourier expansion given by

$$f_{k,m}^{(N)}(z) = q^{-m} + \sum_{n>n_0} a_k^{(N)}(m,n)q^n,$$

so that the gap between q^{-m} and the next nonzero Fourier coefficient is as large as possible. These bases were explicitly constructed for prime and prime power levels N of genus zero in [3, 5, 7, 8, 9].

Let $S_k^\sharp(N)$ be the subspace of $M_k^\sharp(N)$ consisting of forms which vanish at all cusps except possibly at ∞, and let n_1 be the maximal order of vanishing at the cusp at ∞ for a modular form in this space. As we did for $M_k^\sharp(N)$, we may construct a canonical basis $\{g_{k,m}^{(N)}(z)\}_{m\geq -n_1}$ for $S_k^\sharp(N)$, where $g_{k,m}^{(N)}(z)$ is the unique form in $S_k^\sharp(N)$ with Fourier expansion

$$g_{k,m}^{(N)}(z) = q^{-m} + \sum_{n>n_1} b_k^{(N)}(m,n)q^n;$$

again, the gap between q^{-m} and the next nonzero Fourier coefficient is as large as possible.

The first theorem of this paper is that, just as in prime power genus zero levels, the Fourier coefficients of these canonical bases satisfy Zagier duality.

Theorem 1. *For $N = 6, 10, 12, 18$, all even integers k, and all integers m, n, we have*

$$a_k^{(N)}(m,n) = -b_{2-k}^{(N)}(n,m).$$

Additionally, we have the following formula for generating functions for genus zero level basis elements, generalizing the results in [3, 5, 7, 8] to all genus zero levels.

Theorem 2. *For every genus zero level N we have*

$$\sum_{m=-n_0}^{\infty} f_{k,m}^{(N)}(\tau)q^m = \frac{f_{k,-n_0}^{(N)}(\tau)g_{2-k,n_0+1}^{(N)}(z)}{f_{0,1}^{(N)}(z) - f_{0,1}^{(N)}(\tau)}.$$

The principal results of this paper give divisibility results for coefficients for weight zero basis elements for levels 6, 10, 12 and 18 by primes dividing the level. The following congruences hold in level 6.

Theorem 3. *Let r, s be integers relatively prime to 2. Then*

$$a_0^{(6)}(2^a r, 2^b s) \equiv 0 \ (mod\ 2^{a-b+2}) \qquad \text{if } a > b, \tag{1}$$
$$a_0^{(6)}(2^a r, 2^b s) \equiv 0 \ (mod\ 2^2) \qquad \text{if } b > a. \tag{2}$$

Now let r, s be relatively prime to 3. Then

$$a_0^{(6)}(3^a r, 3^b s) \equiv 0 \ (mod\ 3^{a-b+1}) \qquad \text{if } a > b, \tag{3}$$
$$a_0^{(6)}(3^a r, 3^b s) \equiv 0 \ (mod\ 3^1) \qquad \text{if } b > a. \tag{4}$$

Using the U_2 and U_3 operators, these congruences in level 6 lead to very similar congruences for levels 12 and 18.

Theorem 4. *For $N = 12, 18$, let r, s be integers relatively prime to 2. If $N = 18$, assume that $3|r$. Then*

$$a_0^{(N)}(2^a r, 2^b s) \equiv 0 \ (mod\ 2^{a-b+2}) \qquad \text{if } a > b, \tag{5}$$
$$a_0^{(N)}(2^a r, 2^b s) \equiv 0 \ (mod\ 2^2) \qquad \text{if } b > a. \tag{6}$$

Now let r, s be relatively prime to 3. If $N = 12$, assume that $2|r$. Then

$$a_0^{(N)}(3^a r, 3^b s) \equiv 0 \ (mod\ 3^{a-b+1}) \qquad \text{if } a > b, \tag{7}$$
$$a_0^{(N)}(3^a r, 3^b s) \equiv 0 \ (mod\ 3^1) \qquad \text{if } b > a. \tag{8}$$

If $N = 18$ and $3 \nmid r$, or if $N = 12$ and $2 \nmid r$, we have the weaker divisibility results

$$a_0^{(18)}(2^a r, 2^b s) \equiv 0 \pmod{2^{a-b}} \qquad \text{if } a > b, \tag{9}$$
$$a_0^{(12)}(3^a r, 3^b s) \equiv 0 \pmod{3^{a-b}} \qquad \text{if } a > b. \tag{10}$$

Similar congruences hold for level 10, the last genus zero level.

Theorem 5. *If r, s are relatively prime to 2, then*

$$a_0^{(10)}(2^a r, 2^b s) \equiv 0 \ (mod\ 2^{a-b+1}) \qquad \text{if } a > b, \tag{11}$$
$$a_0^{(10)}(2^a r, 2^b s) \equiv 0 \ (mod\ 2^1) \qquad \text{if } b > a. \tag{12}$$

If r, s are relatively prime to 5, then

$$a_0^{(10)}(5^a r, 5^b s) \equiv 0 \ (mod\ 5^{a-b}) \qquad \text{if } a > b. \tag{13}$$

One obvious difference in level 10 is that there is no $b > a$ congruence given modulo powers of 5. This behavior is similar to the congruences for Fourier coefficients in level 13, as described in [9].

We note that because of the duality given by Theorem 1, all divisibility results in weight 0 immediately imply similar results for the coefficients of the weight 2 basis elements. In higher weights, such divisibility results are not expected due to the presence of cusp forms.

In Section 2 of this paper, we construct the canonical bases in levels 6, 10, 12, and 18. In Section 3, we prove Theorems 1 and 2. Sections 4, 5, and 6 contain the proofs of Theorems 3, 4, and 5.

2. Canonical Basis Construction

We now explicitly construct the canonical bases for $M_k^\sharp(N)$ and $S_k^\sharp(N)$ for $N = 6, 10, 12, 18$. We will give explicit explanations for the construction in level 6 and more general outlines for the other levels, as all of the constructions are similar.

The congruence subgroup $\Gamma_0(6)$ has four cusps: ∞, 0, $\frac{1}{2}$ and $\frac{1}{3}$. For this level, we choose the Hauptmodul

$$\psi^{(6)}(z) = f_{0,1}^{(6)}(z) = \frac{\eta(2z)^8\eta(3z)^4}{\eta(z)^4\eta(6z)^8} - 4 = q^{-1} + 6q + 4q^2 - 3q^3 + O(q^4) \in M_0^\sharp(6),$$

which has a pole of order 1 at ∞. We also use the form

$$f_{2,-2}^{(6)}(z) = \frac{\eta(z)^2\eta(6z)^{12}}{\eta(2z)^4\eta(3z)^6} = q^2 - 2q^3 + 3q^4 - q^6 + 7q^8 + O(q^9) \in M_2(6),$$

which has order of vanishing 2 at ∞ and has no other zeros or poles.

For nonnegative even integers k, the dimension of the space $M_k(6)$ of holomorphic modular forms is $k+1$, so increasing the weight by 2 increases the dimension by 2. Note that multiplying or dividing a form in $M_k^\sharp(6)$ by $f_{2,-2}^{(6)}(z)$ increases or decreases both the form's weight and order of vanishing at ∞ by 2, and can only introduce poles at the cusp at ∞. Thus, in weight k we have the initial basis element $f_{k,-k}^{(6)}(z) = \left(f_{2,-2}^{(6)}(z)\right)^{\frac{k}{2}}$, which has maximal order of vanishing at ∞ for that space. We may recursively construct additional basis elements by multiplying by the Hauptmodul $\psi^{(6)}(z)$ and subtracting earlier basis elements to obtain the largest possible gap between the leading power of q and the next nonzero coefficient. Therefore, for all $m \geq -k$, the basis element $f_{k,m}^{(6)}(z)$ has the form

$$f_{k,m}^{(6)}(z) = q^{-m} + \sum_{n=k+1}^{\infty} a_k^{(6)}(m,n)q^n = \left(f_{2,-2}^{(6)}(z)\right)^{\frac{k}{2}} P\left(\psi^{(6)}(z)\right),$$

where $P(x)$ is a polynomial in x of degree $m+k$. We note that because $\psi^{(6)}(z)$ and $f_{2,-2}^{(6)}(z)$ have integral Fourier coefficients, the Fourier coefficients $a_k^{(6)}(m,n)$ of these basis elements and the coefficients of the polynomial $P(x)$ are also all integers. Additionally, we have $n_0 = k$ for the largest possible order of vanishing at ∞ of a form in $M_k^\sharp(6)$.

The congruence subgroup $\Gamma_0(12)$ has six cusps: ∞, 0, $\frac{1}{2}$, $\frac{1}{3}$, $\frac{1}{4}$ and $\frac{1}{6}$. For this level, we choose the Hauptmodul

$$\psi^{(12)}(z) = f_{0,1}^{(12)}(z) = \frac{\eta(4z)^4\eta(6z)^2}{\eta(2z)^2\eta(12z)^4} = q^{-1} + 2q + q^3 + O(q^7) \in M_0^\sharp(12),$$

which may be used to generate the basis for $M_0^\sharp(12)$.

Increasing a nonnegative even weight k by 2 increases the dimension of $M_k(12)$ from $2k$ to $2k+4$. The form

$$\begin{aligned} f_{2,-4}^{(12)}(z) =& \frac{1}{27}\frac{\eta(z)^{10}\eta(4z)\eta(6z)^9}{\eta(2z)^7\eta(3z)^6\eta(12z)^3} + \frac{11}{72}\frac{\eta(z)^7\eta(4z)^4\eta(6z)^9}{\eta(2z)^7\eta(3z)^5\eta(12z)^4} \\ & - \frac{1}{12}\frac{\eta(z)^4\eta(4z)^7\eta(6z)^9}{\eta(2z)^7\eta(3z)^4\eta(12z)^5} + \frac{1}{54}\frac{\eta(z)\eta(4z)^{10}\eta(6z)^9}{\eta(2z)^7\eta(3z)^3\eta(12z)} \\ & - \frac{1}{8}\frac{\eta(z)^9\eta(4z)^3\eta(6z)^2}{\eta(2z)^6\eta(3z)^3\eta(12z)} \\ =& q^4 + O(q^5) \in M_2(12) \end{aligned}$$

has a zero of order 4 at ∞. We compute that there are no other zeros or poles at any other cusp, and note that as eta-quotients have no poles on the upper half plane, the valence formula implies that there are no other zeros. Thus, the first basis element of $M_k^\sharp(12)$ must be $\left(f_{2,-4}^{(12)}(z)\right)^{\frac{k}{2}}$. Multiplying by $\psi^{(12)}(z)$ and row reducing as before, we construct the basis elements $f_{k,m}^{(12)}(z)$ for all $m \geq -2k$, and note that $n_0 = 2k$ in level 12. Again, all Fourier coefficients $a_k^{(12)}(m,n)$ are integers, because $\psi^{(12)}(z)$ and $f_{2,-4}^{(12)}(z)$ have integral Fourier coefficients by the p-adic Sturm bound.

The congruence subgroup $\Gamma_0(18)$ has six cusps: ∞, 0, $\frac{1}{2}$, $\frac{1}{3}$, $\frac{1}{6}$ and $\frac{1}{9}$. For this level, we choose the Hauptmodul

$$\psi^{(18)}(z) = f_{0,1}^{(18)}(z) = \frac{\eta(6z)\eta(9z)^3}{\eta(3z)\eta(18z)^3}$$

and raise or lower the weight by multiplying by the form

$$\begin{aligned}f_{2,-6}^{(18)}(z) =& \frac{25}{216}\frac{\eta(z)^8\eta(6z)^2\eta(9z)^4}{\eta(2z)^4\eta(3z)^4\eta(18z)^2} - \frac{11}{144}\frac{\eta(z)^3\eta(6z)^8\eta(9z)^7}{\eta(2z)^3\eta(3z)^6\eta(18z)^5}\\ &- \frac{121}{972}\frac{\eta(z)^6\eta(6z)^7\eta(9z)}{\eta(2z)^3\eta(3z)^5\eta(18z)^2} - \frac{41}{144}\frac{\eta(z)^6\eta(6z)^2\eta(9z)^6}{\eta(2z)^3\eta(3z)^4\eta(18z)^3}\\ &+ \frac{67}{144}\frac{\eta(z)^4\eta(6z)^7\eta(9z)^3}{\eta(2z)^2\eta(3z)^5\eta(18z)^3} + \frac{1}{972}\frac{\eta(2z)^9\eta(3z)^8\eta(12z)}{\eta(z)^6\eta(6z)^6\eta(9z)^2}\\ &- \frac{125}{1296}\frac{\eta(z)\eta(2z)^4\eta(9z)^2}{\eta(3z)\eta(6z)\eta(18z)}\\ =& q^6 + O(q^7) \in M_2(18).\end{aligned}$$

Each basis element in $M_k^\sharp(18)$ will be $\left(f_{2,-6}^{(18)}(z)\right)^{\frac{k}{2}}$ multiplied by a polynomial in $\psi^{(18)}(z)$, and we have $n_0 = 3k$. Again, all Fourier coefficients $a_k^{(18)}(m,n)$ are integers.

The congruence subgroup $\Gamma_0(10)$ has four cusps: ∞, 0, $\frac{1}{2}$ and $\frac{1}{5}$. For this level, we use the Hauptmodul

$$\psi^{(10)}(z) = f_{0,1}^{(10)}(z) = \frac{\eta(2z)\eta(5z)^5}{\eta(z)\eta(10z)^5} - 1.$$

Increasing the weight k by 4 increases the dimension of $M_k(10)$ by 6, with the dimension of $M_0(10)$ equal to 1 and the dimension of $M_2(10)$ equal to 3. Writing $k = 4\ell + k'$ with $\ell \in \mathbb{Z}$ and $k' \in \{0,2\}$, we use the forms

$$f_{2,-2}^{(10)} = \frac{\eta(10z)^5}{\eta(2z)} = q^3 + O(q^4) \in M_2(10),$$

$$\begin{aligned}f_{4,-6}^{(10)} =& \frac{27}{10000}\frac{\eta(2z)^{14}\eta(5z)^8}{\eta(z)^8\eta(10z)^6} - \frac{13}{4000}\frac{\eta(2z)^{11}\eta(5z)^{11}}{\eta(z)^7\eta(10z)^7} - \frac{14}{125}\frac{\eta(z)^3\eta(2z)\eta(5z)^9}{\eta(10z)^5}\\ &+ \frac{167}{625}\frac{\eta(z)^2\eta(2z)^4\eta(5z)^6}{\eta(10z)^4} + \frac{1}{625}\frac{\eta(z)^{12}\eta(5z)^4}{\eta(2z)^6\eta(10z)^2}\\ &- \frac{7}{32}\frac{\eta(z)\eta(2z)^7\eta(5z)^3}{\eta(10z)^3} + \frac{1}{16}\frac{\eta(2z)^{10}}{\eta(10z)^2}\\ =& q^6 + O(q^7) \in M_4(10)\end{aligned}$$

to see that the first basis element in weight $k = 4\ell + k'$ will be

$$\left(f_{4,-6}^{(10)}(z)\right)^{\ell}\left(f_{2,-2}^{(10)}(z)\right)^{\frac{k'}{2}}.$$

Further basis elements are obtained, as before, by multiplying by a polynomial in $\psi^{(10)}(z)$ to get the correct Fourier expansion, with $n_0 = 6\ell + k'$ and $a_k^{(10)}(m,n) \in \mathbb{Z}$.

To construct the basis elements $g_{k,m}^{(N)}(z)$ for the spaces $S_k^\sharp(N)$, we follow the same general process of finding the basis element with the largest leading power of q, then multiplying by the Hauptmodul and row reducing with previous basis elements. The first basis element $g_{k,-n_1}^{(N)}(z)$ can be constructed as

$$g_{k,-n_1}^{(N)}(z) = f_{k,-n_0}^{(N)}(z) \prod_{\frac{a}{b}} \left(\psi^{(N)}(z) - c_{\frac{a}{b}}\right),$$

where $f_{k,-n_0}^{(N)}(z)$ is the first basis element for $M_k^\sharp(N)$, the product is over all non-∞ cusps of $\Gamma_0(N)$, and $c_{\frac{a}{b}}$ is the value of $\psi^{(N)}(z)$ at the cusp $\frac{a}{b}$. It is clear that this will be a form of the correct weight which vanishes at all non-∞ cusps, and it must have maximal order of vanishing at ∞ in $S_k^\sharp(N)$ because $f_{k,-n_0}^{(N)}(z)$ does in $M_k^\sharp(N)$, and the existence of a Hauptmodul for $\Gamma_0(N)$ means that $f_{k,-n_0}^{(N)}(z)$ cannot vanish at any cusp other than ∞. Therefore, we have $n_1 = n_0 - c(N) + 1$, where $c(N)$ is the number of cusps of $\Gamma_0(N)$.

In level 6, the Hauptmodul has the values 0, 1, 9 at the three non-infinity cusps. This may be computed using Ligozat's theorem (see [14]) to find that the form

$$g_{2,1}^{(6)}(z) = \frac{\eta(2z)^6\eta(3z)^8}{\eta(6z)^{10}} \in M_2^\sharp(6)$$

vanishes at all non-∞ cusps, and then writing the quotient $g_{2,1}^{(6)}(z)/f_{2,-2}^{(6)}(z)$ as a polynomial in $\psi^{(6)}(z)$. The roots of the polynomial are the values at the cusps. Thus, the first basis element of $S_k^\sharp(6)$ will be

$$g_{k,-k+3}^{(6)}(z) = \left(f_{2,-2}^{(6)}(z)\right)^{\frac{k}{2}} \psi^{(6)}(z)(\psi^{(6)}(z) - 1)(\psi^{(6)}(z) - 9).$$

In level 12, the polynomial

$$(\psi^{(12)}(z))(\psi^{(12)}(z) - 1)(\psi^{(12)}(z) + 1)(\psi^{(12)}(z) - 3)(\psi^{(12)}(z) + 3)$$

vanishes at all non-∞ cusps, while levels 18 and 10 have the polynomials

$$(\psi^{(18)}(z))^7 - 7(\psi^{(18)}(z))^4 - 8(\psi^{(18)}(z)),$$
$$(\psi^{(10)}(z) + 2)(\psi^{(10)}(z) + 1)(\psi^{(10)}(z) - 3),$$

which vanish at all cusps away from ∞.

3. Duality and Generating Functions

The proof of Zagier duality for the coefficients of these canonical bases (Theorem 1) follows the argument in [8] for prime power levels, which is itself a generalization of an argument credited to Kaneko by Zagier [15].

Recall that Ramanujan's θ-operator acts on a weakly holomorphic modular form f by $\theta f = q\frac{d}{dq} f = \frac{1}{2\pi i}\frac{d}{dz} f$ and increases the weight by 2, preserving modularity if f is weight 0. With the usual slash operator $f|_k\gamma = (\det\gamma)^{k/2}(cz+d)^{-k} f\left(\frac{az+b}{cz+d}\right)$ for a matrix $\gamma = \left(\begin{smallmatrix} a & b \\ c & d\end{smallmatrix}\right)$, the θ-operator satisfies the intertwining relation

$$\theta(f|_0\gamma) = (\theta f)|_2\gamma.$$

For a form $f \in M_0^\sharp(N)$, we know that $f|_0\gamma$ is holomorphic for any matrix $\gamma \in \mathrm{SL}_2(\mathbb{Z})$ and not in $\Gamma_0(N)$, so the intertwining relation and the fact that derivatives annihilate constants imply that θ actually maps forms in $M_0^\sharp(N)$ to forms in $S_2^\sharp(N)$. Specifically, by examining Fourier expansions we find that

$$\theta\left(f_{0,m}^{(N)}(z)\right) = -mg_{2,m}^{(N)}(z). \tag{14}$$

In weight 2, the maximal order of vanishing n_1 for a form in $S_2^\sharp(N)$ is equal to -1 for $N = 6, 10, 12, 18$, so the functions $\theta\left(f_{0,m}^{(N)}\right) = -mg_{2,m}^{(N)}$ span the space $S_2^\sharp(N)$, and every form in $S_2^\sharp(N)$ must have no constant term in its Fourier expansion.

Now consider the product $f_{k,m}^{(N)}g_{2-k,m}^{(N)}$, which must be in the space $S_2^\sharp(N)$. By examining the Fourier expansions of the forms in the product, we find that its constant term is $a_k^{(N)}(m,n) + b_{2-k}^{(N)}(n,m)$, which must be zero, and Theorem 1 follows.

We next prove Theorem 2; the proof is similar to that in [4]. Fix an even integer weight k and a genus zero level N. Let n_0, N_1 be the maximal orders of vanishing at ∞ for forms in $M_k^\sharp(N)$, $S_{2-k}^\sharp(N)$ respectively. Thus, we have

$$f_{k,m}^{(N)}(z) = q^{-m} + \sum_{n>n_0} a_k(m,n)q^n$$

for every integer $m \geq -n_0$, and

$$g_{2-k,m}^{(N)}(z) = q^{-m} + \sum_{n>N_1} b_{2-k}(m,n)q^n$$

for every integer $m \geq -N_1$. Let

$$F_k(z,\tau) = \sum_{m=-n_0}^{\infty} f_{k,m}^{(N)}(\tau)q^m$$

be the generating function for the $f_{k,m}^{(N)}(z)$. In what follows, we often write $f_{k,m}(z)$ for $f_{k,m}^{(N)}(z)$ for simplicity.

Multiply the Fourier expansions for $f_{0,1}(z)$ and $f_{k,m}(z)$ and write the result, which is a form in $M_k^\sharp(N)$, in terms of basis elements, remembering that the powers of q less than or equal to n_0 completely determine the form. Doing this, we obtain

$$\begin{aligned} f_{0,1}(z)f_{k,m}(z) =& \left(q^{-1} + \sum_{n=1}^{\infty} a_0(1,n)q^n\right)\left(q^{-m} + \sum_{n>n_0} a_k(m,n)q^n\right) \\ =& q^{-m-1} + \sum_{n=1}^{\infty} a_0(1,n)q^{n-m} + \sum_{n>n_0} a_k(m,n)q^{n-1} \\ &+ \sum_{n=1}^{\infty}\sum_{N>n_0} a_0(1,n)a_k(m,N)q^{N+n} \\ =& f_{k,m+1}(z) + a_k(m,n_0+1)f_{k,-n_0}(z) + a_0(1,n_0+m)f_{k,-n_0}(z) \\ &+ \sum_{n=1}^{n_0+m-1} a_0(1,n)f_{k,m-n}(z) \end{aligned}$$

for any $m \geq -n_0$, where the last sum appears only if $m \geq -n_0+2$ and the second to last term appears only if $m \geq -n_0+1$. This gives us a recurrence relation for $f_{k,m+1}(z)$ in terms of forms $f_{k,i}(z)$ with $i \leq m$.

Using this recurrence, we may write

$$\begin{aligned} F_k(z,\tau) =& f_{k,-n_0}(\tau)q^{-n_0} + \sum_{m=-n_0+1}^{\infty} f_{k,m}(\tau)q^m \\ =& f_{k,-n_0}(z)q^{-n_0} + \sum_{m=-n_0}^{\infty} f_{k,m+1}(z)q^{m+1} \\ =& f_{k,-n_0}(\tau)q^{-n_0} + \sum_{m=-n_0}^{\infty} (f_{0,1}(\tau)f_{k,m}(\tau)q^{m+1} - a_k(m,n_0+1)f_{k,-n_0}(\tau)q^{m+1} \\ &- a_0(1,n_0+m)f_{k,-n_0}(\tau)q^{m+1} - \sum_{n=1}^{n_0+m-1} a_0(1,n)f_{k,m-n}(\tau)q^{m+1}). \end{aligned}$$

Using the duality $a_k(m,n_0+1) = -b_{2-k}(n_0+1,m)$, we note that the term

$$-\sum_{m=-n_0}^{\infty} a_k(m,n_0+1)f_{k,-n_0}(\tau)q^{m+1}$$

may, if $N_1 \geq -n_0-1$, be rewritten as $q\cdot f_{k,-n_0}(\tau)(g_{2-k,n_0+1}(z) - q^{-n_0-1})$. We note also that this inequality is true (and is, in fact, an equality) for all genus zero levels. Reversing the order of the double summation, we rewrite other terms similarly to obtain

$$\begin{aligned} F_k(z,\tau) =& f_{k,-n_0}(\tau)q^{-n_0} + qf_{0,1}(\tau)F_k(z,\tau) \\ &+ f_{k,-n_0}(\tau)q(g_{2-k,n_0+1}(z) - q^{-n_0-1}) - f_{k,-n_0}(\tau)q^{1-n_0}(f_{0,1}(z) - q^{-1}) \\ &- q(f_{0,1}(z) - q^{-1})(F_k(z,\tau) - f_{k,-n_0}(\tau)q^{-n_0}), \end{aligned}$$

which simplifies to

$$0 = qf_{0,1}(\tau)F_k(z,\tau) + qf_{k,-n_0}(\tau)g_{2-k,n_0+1}(z) - qf_{0,1}(z)F_k(z,\tau).$$

Therefore, we have the generating function

$$F_k(z,\tau) = \frac{f_{k,-n_0}(\tau)g_{2-k,n_0+1}(z)}{f_{0,1}(z) - f_{0,1}(\tau)}.$$

4. Level 6 Congruences

We will now prove Theorem 3 by first proving the congruence (2), and then showing that (2) implies (1).

We use the usual slash operator defined at the beginning of Section 3, omitting the subscript k for brevity. If $f \in M_0^\sharp(6)$ and $\gamma \in \mathrm{SL}_2(\mathbb{Z})$, this becomes $f|\gamma = f\left(\frac{az+b}{cz+d}\right)$. Recall that if $p|N$, the U_p operator

$$U_p f = \frac{1}{p}\sum_{i=0}^{p-1} f|\left(\begin{smallmatrix}1 & i\\ 0 & p\end{smallmatrix}\right)$$

sends forms in $M_k^!(N)$ to forms in $M_k^!(N)$, while if $p^2|N$, the U_p operator sends forms in $M_k^!(N)$ to forms in $M_k^!(\frac{N}{p})$. Its action on Fourier expansions is given by $U_p \sum a(n)q^n = \sum a(pn)q^n$. Thus, the U_2 operator on forms in $M_0^\sharp(6)$ can be written as

$$2U_2 f(z) = f|\left(\begin{smallmatrix}1 & 0\\ 0 & 2\end{smallmatrix}\right) + f|\left(\begin{smallmatrix}1 & 1\\ 0 & 2\end{smallmatrix}\right).$$

We are interested in the order of vanishing of $U_2 f(z)$ at the other cusps of $\Gamma_0(6)$, which are $0, \frac{1}{2}$, and $\frac{1}{3}$. Computing, we find that

$$\begin{aligned}
2U_2 f|\left(\begin{smallmatrix}0 & -1\\ 1 & 0\end{smallmatrix}\right) &= f|\left(\begin{smallmatrix}0 & -1\\ 1 & 0\end{smallmatrix}\right)|\left(\begin{smallmatrix}2 & 0\\ 0 & 1\end{smallmatrix}\right) + f|\left(\begin{smallmatrix}1 & 0\\ 2 & 1\end{smallmatrix}\right)|\left(\begin{smallmatrix}1 & -1\\ 0 & 2\end{smallmatrix}\right),\\
2U_2 f|\left(\begin{smallmatrix}1 & 0\\ 2 & 1\end{smallmatrix}\right) &= f|\left(\begin{smallmatrix}1 & 0\\ 4 & 1\end{smallmatrix}\right)|\left(\begin{smallmatrix}1 & 0\\ 0 & 2\end{smallmatrix}\right) + f|\left(\begin{smallmatrix}3 & -1\\ 4 & -1\end{smallmatrix}\right)|\left(\begin{smallmatrix}1 & 1\\ 0 & 2\end{smallmatrix}\right),\\
2U_2 f|\left(\begin{smallmatrix}1 & 0\\ 3 & 1\end{smallmatrix}\right) &= f|\left(\begin{smallmatrix}1 & 0\\ 6 & 1\end{smallmatrix}\right)|\left(\begin{smallmatrix}1 & 0\\ 0 & 2\end{smallmatrix}\right) + f|\left(\begin{smallmatrix}2 & -1\\ 3 & -1\end{smallmatrix}\right)|\left(\begin{smallmatrix}2 & 1\\ 0 & 1\end{smallmatrix}\right).
\end{aligned}$$

Under the action of $\Gamma_0(6)$, we have $\frac{1}{4}$ equivalent to $\frac{1}{2}$, and $\frac{3}{4}$ also equivalent to $\frac{1}{2}$. Since replacing z with $\frac{az+b}{d}$ does not change holomorphicity at a cusp, it follows that if $f \in M_0^!(6)$ is holomorphic at 0 and $\frac{1}{2}$, then $U_2 f$ is as well. Therefore, applying U_2 repeatedly to a form in $M_0^\sharp(6)$ will give a form in $M_0^!(6)$ with poles that may occur only at the cusps at ∞ and $\frac{1}{3}$.

We note that for a form in $M_0^!(6)$, slashing with the matrix $\left(\begin{smallmatrix}2 & -1\\ 6 & -2\end{smallmatrix}\right)$ is an Atkin-Lehner involution which takes a form in $M_0^!(6)$ to a form in $M_0^!(6)$. Computing as before, we find that

$$f|\left(\begin{smallmatrix}2 & -1\\ 6 & -2\end{smallmatrix}\right)|\left(\begin{smallmatrix}0 & -1\\ 1 & 0\end{smallmatrix}\right) = f|\left(\begin{smallmatrix}-1 & -2\\ -2 & -6\end{smallmatrix}\right) = f|\left(\begin{smallmatrix}-1 & 0\\ -2 & -1\end{smallmatrix}\right)|\left(\begin{smallmatrix}1 & 2\\ 0 & 2\end{smallmatrix}\right),$$

$$f|\left(\begin{smallmatrix}2&-1\\6&-2\end{smallmatrix}\right)|\left(\begin{smallmatrix}1&0\\2&1\end{smallmatrix}\right)=f|\left(\begin{smallmatrix}0&-1\\2&-2\end{smallmatrix}\right)=f|\left(\begin{smallmatrix}0&1\\-1&1\end{smallmatrix}\right)|\left(\begin{smallmatrix}-2&1\\0&-1\end{smallmatrix}\right),$$
$$f|\left(\begin{smallmatrix}2&-1\\6&-2\end{smallmatrix}\right)|\left(\begin{smallmatrix}1&0\\3&1\end{smallmatrix}\right)=f|\left(\begin{smallmatrix}-1&-1\\0&-2\end{smallmatrix}\right).$$

These calculations show that if f is holomorphic at ∞, then $f|\left(\begin{smallmatrix}2&-1\\6&-2\end{smallmatrix}\right)$ is holomorphic at $\frac{1}{3}$; similarly, if f is holomorphic at $0, \frac{1}{2}$, or $\frac{1}{3}$, then $f|\left(\begin{smallmatrix}2&-1\\6&-2\end{smallmatrix}\right)$ is holomorphic at $\frac{1}{2}, 0$, or ∞ respectively.

We now show that if $b > a$ and r, s are odd integers, then the coefficient $a_0^{(6)}(2^a r, 2^b s)$ is divisible by 4, proving equation (2). Suppose first that $a = 0$. Applying the U_2 operator to the form $f_{0,r}^{(6)}(z) = q^{-r} + \sum_{n\geq 1} a_0^{(6)}(r,n)q^n$ gives us a modular form in $M_0^!(6)$ with Fourier expansion $\sum_{n\geq 1} a_0^{(6)}(r,2n)q^n$ which is clearly holomorphic at ∞; from the discussion above, it must also be holomorphic at 0 and $\frac{1}{2}$, since $f_{0,r}^{(6)}$ is as well. Thus, $U_2 f_{0,r}^{(6)}(z)$ has poles only at $\frac{1}{3}$.

Replacing z with $\frac{2z-1}{6z-2}$ in the formula for $U_2 f(z)$ above, we compute that

$$2U_2 f_{0,r}^{(6)}\left(\frac{2z-1}{6z-2}\right) = f_{0,r}^{(6)}\left(z-\frac{1}{2}\right) + f_{0,r}^{(6)}\left(\frac{8z-3}{12z-4}\right).$$

The first term gives $-q^{-r}+O(1)$ in the Fourier expansion since r is odd. The second term is holomorphic at ∞ because $\left(\begin{smallmatrix}8&-3\\12&-4\end{smallmatrix}\right) = \left(\begin{smallmatrix}2&-3\\3&-4\end{smallmatrix}\right)\left(\begin{smallmatrix}4&0\\0&1\end{smallmatrix}\right)$ and $f_{0,r}^{(6)}|\gamma$ is holomorphic at ∞ for any $\gamma \in \mathrm{SL}_2(\mathbb{Z})$ not in $\Gamma_0(6)$. Slashing with $\left(\begin{smallmatrix}0&-1\\1&0\end{smallmatrix}\right), \left(\begin{smallmatrix}1&0\\2&1\end{smallmatrix}\right), \left(\begin{smallmatrix}1&0\\3&1\end{smallmatrix}\right)$ as above, we find that this form is holomorphic at 0 and $\frac{1}{2}$, and that at the cusp at $\frac{1}{3}$ it is equal to $f_{0,r}^{(6)}(\frac{z+1}{4}) + f_{0,r}^{(6)}(\frac{z+3}{4})$. Replacing z by $\frac{z+1}{4}$ and $\frac{z+3}{4}$ in the Fourier expansion of $f_{0,r}^{(6)}(z)$, we find that the negative powers of q cancel because r is odd, and conclude that the form is also holomorphic at $\frac{1}{3}$. Since its Fourier expansion begins $-q^{-r}$, it must be $-f_{0,r}^{(6)}(z)$.

Let $\psi(z)$ be the Hauptmodul $\psi(z) = \frac{\eta(z)^5\eta(3z)}{\eta(2z)\eta(6z)^5} \in M_0^\sharp(6)$; we note that $\psi(z)$ has a pole of order 1 at ∞ and vanishes at the cusp at 0. From the construction of the basis elements $f_{0,m}^{(6)}(z)$, we may write each basis element as an integral polynomial in powers of $\psi(z)$, giving us

$$2U_2 f_{0,r}^{(6)}\left(\frac{2z-1}{6z-2}\right) = -f_{0,r}^{(6)}(z) = \sum_{i=0}^{r} c_i\psi^i(z),$$

with the coefficients $c_i \in \mathbb{Z}$.

Using the equalities $\eta(z+1) = e^{2\pi i/24}\eta(z)$ and $\eta(-1/z) = \sqrt{z/i}\,\eta(z)$, we compute that

$$\psi|\left(\begin{smallmatrix}2&-1\\6&2\end{smallmatrix}\right) = \frac{-2^3\eta(2z)^5\eta(6z)}{\eta(3z)^5\eta(z)}.$$

We note that $\psi_{\frac{1}{3}}(z) = \frac{\eta(2z)^5\eta(6z)}{\eta(3z)^5\eta(z)}$ is a modular form in $M_0^!(6)$ with integer coefficients; it is holomorphic at ∞ and 0, vanishes at $\frac{1}{2}$, and has a pole of order 1 at $\frac{1}{3}$.

Replacing z with $\frac{2z-1}{6z-2}$ in our expression for $2U_2 f_{0,r}^{(6)}$ above, we find that

$$2U_2 f_{0,r}^{(6)}(z) = \sum_{i=0}^{r} c_i 2^{3i} \psi_{\frac{1}{3}}(z)^i.$$

Because the c_i and all Fourier coefficients of $\psi_{\frac{1}{3}}$ are integers, we conclude that for $n \geq 1$, all of the Fourier coefficients $a_0^{(6)}(r, 2n)$ of $U_2 f_{0,r}^{(6)}(z)$ must be divisible by 4.

Now suppose that $a \geq 1$ and that the coefficients $a_0^{(6)}(2^\alpha r', 2^\beta s')$ are divisible by 4 for every nonnegative integer $\alpha < a$ and every $\beta > \alpha$ and odd r', s'. We will show that $a_0^{(6)}(2^a r, 2^b s)$ is divisible by 4 for every integer $b > a$, proving the theorem by induction.

Define $G(z)$ to be the form $U_2 f_{0,2^a r}^{(6)}(z) - f_{0,2^{a-1}r}^{(6)}(z)$. We note that $G(z)$ is holomorphic at ∞ because the $q^{2^{a-1}r}$ terms cancel, that it is holomorphic at 0 and $\frac{1}{2}$ and meromorphic at $\frac{1}{3}$, and that its nth Fourier coefficient is $a_0^{(6)}(2^a r, 2n) - a_0^{(6)}(2^{a-1}r, n)$. Replacing z by $\frac{2z-1}{6z-2}$ as before, we have the form

$$2G\left(\frac{2z-1}{6z-2}\right) = f_{0,2^a r}^{(6)}\left(z - \frac{1}{2}\right) + f_{0,2^a r}^{(6)}\left(\frac{8z-3}{12z-4}\right) - 2 f_{0,2^{a-1}r}^{(6)}\left(\frac{2z-1}{6z-2}\right),$$

which is holomorphic away from ∞. The second two terms are holomorphic at ∞, and the first term gives a $q^{-2^a r}$, so this must be $f_{0,2^a r}(z)$, which has integer coefficients. Thus, we have

$$2G\left(\frac{2z-1}{6z-2}\right) = \sum_{i=0}^{2^a r} c_i \psi(z)^i,$$

with the coefficients $c_i \in \mathbb{Z}$. Replacing z with $\frac{2z-1}{6z-2}$ as before, we get

$$2G(z) = \sum_{i=0}^{2^a r} c_i 2^{3i-1} \psi_{\frac{1}{3}}(z)^i. \tag{15}$$

Applying the U_2 operator a times to this equation, we get a form $U_2^a G(z) \in M_0^!(6)$ with poles only at the cusp at $\frac{1}{3}$ and with nth Fourier coefficient given by

$$a_0^{(6)}(2^a r, 2^{a+1} n) - a_0^{(6)}(2^{a-1} r, 2^a n).$$

By our inductive hypothesis, the second term in this expression is divisible by 4. Since each positive power of q^n in (15) has a coefficient divisible by 4 as well, we conclude that $a_0^{(6)}(2^a r, 2^{a+1} n)$ is divisible by 4, proving congruence (2).

Now consider the coefficient $a_0(2^a r, 2^b s)$ with $a > b$. By duality and equation (14), we have

$$a_0(2^a r, 2^b s) = \frac{-2^a r}{2^b s} b_2(2^a r, 2^b s) = \frac{2^a r}{2^b s} a_0(2^b r, 2^a s).$$

Note that $\frac{r}{s}a_0(2^b r, 2^a s)$ must be an integer because all basis elements have integer coefficients in level 6. Then by the congruence just proved, since $a > b$, we have $2^2 | a_0(2^b r, 2^a s)$, and it follows that $2^{a-b+2} | a_0(2^a r, 2^b s)$.

The proof of congruences for powers of 3 in level 6 is very similar; we will note explicit calculations that differ. The U_3 operator on forms $f \in M_0^\sharp(6)$ can be written as

$$3U_3 f(z) = f | \left(\begin{smallmatrix} 1 & 0 \\ 0 & 3 \end{smallmatrix}\right) + f | \left(\begin{smallmatrix} 1 & 1 \\ 0 & 3 \end{smallmatrix}\right) + f | \left(\begin{smallmatrix} 1 & 2 \\ 0 & 3 \end{smallmatrix}\right).$$

To check the order of vanishing of $U_3 f(z)$ at the cusps 0, $\frac{1}{2}$ and $\frac{1}{3}$ of $\Gamma_0(6)$, we compute that

$$3U_3 f | \left(\begin{smallmatrix} 0 & -1 \\ 1 & 0 \end{smallmatrix}\right) = f | \left(\begin{smallmatrix} 0 & -1 \\ 1 & 0 \end{smallmatrix}\right) | \left(\begin{smallmatrix} 3 & 0 \\ 0 & 1 \end{smallmatrix}\right) + f | \left(\begin{smallmatrix} 1 & 0 \\ 3 & 1 \end{smallmatrix}\right) | \left(\begin{smallmatrix} 1 & -1 \\ 0 & 3 \end{smallmatrix}\right) + f | \left(\begin{smallmatrix} 2 & -1 \\ 3 & -1 \end{smallmatrix}\right) | \left(\begin{smallmatrix} 1 & 1 \\ 0 & 3 \end{smallmatrix}\right),$$

$$3U_3 f | \left(\begin{smallmatrix} 1 & 0 \\ 3 & 1 \end{smallmatrix}\right) = f | \left(\begin{smallmatrix} 1 & 0 \\ 9 & 1 \end{smallmatrix}\right) | \left(\begin{smallmatrix} 1 & 0 \\ 0 & 3 \end{smallmatrix}\right) + f | \left(\begin{smallmatrix} 4 & -1 \\ 9 & -2 \end{smallmatrix}\right) | \left(\begin{smallmatrix} 1 & 1 \\ 0 & 3 \end{smallmatrix}\right) + f | \left(\begin{smallmatrix} 7 & 3 \\ 9 & 4 \end{smallmatrix}\right) | \left(\begin{smallmatrix} 1 & -1 \\ 0 & 3 \end{smallmatrix}\right),$$

$$3U_3 f | \left(\begin{smallmatrix} 1 & 0 \\ 2 & 1 \end{smallmatrix}\right) = f | \left(\begin{smallmatrix} 1 & 0 \\ 6 & 1 \end{smallmatrix}\right) | \left(\begin{smallmatrix} 1 & 0 \\ 0 & 3 \end{smallmatrix}\right) + f | \left(\begin{smallmatrix} 1 & 0 \\ 2 & 1 \end{smallmatrix}\right) | \left(\begin{smallmatrix} 3 & 1 \\ 0 & 1 \end{smallmatrix}\right) + f | \left(\begin{smallmatrix} 5 & -1 \\ 6 & -1 \end{smallmatrix}\right) | \left(\begin{smallmatrix} 1 & 1 \\ 0 & 3 \end{smallmatrix}\right).$$

Since the cusps at $\frac{2}{3}$, $\frac{1}{9}$, $\frac{4}{9}$, and $\frac{7}{9}$ are equivalent to the cusp at $\frac{1}{3}$, if $f(z)$ is holomorphic at 0 and $\frac{1}{3}$, then $U_3 f(z)$ will be also. Thus, applying U_3 repeatedly to a form in $M_0^\sharp(6)$ gives a form with poles only at $\frac{1}{2}$ and ∞.

We again use an Atkin-Lehner involution $\left(\begin{smallmatrix} 3 & 1 \\ 6 & 3 \end{smallmatrix}\right)$. Computing as before, we find that

$$f | \left(\begin{smallmatrix} 3 & 1 \\ 6 & 3 \end{smallmatrix}\right) | \left(\begin{smallmatrix} 0 & -1 \\ 1 & 0 \end{smallmatrix}\right) = f | \left(\begin{smallmatrix} 1 & -3 \\ 3 & -6 \end{smallmatrix}\right) = f | \left(\begin{smallmatrix} 1 & 0 \\ 3 & 1 \end{smallmatrix}\right) | \left(\begin{smallmatrix} 1 & -3 \\ 0 & 3 \end{smallmatrix}\right),$$

$$f | \left(\begin{smallmatrix} 3 & 1 \\ 6 & 3 \end{smallmatrix}\right) | \left(\begin{smallmatrix} 1 & 0 \\ 2 & 1 \end{smallmatrix}\right) = f | \left(\begin{smallmatrix} 5 & 1 \\ 12 & 3 \end{smallmatrix}\right) = f | \left(\begin{smallmatrix} 5 & 2 \\ 12 & 5 \end{smallmatrix}\right) | \left(\begin{smallmatrix} 1 & -1 \\ 0 & 3 \end{smallmatrix}\right) = f | \left(\begin{smallmatrix} 1 & -1 \\ 0 & 3 \end{smallmatrix}\right),$$

$$f | \left(\begin{smallmatrix} 3 & 1 \\ 6 & 3 \end{smallmatrix}\right) | \left(\begin{smallmatrix} 1 & 0 \\ 3 & 1 \end{smallmatrix}\right) = f | \left(\begin{smallmatrix} 6 & 1 \\ 15 & 3 \end{smallmatrix}\right) = f | \left(\begin{smallmatrix} 2 & 3 \\ 5 & 8 \end{smallmatrix}\right) | \left(\begin{smallmatrix} 3 & -1 \\ 0 & 1 \end{smallmatrix}\right)$$

for any $f \in M_0^!(6)$.

Suppose now that $b > a$ and that r, s are not divisible by 3. In the $a = 0$ case, replacing z with $\left(\begin{smallmatrix} 3 & 1 \\ 6 & 3 \end{smallmatrix}\right)$ in the formula for $U_3 f(z)$, we compute that

$$3U_3 f_{0,r}^{(6)}\left(\frac{3z+1}{6z+3}\right) = f_{0,r}^{(6)}\left(z + \frac{1}{3}\right) + f_{0,r}^{(6)}\left(\frac{9z+4}{18z+9}\right) + f_{0,r}^{(6)}\left(z + \frac{2}{3}\right).$$

The second term is holomorphic at ∞ since $\left(\begin{smallmatrix} 9 & 4 \\ 18 & 9 \end{smallmatrix}\right) = \left(\begin{smallmatrix} 1 & 2 \\ 4 & 9 \end{smallmatrix}\right)\left(\begin{smallmatrix} 9 & 0 \\ 0 & 1 \end{smallmatrix}\right)$. The first and last terms give $-q^{-r} + O(1)$ because $r \not\equiv 0 \pmod 3$. Slashing with $\left(\begin{smallmatrix} 0 & -1 \\ 1 & 0 \end{smallmatrix}\right)$ and $\left(\begin{smallmatrix} 1 & 0 \\ 3 & 1 \end{smallmatrix}\right)$, we find that this form is holomorphic at 0 and $\frac{1}{3}$. Slashing with $\left(\begin{smallmatrix} 1 & 0 \\ 2 & 1 \end{smallmatrix}\right)$ gives $f_{0,r}^{(6)}(\frac{z+2}{9}) + f_{0,r}^{(6)}(\frac{z+8}{9}) + f_{0,r}^{(6)}(\frac{z+5}{9})$, and all negative powers of q cancel out since $r \not\equiv 0 \pmod 3$. We conclude that $3U_3 f_{0,r}^{(6)}\left(\frac{3z+1}{6z+3}\right) = -f_{0,r}^{(6)}(z)$.

Now let $\psi(z)$ be the level 6 Hauptmodul $\psi(z) = \frac{\eta(z)^5 \eta(3z)}{\eta(2z)\eta(6z)^5}$. Because powers of ψ form a basis for $M_0^\sharp(6)$, we can write

$$3U_3 f_{0,r}^{(6)}\left(\frac{3z+1}{6z+3}\right) = -f_{0,r}^{(10)}(z) = \sum_{i=0}^{r} c_i \psi(z)^i,$$

with the coefficients $c_i \in \mathbb{Z}$. We compute that

$$\psi|\left(\begin{smallmatrix}3&1\\6&3\end{smallmatrix}\right) = \frac{3\eta(3z)^5\eta(z)}{\eta(6z)\eta(2z)^5}.$$

Note that $\psi_{\frac{1}{2}}(z) = \frac{\eta(3z)^5\eta(z)}{\eta(6z)\eta(2z)^5}$ is a modular form in $M_0^!(6)$ with integer coefficients that is holomorphic at ∞ and 0, vanishes at $\frac{1}{3}$, and has a pole of order 1 at $\frac{1}{2}$.

We then use near-identical arguments for the base case and inductive step to show that the stated congruences hold for powers of 3.

5. Congruences in Levels 12 and 18

To extend the level 6 congruences to levels 12 and 18, we first relate basis elements in these levels to basis elements in level 6, using the usual U_p and V_p operators with $p = 2, 3$. Recall that $V_p(f(z)) = f(pz)$, and that if $f \in M_k^!(N)$, then $V_p f \in M_k^!(pN)$.

Lemma 1. *For any nonnegative integer m, we have*

$$U_2\left(f_{0,m}^{(12)}\right) = f_{0,\frac{m}{2}}^{(6)} \text{ if } 2|m,$$

$$U_3\left(f_{0,m}^{(18)}\right) = f_{0,\frac{m}{3}}^{(6)} \text{ if } 3|m.$$

If m is odd, then $U_2(f_{0,m}^{(12)}) = 0$, and if $3 \nmid m$, then $U_3(f_{0,m}^{(18)}) = 0$.

To see this for $p = 2$, let $f_{0,2m} \in M_0^\sharp(12)$. We compute that V_2U_2 preserves the space $M_0^\sharp(12)$ as in section 3 of [7]. The first term of $f_{0,2m}$ will be q^{-2m}, as will the first term of $V_2 f_{0,m}^{(6)}$. Then in the expression

$$V_2\left(U_2\left(f_{0,2m}^{(12)}(z)\right) - f_{0,m}^{(6)}(z)\right),$$

the first terms will cancel and the resulting form will have only nonnegative powers of q. Since it is in the space $M_0^\sharp(12)$, it must be identically zero. If 2 does not divide m, then $U_2 f_{0,m}^{(12)}$ will be a form in $M_0^\sharp(6)$ which vanishes at ∞, and must be zero; thus, when m is odd, the form $f_{0,m}^{(12)}(z)$ will have only odd powers of q in its Fourier expansion. The proof of the case for level 18 is similar.

Using Lemma 1, we can prove the congruences for levels 12 and 18. Consider $f_{0,m}^{(12)} \in M_0^\sharp(12)$, and assume that $m = 2^a r$ for some $a \geq 0$ and odd r. Then by Lemma 1, we have $U_2(f_{0,m}^{(12)}) = f_{0,2^{a-1}r}^{(6)}$ if $a \geq 1$ and $U_2(f_{0,m}^{(12)}) = 0$ if $a = 0$. If $a \geq 1$, then for some $b \geq 0$ and some odd s, we have the relation

$$a_0^{(12)}(2^a r, 2^b s) = a_0^{(6)}(2^{a-1} r, 2^{b-1} s).$$

Note that if $b=0$, then $a_0^{(12)}(2^a r, 2^b s)=0$. Suppose that $a>b$. If $b=0$, then the congruence holds vacuously. If $b>0$, then $a-1>b-1$ and the congruence in level 12 follows from the congruence in level 6. If instead $b>a$, then if $a=0$, we have $U_2 f_{0,2^a r}^{(12)}=0$ and the congruence vacuously holds. Otherwise, $b-1>a-1$ and the congruence again follows from the congruence in level 6.

For the congruences modulo powers of 3, write $m=2^a 3^c r$, with $a \geq 1$ and r' not divisible by 2 or 3. Using the U_2 operator again, we obtain the relation

$$a_0^{(12)}(2^a 3^c r, 2^b 3^d s) = a_0^{(6)}(2^{a-1} 3^c r, 2^{b-1} 3^d s')$$

for appropriate b, d, s' with s' not divisible by 2 and 3. Again, the congruences follow directly from the congruences in level 6. The weaker congruence (10) modulo powers of 3 when 2 does not divide the order of the pole can be proved in a manner similar to the way in which (2) implies (1), using duality, equation (14), and the fact that all coefficients $a_0^{(12)}(m,n)$ are integers.

The proof for the level 18 congruences is similar.

6. Congruences in Level 10

The proof of congruences (11) and (12) for powers of 2 in level 10 is very similar to the proof for level 6 in section 4. We will make note of explicit calculations that differ.

Let $f \in M_0^\sharp(10)$. Again, we want to find the behavior at the cusps under the U_2 operator. Level 10 has the three non-infinity cusps 0, $\frac{1}{2}$, and $\frac{1}{5}$. The calculations showing order of vanishing for the cusps at 0 and $\frac{1}{2}$ are similar to the level 6 calculations. For the cusp at $\frac{1}{5}$, we find that

$$2U_2 f \big| \left(\begin{smallmatrix} 1 & 0 \\ 5 & 1 \end{smallmatrix}\right) = f \big| \left(\begin{smallmatrix} 1 & 0 \\ 10 & 1 \end{smallmatrix}\right) \big| \left(\begin{smallmatrix} 1 & 0 \\ 0 & 2 \end{smallmatrix}\right) + f \big| \left(\begin{smallmatrix} 3 & -2 \\ 5 & -3 \end{smallmatrix}\right) \big| \left(\begin{smallmatrix} 2 & 1 \\ 0 & 1 \end{smallmatrix}\right).$$

It follows that applying U_2 repeatedly to a form in $M_0^\sharp(10)$ gives forms with poles only at $\frac{1}{5}$ and ∞.

For the space $M_0^!(10)$, the matrix $\left(\begin{smallmatrix} 4 & -1 \\ 10 & -2 \end{smallmatrix}\right)$ is an Atkin-Lehner involution. Computing as before,

$$f \big| \left(\begin{smallmatrix} 4 & -1 \\ 10 & -2 \end{smallmatrix}\right) \big| \left(\begin{smallmatrix} 0 & -1 \\ 1 & 0 \end{smallmatrix}\right) = f \big| \left(\begin{smallmatrix} -1 & -4 \\ -2 & -10 \end{smallmatrix}\right) = f \big| \left(\begin{smallmatrix} 1 & 0 \\ 2 & 1 \end{smallmatrix}\right) \big| \left(\begin{smallmatrix} -1 & -4 \\ 0 & -2 \end{smallmatrix}\right),$$

$$f \big| \left(\begin{smallmatrix} 4 & -1 \\ 10 & -2 \end{smallmatrix}\right) \big| \left(\begin{smallmatrix} 1 & 0 \\ 2 & 1 \end{smallmatrix}\right) = f \big| \left(\begin{smallmatrix} 2 & -1 \\ 6 & -2 \end{smallmatrix}\right) = f \big| \left(\begin{smallmatrix} 1 & 0 \\ 3 & 1 \end{smallmatrix}\right) \big| \left(\begin{smallmatrix} 2 & -1 \\ 0 & -1 \end{smallmatrix}\right),$$

$$f \big| \left(\begin{smallmatrix} 4 & -1 \\ 10 & -2 \end{smallmatrix}\right) \big| \left(\begin{smallmatrix} 1 & 0 \\ 5 & 1 \end{smallmatrix}\right) = f \big| \left(\begin{smallmatrix} -1 & -1 \\ 0 & -2 \end{smallmatrix}\right).$$

Now if we replace z with $\left(\begin{smallmatrix} 4 & -1 \\ 10 & -2 \end{smallmatrix}\right)$ in the formula for $U_2 f(z)$, we compute that

$$2U_2 f_{0,r}^{(10)}\left(\frac{4z-1}{10z-2}\right) = f_{0,r}^{(10)}\left(\frac{4z-1}{20z-4}\right) + f_{0,r}^{(10)}\left(z-\frac{1}{2}\right).$$

Note that the first term here is holomorphic at ∞ since $\left(\begin{smallmatrix} 4 & -1 \\ 20 & -4 \end{smallmatrix}\right) = \left(\begin{smallmatrix} 1 & -1 \\ 5 & -4 \end{smallmatrix}\right)\left(\begin{smallmatrix} 4 & 0 \\ 0 & 1 \end{smallmatrix}\right)$ and that the second term will give $-q^{-r} + O(1)$. After checking that this is holomorphic at the other cusps, we conclude that $2U_2 f_{0,r}^{(10)}\left(\frac{4z-1}{10z-2}\right) = -f_{0,r}^{(10)}(z)$.

Now let $\psi(z)$ be the level 10 Hauptmodul $\psi(z) = \frac{\eta(2z)\eta(5z)^5}{\eta(z)\eta(10z)^5}$. By the construction of basis elements, we can write

$$2U_2 f_{0,r}^{(10)}\left(\frac{4z-1}{10z-2}\right) = -f_{0,r}^{(10)}(z) = \sum_{i=0}^{r} c_i \psi^i(z),$$

with the coefficients $c_i \in \mathbb{Z}$. We compute that

$$\psi|\left(\begin{smallmatrix} 4 & -1 \\ 10 & 2 \end{smallmatrix}\right) = \frac{-2^2\eta(z)\eta(10z)^5}{\eta(2z)\eta(5z)^5}.$$

Note that $\psi_{\frac{1}{5}}(z) = \frac{\eta(z)\eta(10z)^5}{\eta(2z)\eta(5z)^5}$ is a modular form in $M_0^!(10)$ with integer coefficients that is holomorphic at ∞, vanishes at $\frac{1}{2}$ and 0, and has a pole of order 1 at $\frac{1}{5}$.

After these explicit calculations, we can use near-identical arguments for the base case and inductive step to show that congruence (12) holds. Congruence (11) follows as before using duality and equation (14); the congruence (13) modulo powers of 5 is proved in the same way.

References

[1] H.-F. Aas, *Congruences for the coefficients of the modular invariant $j(\tau)$*, Math. Scand. **14** (1964), 185–192.

[2] N. Andersen and P. Jenkins, *Divisibility properties of coefficients of level p modular functions for genus zero primes*, Proc. Amer. Math. Soc. **141** (2013), no. 1, 41–53.

[3] W. Duke and P. Jenkins, *On the zeros and coefficients of certain weakly holomorphic modular forms*, Pure Appl. Math. Q. **4** (2008), no. 4, Special Issue: In honor of Jean-Pierre Serre. Part 1, 1327–1340.

[4] A. El-Guindy, *Fourier expansions with modular form coefficients*, Int. J. Number Theory **5** (2009), no. 8, 1433–1446.

[5] S. Garthwaite and P. Jenkins, *Zeros of weakly holomorphic modular forms of levels 2 and 3*, Math. Res. Lett. **20** (2013), no. 4, 657–674.

[6] M. Griffin, *Divisibility properties of coefficients of weight 0 weakly holomorphic modular forms*, Int. J. Number Theory **7** (2011), no. 4, 933–941.

[7] A. Haddock and P. Jenkins, *Zeros of weakly holomorphic modular forms of level 4*, Int. J. Number Theory **10** (2014), no. 2, 455–470.

[8] P. Jenkins and DJ Thornton, *Weakly holomorphic modular forms in prime power levels of genus zero*, preprint, arXiv:1703.08145v1 [math.NT].

[9] P. Jenkins and DJ Thornton, *Congruences for coefficients of modular functions*, Ramanujan J. **38** (2015), no. 3, 619–628.

[10] O. Kolberg, *The coefficients of $j(\tau)$ modulo powers of* 3, Arbok Univ. Bergen Mat.-Natur. Ser. **1962** (1962), no. 16, 7.

[11] O. Kolberg, *Congruences for the coefficients of the modular invariant* $j(\tau)$., Math. Scand. **10** (1962), 173–181.

[12] J. Lehner, *Divisibility properties of the Fourier coefficients of the modular invariant* $j(\tau)$, Amer. J. Math. **71** (1949), 136–148.

[13] J. Lehner, *Further congruence properties of the Fourier coefficients of the modular invariant* $j(\tau)$, Amer. J. Math. **71** (1949), 373–386.

[14] G. Ligozat, *Courbes modulaires de genre* 1, Société Mathématique de France, Paris, 1975, Bull. Soc. Math. France, Mém. 43, Supplément au Bull. Soc. Math. France Tome 103, no. 3.

[15] D. Zagier, *Traces of singular moduli*, Motives, polylogarithms and Hodge theory, Part I (Irvine, CA, 1998), Int. Press Lect. Ser., vol. 3, Int. Press, Somerville, MA, 2002, pp. 211–244.

15

CYCLES IN THE COPRIME HYPERGRAPH OF INTEGERS

Jan-Hendrik de Wiljes
Institute of Mathematics and Applied Computer Science, University of Hildesheim, Hildesheim, Germany
wiljes@uni-hildesheim.de

Abstract

In this paper we investigate the existence of different kinds of cycles in the coprime hypergraph of integers CHI_k and its induced subgraphs. We obtain conditions for the existence of cycles as well as examples which show that the results cannot be improved. We will also briefly discuss different forms of acyclicity of coprime hypergraphs. Our results generalize similar results of Erdős and Sárközy on cycles in the coprime graph of integers CHI_1, although not in the sense that they imply their findings (since we require $k \geq 2$).

1. Introduction

1.1. Definitions and Notations

We start by explaining the necessary notation. Let $[n]$ denote the set of the first n positive integers; similarly we define $[m, n] := [n] \setminus [m-1]$ for positive integers m, n with $n > m$. The number of integers from $[n]$, which have a divisor among the first j primes, is denoted by $f(n, j)$. For $A \subseteq \mathbb{N}$ and non-negative integers s, t, we define $A_{(s,t)} := \{a \in A : a \equiv t \mod s\}$. Further, $(a_1, a_2, \ldots, a_j)$ denotes the *greatest common divisor* of $a_1, a_2, \ldots, a_j \in \mathbb{Z}$. A set of size l is called an *l-set*, and $\mathcal{P}(A)$ denotes the *power set* of the set A. Finally, A^l denotes the set and $\binom{|A|}{l}$ the number of l-subsets of the finite set A.

A *hypergraph* H is a pair (V, E) consisting of a set $V \neq \emptyset$ of *vertices* and a set E of subsets of V which are called *(hyper)edges*. If all the elements of E have the same cardinality k, the hypergraph is called *k-uniform*. In this case we also call H a *k-graph*, in particular a 2-graph is just a *graph*. We can associate a graph $G(H)$ to H, its so called *primal graph*, which has vertex set V and contains the edge $\{u, v\}$ if and only if there exists some edge in E that contains both, u and v.

A *clique* in a hypergraph $H = (V, E)$ is a subset W of V such that, for any two elements u, v of W, there exists some edge in E that contains both, u and v. We call H *chordal* if it is *conformal*, i. e., every clique of H is contained in an edge of

H, and if $G(H)$ is *chordal*, i. e., $G(H)$ does not contain induced (normal graph) cycles of length greater than 3 (see [1, 2]).

Different types of cycles for hypergraphs have been introduced in the past. We will restrict ourselves to the following concepts (exemplary literature where those concepts can be found is given after every definition).

A *Berge cycle* of length l in a hypergraph G is a sequence $v_1, e_1, v_2, e_2, \ldots, v_l, e_l$ of distinct vertices $v_1, v_2, \ldots, v_l$ and distinct edges $e_1, e_2, \ldots, e_l$ of G, where $v_1, v_l \in e_l$ and $v_i, v_{i+1} \in e_i$ for every $i \in [l-1]$ (see [2]).

A k-graph $C = (V, E)$ is a *t-cycle* if the elements of V can be ordered, say $v_1, v_2, \ldots, v_n$, such that E consists of

$$\{v_1, v_2, \ldots, v_k\}, \{v_{k-t+1}, v_{k-t+2}, \ldots, v_{2k-t}\}, \ldots,$$
$$\ldots, \{v_{n-k+t+1}, v_{n-k+t+2}, \ldots, v_n, v_1, v_2, \ldots, v_t\},$$

i. e., consecutive edges intersect in precisely t vertices.[1] The number of edges of C is called its *length*. We say that a hypergraph contains a t-cycle of length r if it has a subhypergraph which is a t-cycle of length r (see [12, 13, 14]). Only the cases $t = 1$ and $t = k-1$, which are also known as *loose* and *tight cycles* respectively, will be discussed in detail in this article.

The following cycle type is quite different from those considered above, but we find it interesting to study. It is motivated by a definition of a graph cycle using matroids (see [2]).

A nonempty set $S \subseteq \mathcal{P}(B^k)$, where B is a finite set, is called *dependent* if every set $A \subseteq B^{k-1}$ is a subset of an even number of elements from S, and it is called *reduced* if no proper nonempty subset of S is dependent. For a k-graph H, we call a nonempty l-set C of edges, which is dependent and reduced, a *matroid cycle* of H of length l (see [8]).

1.2. The Coprime Hypergraph of Integers

The object of interest, as introduced in [4], is the $(k+1)$-uniform *coprime hypergraph of integers* CHI_k, which has vertex set $\mathbb{Z}$ and $k+1$ elements of $\mathbb{Z}$ form a hyperedge if their greatest common divisor equals 1. Let A be a set of positive integers. The coprime hypergraph of A, denoted by $CHI_k(A)$, is the induced subgraph in CHI_k by A. The set of edges of $CHI_k(A)$ will be denoted by $E_k(A)$.

In [4, 5, 6] the maximal size of complete subgraphs of $CHI_k([n])$ was investigated. We will consider another natural question, namely the existence of cycles in CHI_k and its induced subgraphs. For $CHI_1(A)$, the existence of cycles depends on the size of $A \subseteq [n]$ compared to $f(n,1)$ for even cycles and to $f(n,2)$ for odd cycles,

[1] To avoid unnecessarily long notation we sometimes say that $v_1, v_2, \ldots, v_n$ yields or induces a t-cycle.

see [10]. It turns out that for $k \geq 2$ the existence of (certain types of) cycles in $CHI_k(A)$ only depends on $f(n, 1)$.

It should be mentioned that our results generalize the ones from [10] to coprime hypergraphs, but not in the sense that they imply the results from [10] (since they only hold for $k \geq 2$).

The following lemma from [9] will be used extensively.

Lemma ESS. (Erdős, Sárközy & Szemerédi 1980). *There exists some constant n_0 such that if $n \geq n_0$, $A \subseteq [n]$ with $|A| > f(n, 1)$, and $r = \min\{|A_{(2,1)}|, \lfloor \frac{1}{10} \log\log n \rfloor\}$, then $CHI_1(A)$ contains a complete bipartite graph on $2r$ vertices. In particular, for every $l \in [r]$ we find elements $v_1, v_2, \ldots, v_{2l}$ in A with $(v_i, v_{i+1}) = (v_1, v_{2l}) = 1$ for all $i \in [2l-1]$.*

2. Different Types of Cycles in Coprime Hypergraphs

2.1. Berge Cycles

By filling up edges of $CHI_1([n])$ with arbitrary elements we get $(k+1)$-sets, which are edges of $CHI_k([n])$. Thus, we can certainly use the results from [10] to find Berge cycles. More precisely, Lemma ESS gives us Berge cycles of length l in $CHI_k([n])$ for every $l \in [3, 2r]$ and $n > k \geq 2$. Although certainly there are examples with longer Berge cycles for $k = 2$ (hence improved results might be possible), we could not find a general way to do better in this case. For $k \geq 3$ on the other hand a much stronger result is possible.

Theorem 1. *Let k, n be positive integers with $f(n, 1) \geq k \geq 3$ and $A \subseteq [n]$. If $|A| > \max\{f(n, 1), k+1\}$ then $CHI_k(A)$ contains a Berge cycle of length l for every $l \in \{2, 3, \ldots, |A| - 2\}$.*

Proof. Using the pigeonhole principle, we find two elements $a, b \in A$ with $(a, b) = 1$. If $l = 2$, we pick two distinct $(k-1)$-subsets S_1, S_2 of $A \setminus \{a, b\}$. Then

$$a, S_1 \cup \{a, b\}, b, S_2 \cup \{a, b\}$$

is the desired Berge cycle of length 2.

So, from now on we require $l \geq 3$. We have to distinguish two cases. If $l \leq k$, we pick k elements $v_1, v_2, \ldots, v_k$ from $A \setminus \{a, b\}$ and define $e_i := (\{a, b\} \cup \{v_j : j \in [k]\}) \setminus \{v_i\}$ for every $i \in [l]$. Then

$$v_1, e_3, v_2, e_4, \ldots, v_{l-2}, e_l, v_{l-1}, e_1, v_l, e_2$$

is the desired Berge cycle of length l.

If $l > k$, we pick l elements $v_1, v_2, \ldots, v_l$ from $A \setminus \{a, b\}$ and define

$$e_i := \{v_i, v_{i+1} \ldots, v_{k+i-2}\} \cup \{a, b\}$$

for $i \in [l]$, where $v_{l+j} = v_j$ for every $j \in \mathbb{N}$. Then $v_1, e_1, v_2, e_2, \ldots, v_l, e_l$ is the desired Berge cycle of length l. □

It is not possible to replace $f(n, 1)$ in the statement of the theorem by a smaller number, since $CHI_k([n]_{(2,0)})$ contains no Berge cycles.

Remark. Different notions for acyclicity of hypergraphs, apart from not containing a Berge cycle, have been introduced in the past (see for example [11]), the weakest[2] one being *α-acyclicity*, which is equivalent to being chordal [1]. One might ask if there exists a set $A \subseteq [n]$ with $|A| > f(n, 1)$ such that $CHI_k(A)$ is α-acyclic. This is not the case if $|A| > k + 1 \geq 4$, since the vertices of the Berge cycle of length $|A| - 2$ given in Theorem 1, together with a and b, obviously form a clique of size at least $k + 2$. Therefore, $CHI_k(A)$ is not conformal.

2.2. Loose Cycles

We can easily see that given $A = [n]_{(2,0)} \cup ([2(r-1)+1]_{(2,1)} \setminus \{1\})$ (a similar set was used in [10]), the vertex set of a loose cycle in $CHI_k(A)$ of length $2r - 1$ has to contain at least r odd integers (one for every two edges), which cannot be the case. This gives us the lower bound $f(n, 1) + r$ for the cardinality of A which guarantees a loose cycle of given length $2r - 1$ (as well as $2r$) in $CHI_k(A)$.

Using results from [9] and [10], one can show that $|A| > f(n, 2)$ guarantees the existence of loose cycles of almost every length in $CHI_k(A)$ for $k \geq 2$ (just fill up the missing vertices in the hyperedges by arbitrarily picked numbers, as will be done in the proof of Theorem 2).

For $k \geq 2$ we can even prove a better bound which matches the lower bound from above (therefore in that sense the following theorem cannot be improved), more precisely we have:

Theorem 2. *Let $k \geq 2$ be an integer. There exists some constant n_k such that if $n \geq n_k$, $A \subseteq [n]$, and $|A| \geq f(n, 1) + r$, where $r = \min\{|A_{(2,1)}|, \lfloor \frac{1}{10} \log\log n \rfloor\}$, then $CHI_k(A)$ contains a loose cycle of length $2l$ and a loose cycle of length $2l - 1$ for every l with $2 \leq l \leq r$.*

Proof. We choose $n_k \geq n_0$ (with n_0 from Lemma ESS) such that $f(n, 1) \geq (2k - 1)\lfloor \frac{1}{10} \log\log n \rfloor$ for every $n \geq n_k$. Then, for every l with $2 \leq l \leq r$, we find $v_1, v_2, \ldots, v_{2l} \in A$ with $(v_i, v_{i+1}) = (v_1, v_{2l}) = 1$ for all $i \in [2l - 1]$ using Lemma ESS. Further, since the choices of A and n_k guarantee

$$|A| \geq f(n, 1) + r \geq (2k - 1) \left\lfloor \frac{1}{10} \log\log n \right\rfloor + r \geq 2rk \geq 2lk,$$

[2]Weakest in the sense that it is implied by Berge acyclicity.

we can pick arbitrary $2l(k-1)$ elements from $A \setminus \{v_1, v_2, \ldots, v_{2l}\}$ (and put them in place of the stars below) to obtain

$$v_1, \underbrace{*, \ldots, *}_{k-1 \text{ elements}}, v_2, \underbrace{*, \ldots, *}_{k-1 \text{ elements}}, v_3, *, \ldots, \ldots, *, v_{2l-1}, \underbrace{*, \ldots, *}_{k-1 \text{ elements}}, v_{2l}, \underbrace{*, \ldots, *}_{k-1 \text{ elements}}$$

and

$$v_1, \underbrace{*, \ldots, *}_{k-1 \text{ elements}}, v_2, \underbrace{*, \ldots, *}_{k-1 \text{ elements}}, v_3, *, \ldots,$$
$$\ldots, *, v_{2l-2}, \underbrace{*, \ldots, *}_{k-1 \text{ elements}}, v_{2l-1}, \underbrace{*, \ldots, *}_{k-2 \text{ elements}}, v_{2l},$$

while using only $2(l-1)(k-1)-1$ of the arbitrarily picked elements in the second case. These sequences yield loose cycles in $CHI_k(A)$ of the desired lengths. □

We can do better for $l = 2$:

Proposition 1. *Let k, n be positive integers with $f(n,1) \geq k \geq 2$ and $A \subseteq [n]$ with $|A| \geq \max\{2k, f(n,1)+1\}$. Then $CHI_k(A)$ contains a loose cycle of length 2.*

Proof. As in the proof of Theorem 1, we find $a, b \in A$ with $(a,b) = 1$. Pick $2k-2$ elements $v_1, v_2, \ldots, v_{2k-2}$ from $A \setminus \{a, b\}$. Then

$$a, v_1, v_2, \ldots, v_{k-1}, b, v_k, v_{k+1}, \ldots, v_{2k-2}$$

induces a loose cycle of length 2 in $CHI_k(A)$. □

2.3. Tight Cycles

Since $\{2,3,4,8,9\}$ is the vertex set of a tight cycle of length 5 in $CHI_2([n])$ for $n \geq 9$, the set $[n]_{(2,0)} \cup [n]_{(3,0)}$ from [10] does not pose problems for the existence of odd cycles if $k \geq 2$. But using the set $A = [n]_{(2,0)} \cup ([2\lceil \frac{r}{k+1} \rceil]_{(2,1)} \setminus \{1\})$ instead, we can see that, since a tight cycle of length l in $CHI_k(A)$ has to contain at least $\lceil \frac{l}{k+1} \rceil$ odd integers, $CHI_k(A)$ cannot contain any tight cycle of length l for $l \geq r$. This gives us the lower bound $f(n,1) + \lceil \frac{r}{k+1} \rceil$ for the cardinality of A which guarantees tight cycles of given length r in $CHI_k(A)$.

Any (normal graph) cycle of length l in $CHI_1(A)$ certainly induces a tight cycle of length l in $CHI_k(A)$ for every $k < l$. Applying the results from [10] would therefore guarantee the existence of tight cycles of almost any length in $CHI_k(A)$ for a set $A \subseteq [n]$ with $|A| > f(n,2)$ (for sufficiently large n).

As in the loose cycle case, the lower bound from above is sufficient to guarantee the existence of tight cycles (under certain additional conditions). A similar argument as in the proof of Theorem 2 yields:

Theorem 3. *Let $k \geq 2$ be an integer. There exists some constant n_k such that if $n \geq n_k \geq k+3$, $A \subseteq [n]$, $|A| \geq f(n,1) + \lceil \frac{r}{k+1} \rceil$, where*

$$\lceil \frac{r}{k+1} \rceil = \min\{|A_{(2,1)}|, \lfloor \frac{1}{10} \log\log n \rfloor\},$$

then $CHI_k(A)$ contains a tight cycle of length l for every integer l with $k+1 < l \leq r$.

Proof. We choose $n_k \geq n_0$ (with n_0 from Lemma ESS) such that $f(n,1) \geq k\lfloor \frac{1}{10} \log\log n \rfloor$ for every $n \geq n_k$. Then we define $l' := \lceil \frac{l}{k+1} \rceil$ and $x := l'(k+1) - l \in [0,k]$. We find $v_1, v_2, \ldots, v_{2l'} \in A$ with $(v_i, v_{i+1}) = (v_1, v_{2l'}) = 1$ for all $i \in [2l'-1]$ using Lemma ESS. Further, since the choices of A and n_k guarantee

$$|A| \geq f(n,1) + \left\lceil \frac{r}{k+1} \right\rceil \geq k \left\lfloor \frac{1}{10} \log\log n \right\rfloor + \left\lceil \frac{r}{k+1} \right\rceil \geq \left\lceil \frac{r}{k+1} \right\rceil (k+1) \geq r \geq l,$$

we can pick arbitrary $l - 2l' = (k-1)l' - x$ elements from $A \setminus \{v_1, v_2, \ldots, v_{2l'}\}$ (and put them in place of the stars below) to obtain

$$v_1, v_2, \underbrace{*, \ldots, *}_{k-1 \text{ elements}}, v_3, v_4, \underbrace{*, \ldots, *}_{k-1 \text{ elements}}, v_5, v_6, *, \ldots, \ldots, *, v_{2l'-1}, v_{2l'}, \underbrace{*, \ldots, *}_{(k-1)-x \text{ elements}}.$$

This sequence yields a tight cycle of length l in $CHI_k(A)$ if $x \in [0, k-1]$. For $x = k$ we need to consider

$$v_1, v_2, \underbrace{*, \ldots, *}_{k-1 \text{ elements}}, v_3, v_4, \underbrace{*, \ldots, *}_{k-1 \text{ elements}}, v_5, v_6, *, \ldots,$$

$$\ldots, *, v_{2l'-3}, v_{2l'-2}, \underbrace{*, \ldots, *}_{k-2 \text{ elements}}, v_{2l'-1}, v_{2l'}.$$

This sequence induces a tight cycle of length l in $CHI_k(A)$. □

Remark. It is possible to extend Theorems 2 and 3 to the existence of t-cycles with $1 < t < k$ of length r in $CHI_k(A)$ depending on the size of A. Since the proof of an analogue statement is rather messy, we will just sketch the idea for finding a cycle for fixed t and r. The vertex set of a t-cycle of length r in $CHI_k(A)$ contains precisely $r(k+1-t)$ elements. Therefore, we require $|A| \geq r(k+1-t)$. Further, we need r coprime elements (Lemma ESS). Then, using the idea from Theorem 2, the sequence

$$v_1, \underbrace{*, \ldots, *}_{k-t \text{ elements}}, v_2, \underbrace{*, \ldots, *}_{k-t \text{ elements}}, v_3, *, \ldots, \ldots, *, v_{r-1}, \underbrace{*, \ldots, *}_{k-t \text{ elements}}, v_r, \underbrace{*, \ldots, *}_{k-t \text{ elements}}$$

induces a t-cycle of length r if $1 < t \leq \frac{k+1}{2}$. If $k > t > \frac{k+1}{2}$ this idea would use unnecessarily many elements v_i per edge, therefore we use the idea of Theorem 3 and just push v_{2i} and v_{2i+1} apart. More precisely, the sequence

$$v_1, \underbrace{*,\dots,*}_{t \text{ elements}}, v_2, \underbrace{*,\dots,*}_{k-t \text{ elements}}, v_3, \underbrace{*,\dots,*}_{t \text{ elements}}, v_4, \underbrace{*,\dots,*}_{k-t \text{ elements}}, v_5, *, \dots,$$

$$\dots, *, v_{2j-1}, \underbrace{*,\dots,*}_{t \text{ elements}}, v_{2j}, \underbrace{*,\dots,*}_{k-t \text{ elements}}$$

induces a t-cycle of length $j \cdot \lceil \frac{k+1}{k+1-t} \rceil$.[3] It should be mentioned that depending on k, r and t and for a given number of v_i (analogous to Theorems 2 and 3), the number of elements between v_{2j-1} and v_{2j} might be different (but still in $[t]$) to obtain a cycle (this is the messy part).

2.4. Matroid Cycles

We start with the construction of matroid cycles for the graph $CHI_k([n])$.

Proposition 2. *Let n, k be positive integers. Then $CHI_k([n])$ does not contain a matroid cycle of length l in the following cases:*

(i) $n \leq k+1$,

(ii) $n > k+1, l \leq k+1$,

(iii) $n > k+1, 2 \nmid (k+1)l$.

Proof. Let C be a subset of $E_k([n])$ of size l. Since for $n \leq k+1$ we have $l \leq 1$, there cannot exist a matroid cycle, therefore (i) is true.

For instance, from [3] we know that circuits of $(k+1)$-simplicial matroids (which correspond to matroid cycles) have lengths greater than $k+1$, implying (ii).

If C is a matroid cycle, every k-subset of some $c \in C$ has to occur an even number of times as a subset of the elements of C. This cannot be the case if in total there are an odd number, namely $(k+1)l$, of subsets (counted with multiplicities). Therefore, there cannot exist any odd matroid cycle in $CHI_k(A)$ if $2 \nmid (k+1)l$, which proves (iii). □

Theorem 4. *Let n, k be positive integers with $n > k+1 \geq 3$. Then $CHI_k([n])$ contains a matroid cycle of length l if $l \in [k+2, (n-k-1)k+2]_{(k,2)}$.*

Proof. The set $C_1 = [k+2]^{k+1}$ has exactly $k+2$ elements and every element of $[k+2]^k$ is a subset of an even number of elements of C_1. Therefore, the elements of

---

[3]The set of edges can be partitioned by the property of containing an element v_{2i-1} and for each such vertex there exist $\lceil \frac{k+1}{k+1-t} \rceil$ corresponding edges.

C_1, which are all edges of $CHI_k([n])$ since each contains two consecutive integers, form a matroid cycle of length $k+2$. Now we need to verify that the set

$$C_i = \underbrace{(C_{i-1} \setminus [i, k+i])}_{A} \cup \underbrace{\{D \cup \{k+i+1\} : D \in [i, k+i]^k\}}_{B}$$

of size $ik+2$ is a matroid cycle in $CHI_k([n])$ if C_{i-1} is one. First of all, since each element of B contains two consecutive integers, we have $C_i \subseteq E_k([n])$.

Let $S \subseteq [k+i+1]^k$. We consider the following four cases. If $S \subseteq [i, k+i]$, then S lies in an odd number of elements of A since C_{i-1} is dependent and exactly in one element of B, namely $S \cup \{k+i+1\}$. If $[i-1] \cap S \neq \emptyset$ and $k+i+1 \in S$, then S is not a subset of any of the elements of C_i. If $[i-1] \cap S \neq \emptyset$ and $k+i+1 \notin S$, then $S \cap B' = \emptyset$ for every $B' \in B$, and S lies in an even number of elements from A since C_{i-1} is dependent. Lastly, if $[i-1] \cap S = \emptyset$ and $k+i+1 \in S$, we have $S \cap A' = \emptyset$ for every $A' \in A$, and S lies in exactly two elements of B (since $|[i, k+i+1] \setminus S| = 2$). Therefore, C_i is dependent.

Suppose, C_i was not reduced and S would be a proper subset of C_i, which is dependent. Note that $S \subseteq B$ is impossible. If $B \not\subseteq S$, then $B \cap S = \emptyset$, since the subsets of the form $D \cup \{k+i+1\}$ with $D \in [i, k+i]^{k-1}$ occur only in elements of B (each exactly twice). But then $S \subsetneq C_{i-1}$ contradicts the fact that C_{i-1} is reduced. Therefore, we must have $B \subseteq S$ and $A \not\subseteq S$. But then $(S \setminus B) \cup [i, k+i]$ is a dependent proper subset of C_{i-1} in contradiction of C_{i-1} being a matroid cycle. □

Remark. Certainly any other element of the existing cycle might be substituted in the given way (using the subsets of this element) to get a new cycle, provided these subsets together with the new element $k+i$ form an edge of $CHI_k([n])$. Therefore, if one finds a matroid cycle C of length l with $(i-1)k+2 < l < ik+2$, the method used in the proof of Theorem 4 can be applied to get a cycle of length $l+yk$, where y is at most the number of elements of $[n]$ which are not used in the elements of C.

In case we only want to know if, for a given l, there exists some n such that $CHI_k([n])$ contains a matroid cycle of length l, the problem becomes purely combinatorial since we can use arbitrarily many primes to solve the only obstacle of the sets being edges.

For arbitrary A, the construction of matroid cycles in $CHI_k(A)$ can be done similarly, although A has to contain suitable elements (such as $k+2$ consecutive integers). It does not seem to be easy to obtain the possible length of a matroid cycle given only the size of A (although certainly the ideas in the proof of Theorem 4 should be helpful). This should be part of future research. Therefore, we settle for sufficient conditions on $|A|$ and $|E_k(A)|$ which guarantee a matroid cycle.

Proposition 3. *Let n, k be positive integers and $A \subseteq [n]$.*

(i) If $|A| \geq \max\{f(n,1)+2, k+2\}$ then $CHI_k(A)$ contains a matroid cycle.

(ii) *The hypergraph $CHI_k([n]_{(2,0)} \cup \{a\})$, where $a \in [n]_{(2,1)}$, does not contain a matroid cycle. In particular this means that $|A| > f(n,1)$ does not imply the existence of matroid cycles in $CHI_k(A)$.*

(iii) *If $|E_k(A)| > \binom{|A|}{k}$ then $CHI_k(A)$ contains a matroid cycle.*

Proof. As in the proof of Theorem 1 we find four distinct integers a, b, c, d in A with $(a, b) = (c, d) = 1$. Let S be a $(k+2)$-subset of A containing a, b, c and d. Then S^{k+1} is a matroid cycle of $CHI_k(A)$ (using the same argument as in the proof of Theorem 4), which settles (i).

All the edges of $CHI_k([n]_{(2,0)} \cup \{a\})$ necessarily contain a. Let e be such an edge. Since $e \setminus \{a\}$ cannot be a subset of any element from $E_k([n]_{(2,0)} \cup \{a\})$, the edge e cannot lie in a matroid cycle of $CHI_k([n]_{(2,0)} \cup \{a\})$. This settles (ii).

We fix an ordering $S_1, S_2, \ldots, S_t$, where $t = \binom{|A|}{k}$, of the k-subsets of A and let $f(e)$ be the characteristic vector in $\mathbb{F}_2^t$ of an edge e with respect to the given ordering, meaning it has entry 1 at coordinate i if $S_i \subseteq e$ and entry 0 everywhere else.[4] From linear algebra we know that $\{f(e) : e \in E_k(A)\}$ is linearly dependent over $\mathbb{F}_2$ since $|E_k(A)| > t$. This means that $E_k(A)$ is dependent and, therefore, contains a matroid cycle. □

3. Future Work

We constructed different types of cycles in $CHI_k(A)$ for different sets $A \subseteq [n]$. Other sets should be investigated and the question of the existence of longer cycles in the considered cases should be addressed. Further, it would be natural to try to count the number of cycles of given length.

Since two consecutive numbers are coprime, the graph $CHI_k([n])$ contains a Hamiltonian cycle (cycle containing every vertex of the graph) of every of the first three discussed types. Therefore, it is natural to ask similar extremal questions, namely what size of $A \subseteq [n]$ guarantees the existence of a Hamiltonian cycle in $CHI_k(A)$. Certainly $|A| > f(n,1)$ is necessary for the existence of loose and tight Hamiltonian cycles, but for small k and large n that is not a sufficient condition: $CHI_2([n]_{(2,0)} \cup \{1\})$ neither contains a loose nor a tight Hamiltonian cycle if $n \geq 10$.

There are other types of cycles in hypergraphs whose existence in $CHI_k(A)$ might be investigated. Examples are: (i) k-cycles in [8], which are derived from a generalization of characterizing graph cycles via cut edges and 2-colourings; (ii) t-tight Berge cycles in [7], which are sequences of vertices and edges like Berge cycles but with the condition that $\{v_i, v_{i+1}, \ldots, v_{i+t-1}\} \subseteq e_i$ holds for $i \in [l]$ (indices mod l).

[4] Actually this way the underlying matroid is defined (see for example [8]).

Acknowledgement. The author is grateful to the referee for very valuable comments and suggestions.

References

[1] C. Beeri, R. Fagin, D. Maier, and M. Yannakakis, On the desirability of acyclic database schemes, *J. ACM* **30** (1983), no. 3, 479–513.

[2] C. Berge, *Hypergraphs: Combinatorics in Finite Sets*, Elsevier, New York, 1989.

[3] R. Cordovil, M. Lemos, and C. L. Sales, Dirac's theorem on simplicial matroids, *Ann. Comb.* **13** (2009), no. 1, 53–63.

[4] J.-H. de Wiljes, Complete subgraphs of the coprime hypergraph of integers I: introduction and bounds, *Eur. J. Math.* **3** (2017), no. 2, 379–386.

[5] J.-H. de Wiljes, Complete subgraphs of the coprime hypergraph of integers II: structural properties, *Eur. J. Math.* **4** (2018), no. 2, 676–686.

[6] J.-H. de Wiljes, Complete subgraphs of the coprime hypergraph of integers III: construction, *Eur. J. Math.*, to appear.

[7] P. Dorbec, S. Gravier, and G. N. Sárközy, Monochromatic Hamiltonian t-tight Berge-cycles in hypergraphs, *J. Graph Theory* **59** (2008), no. 1, 34–44.

[8] R. Duke, Types of cycles in hypergraphs, *Ann. Discrete Math.* **27** (1985), 399–418.

[9] P. Erdős, A. Sárközy, and E. Szemerédi, On some extremal properties of sequences of integers, II, *Publ. Math. Debrecen* **27** (1980), nos. 1–2, 117–125.

[10] P. Erdős and G. N. Sárközy, On cycles in the coprime graph of integers, *Electron. J. Combin.* **4** (1997), no. 2, R 8.

[11] R. Fagin, Degrees of acyclicity for hypergraphs and relational database schemes, *J. ACM* **30** (1983), no. 3, 514–550.

[12] G. H. Katona and H. A. Kierstead, Hamiltonian chains in hypergraphs, *J. Graph Theory* **30** (1999), no. 3, 205–212.

[13] D. Kühn, R. Mycroft, and D. Osthus, Hamilton l-cycles in uniform hypergraphs, *J. Combin. Theory Ser. A* **117** (2010), no. 7, 910–927.

[14] V. Rödl, A. Ruciński, and E. Szemerédi, A Dirac-type theorem for 3-uniform hypergraphs, *Combin. Probab. Comput.* **15** (2006), nos. 1–2, 229–251.

Permissions

All chapters in this book were first published in EJCNT, by Integers; hereby published with permission under the Creative Commons Attribution License or equivalent. Every chapter published in this book has been scrutinized by our experts. Their significance has been extensively debated. The topics covered herein carry significant findings which will fuel the growth of the discipline. They may even be implemented as practical applications or may be referred to as a beginning point for another development.

The contributors of this book come from diverse backgrounds, making this book a truly international effort. This book will bring forth new frontiers with its revolutionizing research information and detailed analysis of the nascent developments around the world.

We would like to thank all the contributing authors for lending their expertise to make the book truly unique. They have played a crucial role in the development of this book. Without their invaluable contributions this book wouldn't have been possible. They have made vital efforts to compile up to date information on the varied aspects of this subject to make this book a valuable addition to the collection of many professionals and students.

This book was conceptualized with the vision of imparting up-to-date information and advanced data in this field. To ensure the same, a matchless editorial board was set up. Every individual on the board went through rigorous rounds of assessment to prove their worth. After which they invested a large part of their time researching and compiling the most relevant data for our readers.

The editorial board has been involved in producing this book since its inception. They have spent rigorous hours researching and exploring the diverse topics which have resulted in the successful publishing of this book. They have passed on their knowledge of decades through this book. To expedite this challenging task, the publisher supported the team at every step. A small team of assistant editors was also appointed to further simplify the editing procedure and attain best results for the readers.

Apart from the editorial board, the designing team has also invested a significant amount of their time in understanding the subject and creating the most relevant covers. They scrutinized every image to scout for the most suitable representation of the subject and create an appropriate cover for the book.

The publishing team has been an ardent support to the editorial, designing and production team. Their endless efforts to recruit the best for this project, has resulted in the accomplishment of this book. They are a veteran in the field of academics and their pool of knowledge is as vast as their experience in printing. Their expertise and guidance has proved useful at every step. Their uncompromising quality standards have made this book an exceptional effort. Their encouragement from time to time has been an inspiration for everyone.

The publisher and the editorial board hope that this book will prove to be a valuable piece of knowledge for researchers, students, practitioners and scholars across the globe.

List of Contributors

Neranga Fernando
Department of Mathematics, Northeastern University, Boston, Massachusetts

Jeske Glenn
Mathematics Department, University of Oregon, Eugene, Oregon

Christopher O'Neill
Mathematics Department, San Diego State University, San Diego, California

Vadim Ponomarenko
Mathematics Department, San Diego State University, San Diego, California

Benjamin Sepanski
Mathematics Department, Baylor University, Waco, Texas

Olivier Rozier
Institut de Physique du Globe de Paris, Paris, France

Dominic Klyve
Dept. of Mathematics, Central Washington University, Ellensburg, Washington

Melissa Pidde
Dept. of Mathematics, Central Washington University, Ellensburg, Washington

Kathryn E. Temple
Dept. of Mathematics, Central Washington University, Ellensburg, Washington

Anthony Zaleski
Department of Mathematics, Rutgers University, Piscataway, New Jersey

Neil Hindman
Department of Mathematics, Howard University, Washington, DC

Dona Strauss
Department of Pure Mathematics, University of Leeds, Leeds, UK

Peng Gao
School of Mathematics and Systems Science, Beihang University, Beijing, China

Liangyi Zhao
School of Mathematics and Statistics, University of New South Wales, Sydney, Australia

Paul Kinlaw
Department of Mathematics, Husson University, Bangor, Maine

Tomohiro Yamada
Center for Japanese Language and Culture Osaka University, Aomatanihigashi, Minoo, Osaka, Japan

Abhishek Juyal
Harish-Chandra Research Institute, HBNI, Allahabad, India

Shiv Datt Kumar
Department of Mathematics, Motilal Nehru National Institute of Technology, Allahabad, India

Dustin Moody
National Institute of Standards and Technology, Gaithersburg, Maryland

Idris Mercer
Department of Mathematics and Statistics, Florida International University, Miami, Florida

Min Zhang
School of Applied Science, Beijing Information Science and Technology University, Beijing, P. R. China

Jinjiang Li
Department of Mathematics, China University of Mining and Technology, Beijing, P. R. China

Darij Grinberg
School of Mathematics, University of Minnesota, Minneapolis, Minnesota

Victoria Iba, Paul Jenkins and Merrill Warnick
Department of Mathematics, Brigham Young University, Provo Utah

Jan-Hendrik de Wiljes
Institute of Mathematics and Applied Computer Science, University of Hildesheim, Hildesheim, Germany

Index

www.ingramcontent.com/pod-product-compliance
Lightning Source LLC
LaVergne TN
LVHW080622210926
848485LV00003B/195
9781639893621